274/9800660
21/1/99 ✳

Drug Misuse

Prevention, harm minimization and treatment

02

JOIN US ON THE INTERNET VIA WWW, GOPHER, FTP OR EMAIL:

WWW: http://www.thomson.com
GOPHER: gopher.thomson.com
FTP: ftp.thomson.com
EMAIL: findit@kiosk.thomson.com

A service of I(T)P®

Drug Misuse

Prevention, harm minimization and treatment

Jan Keene

Senior Lecturer
Applied Psychosocial Studies
University of East Anglia
Norwich

CHAPMAN & HALL
London · Weinheim · New York · Tokyo · Melbourne · Madras

Published by Chapman & Hall, 2–6 Boundary Row, London SE1 8HN, UK

Chapman & Hall, 2–6 Boundary Row, London SE1 8HN, UK

Chapman & Hall GmbH, Pappelallee 3, 69469 Weinheim, Germany

Chapman & Hall USA, 115 Fifth Avenue, New York NY 10003, USA

Chapman & Hall Japan, ITP-Japan, Kyowa Building, 3F, 2-2-1 Hirakawacho, Chiyoda-ku, Tokyo 102, Japan

Chapman & Hall Australia, 102 Dodds Street, South Melbourne, Victoria 3205, Australia

Chapman & Hall India, R. Seshadri, 32 Second Main Road, CIT East, Madras 600 035, India

Distributed in the USA and Canada by Singular Publishing Group Inc., 4284 41st Street, San Diego, California 92105

First edition 1997

© 1997 Jan Keene

Typeset in 10/12 pt Palatino by Mews Photosetting, Beckenham, Kent

Printed in Great Britain by Page Bros (Norwich) Ltd

ISBN 0 412 64280 8

∞ Printed on permanent acid-free text paper, manufactured in accordance with ANSI/NISO Z39.48-1992 and ANSI/NISO Z39.48-1984 (Permanence of Paper).

For Bill, again.

Contents

Acknowledgements

With thanks for data collection and conversation to Roger Duncan, Darren James, Norman Preddy, Helen Trinder and Ifor and colleagues. Most of all, I would like to thank all the drug users who contributed.

Introduction – why social and health care professionals should know about drug misuse

At the time of writing (1996), it is apparent that drug misuse is becoming far more widespread and diverse, among young people especially, than was ever imagined possible. This has profound practical implications for professionals in the health and social care fields. This situation is compounded by extensive media coverage contributing to misunderstanding, and unnecessarily complicating the health and welfare task.

PROFESSIONAL NEEDS

Studies in this field have identified a widespread need for information and skills. There is very little appropriate training for generic professionals in the social or health care fields (Galanter *et al.*, 1989; Farrell, 1990; Harrison, 1992) and such training often lacks a cohesive theoretical approach (Gorman, 1993). Social and health care professionals report a lack of knowledge about drug misuse and the kinds of problems associated with it. They are unsure how far they are equipped to deal with these problems and whether it is their job to do so. They do not know where to go for help and advice or where to refer clients. Despite this lack of confidence, the majority of health care and social welfare workers have (often without realizing it) a basic understanding of the problems associated with drug misuse and the practical skills necessary for working with drug misusers.

This potential can only be utilized, however, if practitioners have an understanding of the different kinds of drug-related problems and knowledge of the relevance of their own skills and the range of back-up services available. This book is therefore designed, first, to give indepth understanding of recreational drug misuse, risky drug misuse and

In this book, the term drug misuse is employed throughout to refer to the non-medical use of any drug that is intended only for use in medical treatment and the use of drugs that have no recognized or accepted medical purpose.

dependence and, second, to inform about prevention, harm minimization, treatment and control, in order that professionals can identify, assess and work with different kinds of drug misusers. It also aims to give information about the wide range of multidisciplinary and specialist professionals who can contribute in this field.

BACKGROUND

The ancient Greek word for drug has three meanings: a cure or remedy, a poison and a magical charm. This book will consider these meanings in the modern sense: drugs as medication and as a solution to problems; drugs as dangerous to health; and drugs as magical and hedonistic.

In the recent past, policy and practice guidelines have often been based on a misunderstanding of the diversity and complexity of drug misuse. Professionals have confused different types of drug misuse and/or attempted to compress all types into one narrowly defined category, i.e. all drugs are good or bad. This is the first mistake: it is preferable to go right back to the ancient Greeks and try to understand the range of different effects of drugs on different people.

Practitioners first of all need to understand the complexity of drug misuse in order to develop realistic concepts and construct useful categories for assessment. After this, they are in a position to identify clearly the main risks and problems in the different categories and so to determine which type of intervention is most appropriate.

They have three main sources of information about drug problems: first, there is a range of statistical information on different drug-misusing populations; second, there are the results of academic research; and third, there is professional and clinical experience. Unfortunately, there are limitations to each of these three expert sources and each produces its own (often conflicting) 'evidence' about drug problems. There is, of course, one other source of information, not often explored and standing in even greater contrast to the other three – the views of the drug misusers themselves. These may be collected from a variety of sources but when they are looked at as a whole it becomes evident that they can be ordered into broad categories using common themes that emerge clearly on analysis. What is often regarded as an undifferentiated form of behaviour, drug taking, and moreover one which defies any form of ordering because of the unpredictable patterns of misuse, can be seen in fact to fall into distinct groupings. It is no accident that these are recognized by the drug misusers when talking about themselves and about other people.

It is not, of course, possible to explore all the different forms of drug misuse with equal thoroughness simply because the research evidence is far from complete. Moreover, there is great variation in the amount of information that we have, very often for straightforward practical reasons. Drug misuse which occurs at raves or which is the reason for attendance at an agency is much easier to identify , describe and analyse than that which takes place as a solitary activity in the privacy of one's home. Our understanding of drug misuse must therefore inevitably be partial and if professionals are to practise in an informed manner, to wait for a fully comprehensive account is unrealistic. As our knowledge of drug misuse enlarges, it will be possible to expand our range of possible solutions to problems. In the meantime, we need to bear in mind constantly that the research evidence provides only a series of glimpses into the enormous and complex area of drug misuse.

There is, then, a need to understand drug misuse itself. This can be approached in different ways. In this book the first way is talking to drug misusers, that is, using qualitative research data to gain insight into, and understanding of, other people. This is then placed in the context of statistical and scientific research in the field. This comprehensive understanding is in turn used as a foundation for building practically useful assessment procedures and appropriate practice methods.

THREE DIFFERENT KINDS OF DRUG MISUSE: UNDERSTANDING THE DIFFERENT REASONS FOR USING DRUGS

1. Why do people experiment and then continue to misuse recreationally (as in starting to drive cars and then learning to drive properly and to enjoy it)? People may initially do something frightening as a result of social pressure and peer influence but then, as with an acquired taste, learn to appreciate its intrinsic qualities.
2. Why do they continue to misuse in a dangerous way, despite the risk (as in driving when tired or incapable or beginning to race cars)? When the costs and risks become higher, the motivation for continuing may be greater or qualitatively different.
3. Why do they continue to misuse when dependent/addicted (as in when cars and driving become the main priority and other things assume less importance)?

There may be several reasons at once, from physical addiction and psychological dependence to a need for some remedy or self-medication for other problems. It is necessary to make a distinction here between problems which cause drug misuse (as a remedy) and problems arising from drug misuse itself, as the solutions will be determined accordingly. However, the abstract argument about which came first is as useful as arguing about chicken or egg: the relationship is best described as circular rather than in unidimensional causal terms, problems and drugs each contributing to some extent to each other.

It is only when we understand the different kinds of drug misuse and the different reasons for each that we can respond appropriately. It is important to note that a client's drug misuse can fall into more than one category, but if the problems can be clearly distinguished it is easier to identify needs and set goals for achieving a sequence of small gains, rather than tackle the confusing mixture of problems as a whole.

THREE DIFFERENT KINDS OF INTERVENTION

There are three different approaches to drug misuse – prevention, harm minimization and treatment.These approaches are often used in a blanket or arbitrary way depending on the policy or philosophy of the provider, so some professionals may try to prevent all drug misuse whereas others may attempt to treat it all. Both of these approaches may be inappropriate for a large proportion of clients, but each professional group will be guided by the philosophy, values and priorities of their profession or agency.

Instead of matching the approach of a particular profession or agency to all the drug misusers encountered, it is suggested that a multi-disciplinary perspective should allow for a comprehensive assessment, followed by matching each distinct type of drug misuse to whichever approach is most appropriate, independently of the particular bias of any one professional standpoint. There are three major kinds of drug misuse and misuser (two of these categories will need to be further divided). The metaphor of the car driver will again be used to illustrate the matching of intervention to problem.

Type of drug user	Type of driver	Intervention
1a Experimental, unsafe	Learner driver	Prevention, education
1b Recreational, safer	Weekend driver	Safety regulations
2 Risky, unsafe	Racing driver	Make as safe as possible
3a Dependent, compulsive	Car fanatic	Abstinence or control
3b Has underlying problems	Escapist	Alternative solutions

Obviously any metaphor will not fit perfectly, as the car fanatic does not illustrate dependent drug misuse as well as, for example, the compulsive eater or gambler. But it does illustrate that we do not respond to all drivers in the same way, i.e. simply by punishing them all or preventing driving as a whole, as we do not respond to all gambling or drinking behaviour in the same way.

We cannot prevent or cure all drug misuse, just as we cannot prevent all driving or 'cure' all risk taking. As with driving, much drug taking serves a purposes and many risks are worth the cost. Instead, we react by limiting the damage: by educating people, taking safety precautions and providing help and resources when needed. Instead of using one approach for all, different approaches are appropriate for different kinds of misuser.

THE IMPORTANCE OF PROCESS AND CHANGE OVER TIME

The advantage of discrete categories for assessment of drug problems is that it allows us to match our client to the most appropriate interventions. It is clear that some clients' problems will not fit very well or may fit into more than one category and it is also important to remember that categories are static, whereas drug misusers, like drinkers, may move in and out of different category as time passes.

The categories form a continuum along which drug misusers travel as if it were a 'career' and, more than this, they may travel in a circular fashion, going through the different stages several times. So all drug misusers will start in the 'experimental phase', many will move from experimental to recreational, a few will move on again to the risky phase and a small group will become dependent. At present we know very little about this progression and the relative stages. We need more information about the longitudinal progression from misuse to dependence and better predictors for understanding and identifying young people at risk at each different stage.

We do know that most misusers do not progress very far and those that do will fluctuate between the more and less serious stages at different points in their lives. In this respect drug misusers are no different from many other social and health care clients. Rather than curing or changing people completely, most social and health care work is concerned with improving and/or maintaining their lifestyles, making sure things do not deteriorate beyond repair; there is no reason why working with drug misusers should be any different.

PART ONE
An Introduction to Drug Misuse: Talking to Drug Misusers

The extent of drug misuse

1

POLICY ISSUES

This is not a book about policy but it is important briefly to clarify the contradictions and constraints inherent within drug policies before deciding how far professionals can use statistics and policy documents.

Seen internationally, drug misuse has expanded dramatically, with many societies of 'low risk' history presenting greatly increased rates, together with the appearance of newer, more harmful patterns of drug misuse (WHO Expert Committee, 1994). Many authors have argued that drug policy and practice are based on inaccurate premises, often perpetuated by government policies, such as media prevention campaigns. As Leitner, Shapland and Wiles (1993) point out, the drug prevention campaigns in Britain resulted in the majority of the population having an exaggerated view of the risks involved in drug misuse. Although this serves a useful deterrent function, it also means that the general public and professionals alike have an unrealistic view of drug misuse and misusers.

This has implications for both policy and practice. Solivetti (1994) suggests that public opinion about the feasibility of the control of drugs is seriously flawed and that this will affect policy. Fraser and George (1988) give some insight into the effects of a misplaced local police strategy that in effect simply moves the drug problem around from one locality to another. O'Malley (1991) and Jensen, Gerber and Babcock (1991) suggest that politically driven policies are based on mistaken understandings of drug misuse.

In terms of practice, Easthope (1993) demonstrates that a range of professionals have different perceptions about the causes of drug misuse. Each of these different professional perspectives implies a different way of working with drug misusers and consequently a different policy emphasis, from public health (Stimson, 1987; Engelsman,

1991; Jonas, 1994; Wodak, 1994) to treatment (Edwards, 1989a) to social control (Howitt, 1990).

It is hardly surprising that generic professionals have been fairly united in turning their backs on academic literature and policy documents alike, as neither offers practical guidance or solutions to the wide range of everyday problems they encounter among drug misusers. Instead they have adapted in a variety of ways, mainly by ignoring the media and literature and taking a commonsense approach by providing straightforward services such as education and a range of different kinds of social and health care support.

However, if each professional uses the particular approach and priorities of their own profession, they may make the mistake of applying this one method to all drug misusers irrespective of the differences between them. Any one professional approach is likely to be effective for some but entirely inappropriate for others.

Specialist drug agencies also have difficulties making sense of the situation and maintaining the balance between policy makers, their funders and their clients. In effect, drug agencies provide a range of primary and social care services together with drug treatment and free drugs, but this commonsense range of services must be presented as 'drug treatment' to maintain funding within the context of policy initiatives.

This is also the case for specialist drug 'prevention' programmes, whether in schools or based in the community. Prevention workers provide a range of generic education programmes concerned with social skills, personal responsibility and decision making but these must be presented as drug prevention campaigns.

There are, of course, consistent calls for clearer, more sensible policy (Edwards, 1989b; Smith, 1990, 1991; Burke, 1992; Gorman, 1993), for closer communication between policy makers and researchers (Gunne, 1990; Bell, 1990) and for clearer roles for professionals (Adger, 1991). But as professionals have to deal with everyday problems in the absence of such cohesive guidance, it is best to go back to basics and consider what information we actually have. Before going any further it will be useful to give a brief review of the extent of drug misuse in terms of the statistics available.

HOW MANY, WHO ARE THEY AND WHAT ARE THEY USING?

In summary, the survey statistics show a great deal of recreational or occasional drug misuse, some risky drug misuse and a small proportion of dependent drug misuse. The survey statistics do not show an equal number of recreational and problematic misusers (agency attenders who seek help for risk or dependence): those attending agencies are only a very small proportion of the total number of drug misusers. They also demonstrate that drug misuse has diversified greatly since the early 1960s. A wide range of drugs is now used by a wide range of people, many using more than one type of drug. In addition, the statistics identify a great deal of criminal drug misuse.

Before considering the available statistics it is important to realize that there are different kinds of survey, focusing on different kinds of people. The type of group included will naturally influence the figures, so crime statistics will identify criminal drug misusers whereas agency figures will reflect only those drug misusers with serious problems or those who have come to the attention of the statutory services.

There are three main types of data recording different kinds of drug misuse among different population groups:

1. general and local population surveys (including schools);
2. police crime figures;
3. treatment and registered addicts.

Figures from all three types of survey are given below, but first it will be useful to identify and explain inconsistencies between the three different types of statistics.

To go back to the driving metaphor, different populations are identified depending on what you are looking for. So population surveys will identify all who drive; the police will identify people who are speeding (a large proportion of the population at some time or other); hospitals and GPs will identify those whose driving is more risky; and drug agencies would identify those whose driving is very risky and/or dependent. The main point is that there are far more recreational misusers than problematic drug misusers or criminal misusers (though there is a closer correlation between crime statistics and the general population than there is between treatment statistics and the general population).

The general surveys record the proportion of the whole population misusing drugs. As might be expected, such surveys show a greater proportion of people misusing drugs in a non-problematic way than those who have problems; of these, only a small proportion will ever attend agencies. The local surveys confirm this picture.

The crime figures record the possession of and dealing in drugs amongst those apprehended by the police and/or dealt with by the courts. The figures identify a majority of people misusing drugs in a non-problematic way as most arrests are for drugs such as cannabis. Only a small proportion of these will ever need help.

Treatment figures record the drug misuse of those requesting help from GPs or other agencies. These data record higher levels of more addictive drugs such as heroin and risky behaviour such as injecting drugs.

There is clearly some overlap between the three sources of information but recent Home Office initiatives, such as cautioning schemes and partnerships between drug agencies and probation, indicate how insignificant this overlap is (see Part Three). The majority of the general population and criminal drug misusers will have no need of drug services, as their drug misuse is recreational rather than dependent. However, some of these people may need health care or clean injecting equipment if they pass through stages when their drug misuse becomes more risky.

All three types of survey have flaws. The police crime figures tend to record levels of police activity rather than extent of drug misuse. Drug treatment figures may simply reflect availability of resources and it is estimated that only a fifth of drug misusers are known to their GPs. The general public may under-report drug misuse and school surveys often exclude truants and drop-outs who may be more likely to misuse drugs.

Perhaps the best contemporary reviews of British statistics are those of the Institute for the Study of Drug Dependence and yet even their 1994 survey of drug misuse in Britain states that 'Assessing the prevalence of drug use in Britain is more like piecing together a jigsaw with most of the pieces missing (and the rest fitting poorly or not fitting at all) than an exercise in statistics' (p. 7). They highlight several problematic issues in contemporary data collection, from the lack of continuity of most population surveys to the specific nature of many studies which are confined to at-risk populations or particular localities.

Nevertheless, despite the many limitations of statistical data at present, it is still possible to gain general pictures of the extent and type of drug misuse by reviewing this literature.

The data from surveys as a whole indicate that young people are likely to experiment with the less risky drugs such as cannabis on a temporary basis. Little is known about these populations and drug misuse is seldom studied in overt categories. There is some literature to indicate that as drug misusers reach their early to middle 20s they appear to reduce or stop misusing (Raveis and Kandel, 1987; Harris, 1988). It is also possible that those more likely to develop continuing problems related to their drug misuse are older, more established misusers (Murray, 1986). It becomes clear from the figures below that different populations are identified in different surveys and that interventions may need to be adapted accordingly.

DRUG MISUSERS IN THE GENERAL AND LOCAL POPULATION

It is interesting that surveys highlight an increase in recreational misuse of the 'less problematic' drugs rather than any change in problematic misuse of drugs such as heroin in the population as a whole. For example, NOP (1982), MORI (1989, 1992) and DPT (Leitner, Shapland and Wiles, 1993) show cannabis use as ranging from 10% to 28%, amphetamine misuse from 2% to 5%, LSD from 2% to 6% and heroin and cocaine at 1%. The earlier OPCS (1969) statistics have similar figures for the 'hard' drugs (this despite the fact that more heroin misuse has come to the attention of GPs and specialized clinics, as can be seen from Home Office registration figures below).

Unfortunately the major surveys of drug misuse do not often include drugs such as Ecstasy or recreational benzodiazepine. It is also of note that the increase in the misuse of LSD is seldom mentioned although some more recent surveys record its growing misuse among the younger age groups.

Leitner, Shapland and Wiles, in their Home Office (DPT) funded review, *Drug Usage and Drugs Prevention: Views and Habits of the General Public* (1993) carried out a survey of a random sample of 1000 people in each of four towns, Nottingham, Bradford, Lewisham and Glasgow. The first section of their questionnaire was based on similar items derived from the British Crime Survey. The numbers in this survey were supplemented by a 'booster sample' of younger people considered 'at risk'. Between 14% and 24% of respondents in this group had misused an unprescribed drug at some time. This comprehensive study indicates that, in line with other surveys, opiate misuse has remained fairly stable at about 1% over the past 30 years, but cannabis and amphetamine misuse has increased. The usage of these drugs is most marked among young people under 25 years.

The recent government Green Paper, *Tackling Drugs Together* (1994), and the subsequent White Paper with the same title (Home Office, 1995) give an up-to-date overview of drug misuse statistics for England.

- In any one year, at least 6% of the population take an illegal drug (some 3 million people).
- Among school children, 3% of 12–13-year-olds and 14% of 14–15-year-olds admit to taking an illegal drug.
- Drug misuse is more prevalent among men; one-third of male survey respondents aged 16–29 report misuse of an illegal drug in the past compared with just over a fifth who were female.
- Of young people living in inner city areas, some 42% aged 16–19 and 44% aged 20–24 have taken drugs at some time, with between 20–30% of the 19–29 age group misusing a drug in the last year.
- Cannabis is the most widely misused illegal drug. For people between 23 and 59, some 14% report experience of its misuse; 24% of people aged 16–29 report long-term cannabis misuse; 18% of the 16–19 age group and 14% of those aged 20–24 report misuse in the past year.

- A feature of drug misuse in the 1990s has been the integration of LSD and Ecstasy into mass youth culture. Among 16–19 year-olds, 11% have tried amphetamines, 9% Ecstasy and 8% LSD..
- Among general population samples, under 1% report ever misusing heroin, cocaine and crack. Samples of inner-city people (aged 16–25) suggest that 1% have misused crack cocaine, 2% heroin and 4% cocaine.
- Rates of reported misuse of solvents for young people are: 12–13 years (3%); 14–15 years (7%) and 15–16 years (6%).

HOME OFFICE CRIME STATISTICS

The number of drug seizures has risen steadily during the 1990s. It has to be remembered, however, that individual seizures in any one year may be unrepresentative. Moreover, the official statistics omit the number of offences reported to the police which have not resulted in a seizure or a conviction. All such enforcement statistics give a particular perspective on drug misuse; at the same time, they have the advantage of regular publication and thus indicate trends.

The number of seizures of controlled drugs in the UK made by police and other authorities under the Misuse of Drugs Act rose, for Class A drugs, as follows:

- for cocaine and crack, from 1410 in 1990 to 1974 in 1992;
- for heroin, from 2321 in 1990 to 2803 in 1992;
- for LSD, from 1772 in 1990 to 2404 in 1992;
- for MDMA, from 374 in 1990 to 2361 in 1992.

Similar increases are reported for Class B drugs, with amphetamines rising from 4490 in 1990 to 10,409 in 1992 and cannabis from 47,863 in 1990 to 53,811 in 1992.

Revealing a similar pattern, the number of persons found guilty, cautioned or dealt with by compounding for drug offences rose from 1990 to 1992. The figures are:

- for cocaine, 860 in 1990 and 913 in 1992;
- for heroin, 1605 in 1990 and 1415 in 1992;
- for LSD, 915 in 1990 and 1428 in 1992;
- for MDMA, 286 in 1990 and 1516 in 1992;
- for amphetamines, 2330 in 1990 and 5653 in 1992;
- for cannabis, 40,194 in 1990 and 41,353 in 1992.

As will be evident, cannabis alone accounts for four-fifths of all convictions but as there are reliable estimates of some 2 million misusers in the UK, only about one in 50 is sentenced.

DRUG MISUSERS IN TREATMENT

Addict notifications

As with enforcement statistics, notifications of addiction give a very partial view of drug misuse. They are derived entirely from doctors' reporting cocaine or opiate addiction to the Home Office. Other types of drugs are not covered and only addicts seen by doctors are included. It is estimated that only one in five opiate addicts is notified. In 1993 over 25,000 addicts in England were notified (41% were new notifications); 60% were in their 20s. Heroin accounts for 68% of all notifications, though this has fallen by 17% since 1990.

An important new source of information has appeared with the development of clinical drug misuse databases in England, Wales and Scotland. The first English national report was published by the Department of Health in 1994, limited to the six months between 1 October 1992 and 31 March 1993; 17,800 drug misusers started treatment either for the first time or after a six month gap (male:female ratio of 3:1). Over half the clients were in their 20s, with 40% under 25; 96% were in the age range 15–44.

Heroin was the primary drug of approximately half of the clients. Fifteen per cent reported methadone as the primary drug; 11% amphetamines; 7% cannabis; 4% benzodiazepines and 3% cocaine.

Thirty-eight per cent of clients reported injecting their main drug, while 60% of heroin misusers were injectors. There was an increase in those known to be injecting benzodiazepines (5% to 10%) and cocaine (11% to 22%), when these were used as a subsidiary drug.

The age distribution of those covered by this first report is of some interest: under 15, 1%; 15–19, 11%; 20–24, 27%; 25–29, 28%; 30–34, 16%; 35–39, 8%; over 40, 16%. The age, gender and type of drug misuse of problematic drug misusers attending agencies are of particular interest as these data can be compared to recreational and criminal misusers to identify differences between these drug-misusing populations.

It is necessary to repeat the caution here against overgeneralizing from local surveys as there are clear variations in type, method and patterns of drug misuse from one locality to another. This is also true of different countries; for example, Wales reports more amphetamine and benzodiazepine misuse than the south of England and further afield, the American increase in crack cocaine has not taken place in Britain.

DISCUSSION

It will be clear from the above statistics that there are different types of drug misusers, from the experimental and recreational to the problematic and dependent. The surveys carried out identify different types of misuser in different environments.

If the statistics concerning general populations are compared with those relating to problematic drug misusers in treatment, it can be seen that these constitute two distinct (though overlapping) populations, the former being a far larger group than the latter. The differences between these populations are explored in the following chapters. The police statistics also highlight a larger population than those attending treatment services. This group is more closely correlated with actual misuse in the general population among young men in this age group (see Chapters 5 and 9).

These statistical data have implications for both policy and practice. It is unfortunately not yet clear what proportion of all drug misusers continue misusing for more than a few years, have problems or end up in treatment. However, it is clear that many only misuse drugs occasionally for short periods in their youth and, whilst they may take risks, do not have serious problems or become dependent.

As we have seen above, there are recent strong indications that the misuse of cannabis, LSD, Ecstasy and amphetamines is becoming widespread among young people. The ISDD Annual National Audit of Drug Misuse (1995) stated that the extensive misuse of cannabis reflected its relative social acceptability in 1992 when 'Britain's enforcement services decided not to prosecute half of all cannabis offenders they apprehended' (p. 10). This reflects the development of cautioning schemes and arrest referral schemes across Britain. The report concluded that, although heroin misuse was growing, it was not on the same scale as other types of drug misuse and that it may be due in part to increasing notification activity by GPs, remaining fairly stable at 100,000 people. The report also notes that cocaine misuse seems more widespread but irregular in character rather than habitual or dependent. Perhaps of greater significance is the apparent rise in amphetamine misuse alongside other recreational drugs such as MDMA (Ecstasy) and LSD. The misuse of these stimulant/hallucinogenic drugs is clearly associated with social activities such as raves and all-night dances.

Fashions in drug misuse change over time: in the late 1980s and early 1990s a change was identified by an ISDD Audit (1991) which concluded that:

> Converging reports from enforcement sources, researchers and drug agencies confirm a new wave of stimulant and psychedelic use by young people, associated with the now hugely popular 'rave' dance culture that grew out of the late 80's dance culture. Already agencies are seeing the effects of amphetamine induced

'burn-out' and the 'bad trips' and confusion sometimes caused by hallucinogens like LSD and Ecstasy.

These reports are confirmed by Newcombe (1991), Pearson *et al.* (1991) in Britain and Solowij, Hall and Lee (1992) in Australia. These drug misusers are variously described as 'A group dedicated to lively enjoyment ... requiring a different treatment and educational approach from opiate addicts' (Gilman, 1991); 'A different type of drug user to the older heroin addicts ... with problems of stimulant and hallucinogen use'; or 'Happy consumers ... There could be at least a thousand and perhaps as many as two thousand at any one time' (Fraser, 1991). Finally, a recent report indicates that drug misuse may be widespread amongst teenagers as well as young adults. Collating data from a sample of 750 young people aged 15–16 in the north of England, Measham, Parker and Newcombe found that 71% had been offered drugs, 41% had taken cannabis and 33% had taken LSD. They reported that 50% of this group had experimented with drugs and 30% were regular misusers (Measham, Newcombe and Parker, 1993; Parker, 1993).

It is possible that the literature focusing on special populations, such as the mentally ill and the prison and probation populations, may be misinterpreted if it is taken out of the context of drug misuse amongst young people as a whole. It is significant that much of the recent literature generally reports an increase of stimulant and 'dance drugs' rather than opiates. Data regarding these particular drugs are not so readily available from Home Office sources.

AMERICAN STATISTICS

The conclusions in the last section are derived from an examination of British survey material, much of it very recent. It appears that similar conclusions can be drawn from the US experience. The National Institute of Drug Misuse (NIDA) relies on three sources of information: a national household survey, data from hospitals on drug-related admissions and surveys of high school seniors (this excludes police and prison figures). As Gustavsson (1991) points out, these figures may well underestimate the extent of drug misuse in particular populations. The NIDA figures for misuse of drugs in the US in 1988 indicate that, while 37% of the population over 12 years old have misused cannabis at some time, less than 1% have used heroin or cocaine. Chasnoff (1989) has estimated drug misuse among expectant mothers in both private and public clinics. Although figures on this group are notoriously inconsistent, he found that 16% and 13% respectively had misused drugs and estimated that 375,000 infants would be exposed to drugs in the US annually.

It is important to point out that, in common with British surveys, the data collected in different surveys can lead to differing conclusions. For example, two major population studies suggest a reduction in drug misuse from 1970, whereas a study of drug-related deaths shows a recent increase (Harrison, 1992). Harrison argues that, despite methodological problems in comparing different survey data, drug misuse as a whole is decreasing in the US; this is, paradoxically, in conjunction with increased public concern in national polls.

SOCIOLOGICAL RESEARCH

Sociological researchers can be very different in their perspective and methodologies, yet provide valuable insights into drug misuse. One of their starting points is that, as several authors have demonstrated, most studies of drug misusers are based on clinical studies or research with misusers or former misusers in treatment programmes (Waldron, 1969). It is clear that these drug misusers are more likely to suffer from problems or dysfunction than those who do not attend programmes (Ellis and Stephens, 1976).

Two types of research have recently helped to offset this imbalance: first, surveys of large populations in Britain and America and second, ethnographic or sociological studies of drug misusers who do not attend services. From this work we can begin to see how inadequate research of clinical populations has been. Survey work and sociological studies indicate that non-problematic or recreational drug misuse is more common than previously thought, particularly among young people. Moreover, drug misusers can be seen as quite normal, educated, employed and married with mortgages.

As long ago as the 1960s, Winick (1991) argued that drug misuse does not necessarily affect the ability to work and can enhance it. He cites research illustrating this, from early work such as Dai (1937), who showed that four-fifths of the addicts he studied were in conventional occupations, to a later study by the National Institute of Mental Health showing that 30% of a population of 12,000 were in employment (Waldron, 1969).

Assessing drug misuse in the general population and among non-therapeutic, non-criminal groups is very difficult. However, several sociological studies and surveys have been carried out recently which make a useful contribution in this challenging area. Frank, Marel and Schmeidler (1984), for example, demonstrated that high-income families misused more drugs than any other income group. They did not find any increase in levels of dysfunction amongst them. Other sociological and ethnographic studies have indicated that the life of drug misusers can require many different skills similar to those encountered in business. For example, Preble and Casey (1969), in their paper 'Taking Care of Business', outlines the difficulties inherent in the lifestyle of an illicit drug user/dealer. Gould *et al.* (1974) describe similar challenges in the life of a drug 'hustler'. In contrast, Stimson (1973), Zinberg (1984) and Cohen (1989) have described the everyday life and tasks of conventional drug misusers, often in legitimate employment, who may misuse drugs to help maintain their lifestyle and jobs. Stimson also describes how drug misusers will usually mature out of drug misuse as they reach middle age and continue conventional lifestyles. These findings are replicated in studies of drug agencies where the average length of drug misuse prior to admission at the agencies is up to nine years, indicating that it may not have been problematic before that time.

Parker, Baker and Newcombe (1988), while finding some American epidemiological studies of drug misuse valuable, consider that much of the sociological and criminological literature in this field has been distinctly unhelpful.

> Surprisingly, British sociology and criminology, routinely adept at providing such frameworks, have not been instantly helpful on this occasion ... Studies of drug use conducted during the 1970s tended to focus on middle-class cannabis use, accurately portrayed as victimless. Societal over-reaction was also documented in relation to 'moral panics' of adults coping with youthful deviance and the social 'misconstruction' and thus mismanagement of subcultural responses by working class youth' (p. 31).

These authors also criticize the medical literature on the grounds that it is based on clinic populations of 'addicts' very different to those drug misusers who do not attend clinics or seek help.

To recapitulate, the information collected from different types of surveys and research demonstrates that there are different types of drug misuser. It is therefore necessary to distinguish these different types before deciding on the best intervention for each. A good way to determine the similarities and differences between types is simply to understand the different reasons why each misuse drugs and the problems associated with them. (NB: All individual drug misusers will, of course, be different in some respects but useful categories have emerged from research which allow us to tailor our interventions, rather than simply treat all drug misusers in the same way.)

METHODOLOGY

The material employed in the next four chapters is taken from a series of research studies using methods taken from social anthropology and the well-established British tradition of ethnographic fieldwork (see Ellen,

1984). The essence of this approach is that the categories used, the concepts deployed and the theories constructed arise from the data themselves rather than being created as abstract theoretical entities which are then applied to the material to see if it fits (Glaser and Strauss, 1970).

Apart from the sociological research mentioned above, the greater part of the research undertaken into drug misuse lies within the ambit of the medical and psychological disciplines. As such it is positivistic in its approach and is based on the fundamental assumption that 'addiction' can be defined as a discrete phenomenon to which the scientific experimental method can be applied. It follows that the nature of drug misuse can be precisely specified, its treatment determined and outcomes measured. This ideal model has driven many research initiatives and, even if such activities have not provided definitive results, energy and resources continue to be devoted to further investigations. There is still much confusion about what exactly drug misuse is and what ought to be done for drug misusers, which is reflected in the lack of agreement about the definition of both the problem and the successful outcome of treatment. In basic terms this means researchers and clinicians cannot agree on what to measure.

The perspective adopted in this book takes it for granted that this scientific approach is not sufficient in itself, that its methodology has significant limitations and that it is by its nature unable to deal with phenomena that are crucial to the full understanding of drug misuse. In particular, the influence of variables such as subjective understandings – of both clients and professionals – and individual developmental processes is seen as central.

In summary, the advantages of this methodology for both clinicians and researchers lie in its potential to explore in depth both individuals and groups without the preconceived categories inherent within scientific studies. It permits the exploration of individual differences, subjective interpretations and is potentially dynamic in its ability to examine the developmental process over time. The methods used include questionnaires and structured and semistructured interviews with drug misusers; non-participant observation in informal settings and 'passive' observation (Spradely, 1980, p. 59). These methods allow an understanding of drug misusers' views in terms of their own concepts and beliefs (Lofland, 1971) and are particularly useful in exploratory research. As Spencer and Dale (1979, p. 697) observe, such methods provide 'many opportunities for developing and refining concepts rather than relying on the assumption that we know the relevant properties and categories in advance.'

The greater part of the information in the following chapters was obtained from interviews, self-completion questionnaires and participant observation. The questionnaires were distributed to subjects by various professionals and volunteers in drug misusers' own homes, in nightclubs and raves, in prisons and in drug agencies. Workers were present to offer advice and information about questionnaire completion. Subjects did not receive payment for completing interviews or questionnaires, which were designed to collect basic information about subject characteristics, drugs misused, patterns of misuse and methods of ingestion, in conjunction with data concerning social context and personal attitudes to drug misuse.

This informal approach used with respondents who have no experience of research and may be suspicious of questionnaires makes it all the more important that their independence and privacy are respected. The quality of the data was assured as far as possible by standardizing the methods of data collection. Information about questionnaires was available to all volunteers and clear instructions to respondents for completing questionnaires were contained within them. Unrecorded responses and non-responses were treated as missing data, as were spoiled or unclear responses.

Self-reports of behaviour and attitudes have formed much of the research into drinking and drug misuse (Williams *et al.*, 1985; Czarnecki *et al.*, 1990; McMurran, Hollin and Bowen, 1990). The self-report measures used in this study are, of course, subject to errors, as are all questionnaire/interview procedures. Previous work has tested the reliability and validity of self-report measures (McMurran, Hollin and Bowen, 1990; Czarnecki *et al.*, 1990; Maden, Swinton and Gunn, 1992; Skog, 1992). Skog points out that there are biases in sampling due to selective non-response and the possibility that respondents do not give accurate reports. Response rates usually fall well short of 100% and often underestimate drug misuse. The frequency of false positives is generally held to be less than false negatives (Single, Kandel and Johnson, 1975; Barnea, Rahav and Teichman, 1987).

There are, of course, many weaknesses in the qualitative approach; its inability to produce testable hypotheses, develop replicable experiments or control variables has to be recognized. It is therefore not usually considered possible to generalize from small groups to larger populations (although Yin (1989) and others refute this assumption and give detailed accounts of analysis of multiple case studies which allow generalization). The intention here is simply to use the qualitative approach as an initial exploratory tool prior to an analysis of the scientific and quantitative research literature in Part Two. Although the methods derived from the ethnographic perspective do not confer scientific credibility, they give an indepth understanding of the phenomenon of drug misuse.

In conclusion, it is worthy of note that, although scientific research has virtually abandoned any interest in qualitative data, they are still of relevance to clinicians. Clinical knowledge and professional experience are themselves based on deep understanding gained from listening to the problems of individual clients. The case studies of psychiatry and clinical psychotherapy are in essence products of subjective interpretation. Diagnosis and treatment are often based as much on individual casework and experience as research findings and many clinicians see their activities as a professional art rather than the unthinking application of scientific procedures. Part One goes some way to explaining why professionals may find understanding of individual experience as useful as the scientific evidence presented in Part Two.

Talking to recreational drug misusers
2

INTRODUCTION

This chapter will examine the views of respondents concerning the normality and relative safety of their own drug misuse, their reasons for misusing drugs and the possible risks and consequences. As we saw in Chapter 1, the growth of a 'dance culture' in Britain during the late 1980s, a development that has accelerated significantly during the 1990s, has transformed the misuse of drugs among many adolescents and young adults. All the reports in recent years confirm that the misuse of a wide variety of drugs at raves is commonplace and taken for granted by those who attend these events. To a lesser extent the nightclub, an older but changing form of entertainment, has witnessed a similar transformation. This chapter analyses data from observation and interviews with drug misusers at nightclubs and raves.

DRUG MISUSE IN RAVES AND NIGHTCLUBS

The original 'house party' or 'pay party' rave has changed in the past few years to become more a part of 'normal' club culture. Some estimate that between 'twenty and thirty thousand young people go to raves every weekend' (Newcombe, 1991; Pearson *et al.*, 1991). Newcombe carried out research into dance drugs in the north-west of England and found that the rave scene had changed since its initial emergence with all-night illegal warehouse parties to a more regular series of nine-to-two dance clubs.

In 1989 the Home Office advised local authorities to adopt the 1967 Private Places of Entertainment (Licensing) Act. This required local councils to license large private events in the same way as public events. The Association of District Councils acted on this advice and as a consequence, local councils could effectively close the legislative loophole that had previously allowed all-night raves to remain legal. Police and local authorities may now respond to unauthorized raves and house music clubs by reference to licensing laws. At the same time the rave has changed its status from a small underground illicit culture to embrace a much wider section of young people, including those mainly interested in the music and those caught up in drug misuse. Many nightclubs now play rave music only or, more commonly, will play particular types of rave music on particular nights. Similarly, raves can be organized as 'one-offs' or as regular annual occurrences.

The overall atmosphere and experience at a rave is constructed by loud captivating music, psychedelic lighting, use of dry ice or smoke, an almost tropical climate and a large mass of dancers. Perhaps this can be best described as the difference between attending a New Year's Eve party and a quiet drink at the local pub. Newcombe (1991) describes the unique setting of raves as 'an essential component to spark off the orgasmic trance dance atmosphere sought by ravers, described as "mental", "happening" or "kicking"'. The house music played continuously at raves is the central feature of the experience. Its notable features include a fast beat, rhythmic baselines, eerie sound effects and nostalgic snatches from old songs. Most house music is either instrumental or has minimal lyrics, the latter often based on repetitive chants that can be construed variously as meaningless, meaningful or mundane.

This chapter describes the misuse of recreational drugs as part of the dance culture, largely through the eyes of the participants. As we have seen, young people may start misusing drugs to gain access to a particular social group and acquire status within it. In time they will learn to appreciate the inherent properties of the drugs and develop their potential within this social context. Some young people may misuse drugs simply for their utility in enhancing sociability, others because the physiological effects are intrinsically satisfying. Whatever the reasons, and they may be complex, the milieu of the rave and the nightclub provides a powerful stimulus for experiment.

The data used here were obtained through a study of 206 young men and women who attended raves and nightclubs in one locality in south Wales in 1993 over a period of seven months. The research was undertaken in 30 regular raves and clubs and a few occasional events. Although there is much information available on those drug misusers who attend agencies (characterized in this book as problematic), there has been relatively little research on recreational misusers who have not developed problems or dependencies.

The material was obtained through interviews, conversations, questionnaires and non-participant observation. Questionnaires were distributed to as many young people as possible by project workers and volunteers in raves and nightclubs and collected back on the same night.

Advice and information about completing the forms were available. The respondents did not receive payment. Interviews were also conducted by project staff and volunteers. The respondents were not a representative cross-section of people attending raves and nightclubs. Nevertheless, the results of the quantitative analysis of the data are very similar to those reported elsewhere (see ISDD, 1994a).

A BRIEF QUANTITATIVE REVIEW

The great majority of subjects (about 70%) had attended a nightclub in the two months before they took part in the survey and about one-third went once a week. More than two-thirds had attended a rave during the previous two months, of whom a quarter attended at least once a week. It also seems that going to both clubs and raves was a frequent and regular occurrence.

Overall there are few differences between different age groups or between rave and nightclub attenders. This regularity of response across variables indicates a consistency in the data as a whole. The demographic analysis indicates that more than a third of the group were women (37%), a finding that is consistent with reports from studies in Scotland (McDermott, Matthews and Bennett, 1992). Men predominate in the agency clientele, constituting some fourth-fifths of those who attend. The mean age of those who participated in the survey was only 19 years and only one in six was over 21, which is substantially younger than found elsewhere. They are also much younger than those who frequent agencies, where the mean age in Wales at the time of the research was 26 years (Welsh Office, 1993).

The data indicate that within these clubs there are people who misuse fairly large quantities of various drugs on a recreational basis (87% of the respondents). Four-fifths of them misused cannabis, about three-quarters misused amphetamine and two-thirds misused Ecstasy and LSD. In contrast, the numbers misusing 'hard' drugs were much smaller, with about a third misusing cocaine and only one in 10 misusing heroin. During attendance at raves and nightclubs, however, the dance drugs Ecstasy, LSD and amphetamine (and to a lesser extent cocaine) were most common. It has been suggested that Ecstasy is almost exclusively a dance drug (see Henry, 1992).

These young people misuse a wide variety of drugs before, during and after visiting a rave or nightclub, often misusing stimulant drugs before going out and depressants such as benzodiazepines or alcohol after a night out. The respondents did not consider their own or others' drug misuse as unusual. Approximately three-quarters of the total group thought it was normal to misuse drugs at a rave and about half that this was normal in a nightclub.

Drugs misused by the rave attenders were different to those of agency clients. Heroin and injecting were far more common among the latter. Although amphetamine was common to all groups, it was more frequently injected by those who attended drug agencies.

Although the difference is fairly small, younger attenders misused proportionally more drugs and the type of drug misuse varied slightly depending on age. The frequency of cannabis and amphetamine is similar across age groups, but LSD and Ecstasy misuse are more common among the 18–20 age group and heroin and cocaine among older respondents. This is also reflected in the small numbers who inject (7%), these being found more commonly among the older respondents.

In summary, the drug profile of recreational misusers briefly illustrated here differs significantly from those of drug agencies' clients in this area during a similar period. Although the most popular drugs misused by agency attenders were also cannabis and amphetamine, LSD, Ecstasy and cocaine were far less common. The number misusing LSD were increasing in the agencies (21%) but this was still much less than the 60% of the recreational misusers. The difference is even more pronounced with Ecstasy and cocaine; agency numbers (4% and 7% respectively) were very low, possibly suggesting that these drugs are less problematic.

As we have seen, by the age of 21 attendance at raves/clubs has reduced markedly and only four people were over 25. There was no-one over 30. There are very few data on this younger group of recreational misusers who have not developed problems or dependencies. However, it should be remembered that some of this group may well continue to misuse drugs elsewhere as they get older. There is no evidence at present as there are no longitudinal or cohort studies of this very recent phenomenon.

QUALITATIVE DATA: THE DRUG MISUSERS SPEAK

The remainder of this chapter is devoted to reporting and analysing the attitudes and opinions expressed by the young people who attended raves and nightclubs. As we shall see, a clear picture of drug misuse emerges which characterizes it as normal, commonplace, fairly safe and far removed from the popular conceptions of addiction.

IS DRUG MISUSE NORMAL AT RAVES AND NIGHTCLUBS?

It is important to bear in mind that most of the young men and women who participated in this study misuse drugs. Their views are therefore very likely to be consistent with a general justification of their own behaviour: they are not on the whole defensive nor do they see themselves as deviant.

Raves

The rave is a social event; attending it is a means of both entry into a social group and establishing one's credentials. This is achieved by what is perceived as behaving normally:

Drug use is normal because it is part of the rave culture.

A lot of people who go to raves take drugs, but it doesn't mean that everybody does. Some go and get high on the music and atmosphere.

It seems that everybody there has taken something, as everybody is so nice and friendly to each other. You don't usually see that in nightclubs. It also helps you appreciate the music, colours, people, setting, etc.

A person does drugs to enjoy themselves, relax and generally feel at one with everybody else. Normally they do that at a rave, but hardcore nightclubs are a drug-must. To do drugs you must essentially be a person who loves socializing and generally enjoying yourself. You cannot be an at-home *Eastenders* addict!

The event, the enjoyment and the misuse of drugs are therefore seen as inseparable.

Nightclubs

The portrait of nightclub attendance is somewhat different. Drug misuse is seen as less widespread, competing with alcohol and with other forms of socializing. It is, as it were, a less sharply focused occasion with a wider clientele:

Yes, people use drugs in most places.

Yes, people also take drugs at nightclubs and pubs, but to a lesser degree than in raves.

Not as much as raves. But rural areas don't have as many raves, so people will therefore go to clubs and do drugs. There seems to be a lot more people doing drugs than there used to be.

Nightclubs are different in their approach and purpose to raves. Many of the people who go to clubs are only looking for nice clubs, cheap drinks and a place to pick up a woman. All these things are not important in raves ... it's the music and company that matters.

TYPES OF DRUGS USED

As we have seen in the quantitative review and from reports of drug misuse in other parts of Britain, stimulant and other dance drugs predominate and this is accompanied by quite clear distinctions between these and other drugs. The nature of the entertainment and the physical demands of prolonged participation militate strongly against opiates and depressants.

> Drugs are very much part of the rave culture and it is quite normal and acceptable for people to do drugs like E's (Ecstasy), Acid (LSD) and cannabis.

> I'd say about 50% of the people I know who go to raves take some drugs which includes Ecstasy, speed, coke or acid. It helps people have a good time.

> Yes, you use drugs because it gets you hyped up and you can dance all night. Mostly E's and LSD ... it helps with the lights and atmosphere.

> It's less common to use drugs at a nightclub, but a lot of my friends will take some speed before going.

> Most ravers only do wiz at a nightclub due to the amazing 'talky head and pulling power'. At a nightclub, a drug-free night is a no-no.

It is important to remember, however, that use of dance drugs at raves or nightclubs may only be a part of total drug misuse:

> I've been using since I was at school (mainly cannabis). Now mainly speed and E's.

> I've been using speed for about two years, but after trying Ecstasy there is no comparison. E is more powerful and puts you more in touch with yourself and everybody around you.

> I've used drugs prescribed by my GP and illegal ones, I've enjoyed taking them, I use in a safe way and therefore it's acceptable, I've not harmed anyone. I use dope (cannabis) regular, I was brought up on it. Also E's occasionally and speed every weekend. I only tried acid once and don't use coke any more.

> Yes, E's, acid, coke, benzos, alcohol, blow (cannabis). I've learnt from my own experiences when I was younger and through experimenting with friends. However, now I know the drugs I like

and stick to these. I have injected in the past (a few times) and I now feel disgusted with myself ... never again. I've learnt that injecting is not for me. I also stopped using so many E's as too many do your head in.

IS DRUG MISUSE SAFE?

The conventional wisdom is that drug misuse is extremely risky, that normal people do not take unnecessary risks and that, however exciting it may be, drug taking is dangerous and may well lead to dependence. This view is, of course, applied to illegal drugs but much less often to tobacco and alcohol. The young people in the survey had clearly given thought to such matters and were aware of potential risks, but on the whole saw them as a problem for the young, inexperienced and ill informed. In this regard they are probably no different from those who think they can drink and drive safely or who believe that the dangers of cigarette smoking do not apply to them. When prompted to enlarge on this issue, however, the element of risk is clearly recognized, only to be dismissed as worth taking and not a major preoccupation:

> If they know how to use safer they probably will. But there is a lot who don't know what they're doing and need to be educated.

> People are usually aware of the risks involved with drug taking, if they don't try to reduce the risks they are stupid.

> I suppose there are risks involved, but when you're young you don't care.

> It is mainly kids who are just starting to experiment who don't know what they're doing, therefore they're likely to use dangerously.

It is worth noting that nearly one in eight of those who participated in the study were aged 18 or less.
More specifically on danger and risk:

> Most people are aware that there are dangers involved with drugs, but there are ways of doing it safely. You're never going to stop people using drugs, the aim should be to encourage people to use safely.

> It is very dangerous if they don't know the risks involved and are stupid about taking too much. They could really mess up and have no-one to go and sort them out. People try and make it safer by doing it in proportion.

> Most ravers are aware of the risks in taking dance drugs, but most are sensible and know about their drugs. OK, you get the odd person who is determined to make a prat of themselves by taking loads of drugs but these are the exceptions.

> People usually test drugs out first to see what happens. They dab speed and if they cope and like it, they'll go on to snort.

> Yes, people are bound to take them safely, no-one wants to catch diseases like AIDS, etc. Only a minority use unsafely, these are heavily addicted and they don't care.

Moreover, many are convinced that much of the risk can be easily avoided:

> There are a lot of risks, but a lot of them can be reduced, e.g. heat-stroke ... plenty of water and chill out.

> There's the obvious risks of overdoing it and also drugs that have been spiked with something else.

> There are risks involved with all drugs but with E's the main ones are dehydration and overheating. All you need to do is avoid booze and chill out regularly.

Or that it is exaggerated:

> You hear about people dying from Ecstasy but in comparison to how many are using every weekend, it's only a very small figure which is blown up by the newspapers. It's mainly the pushers' fault, they're not ravers, they don't care and they want cash, they don't care what they sell.

IS IT WORTH IT IT? WHAT ARE THE BENEFITS?

The statements given above demonstrate an unequivocal awareness that the misuse of drugs is risky and that, although precautions may be taken, there is always some danger – even if that is seen as mainly affecting other people. What is equally clear, however, is that the great majority of these young people believe that the risk is worthwhile:

> People tell you that E's can kill you and are really dangerous. I've taken loads of them and have suffered no side-effects (apart from being knackered), therefore it's definitely worth it.

Like most things in life you've got to weigh up the risks. The risks with taking E's can't be any higher than crossing the road, OK so long as you observe how to do it properly.

Even though I'm aware of the risks I think it is worth it. If I didn't go raving my life would be well boring with nothing to look forward to at weekends.

The list of benefits is, not surprisingly, rather long. Drug taking allows people to relax:

Drugs allow people to be themselves, relaxed and friendly.

I look forward to every weekend so that I can relieve the stress and tensions of the week and go for it.

Gives you a good head, you become very happy and relaxed.

It has a marked effect on sociability, it allows young people to mix freely, instills confidence and creates friendships:

Like alcohol it's a social thing, as all my friends do it and it's not too expensive.

Gives me confidence to go out and mix with other people more freely.

The atmosphere and feelings you get, everyone is so friendly, without any bad feeling toward each other that sometimes comes with drinking. Everyone is wrecked but totally in control.

To escape from mundane reality of everyday. It gives a far better and clean high than alcohol and it is far more social and brings out the best in people not the worst like alcohol can do.

People who are total strangers become your close friends overnight and that is what appeals to me.

I'm probably a nicer person because of E.

Above all, it provides 'a good time':

It just helps you have a better than normal time.

It opens up your mind and body to new experiences and helps you have a good time.

Enjoyment and fun and positive experiences.

Making the night completely amazing.

Enjoyment, fulfilment, a good laugh out with your mates. The music makes it better, the head rushes and the body shivers.

and a sense of great physical exhilaration.Almost all the respondents referred to the 'buzz', 'rushes' and 'out of body experience'.

In contrast to the benefits, the list of costs was much shorter:

Feeling spaced out and like a cabbage the following day.

Comedowns; feeling knackered; flashbacks: irritable; snappy.

Come down the next day, paranoia, lack of sleep.

Feel ill in the morning and whacked on Monday morning.

I'm like a wet rag the morning after.

Taking too much and suffering the consequences.

Felt dizzy.

Paranoia, bad trips.

A false impression of what is going on around you.

Your willy shrinks and you get spots.

Physical exhaustion, which is very easy when using Class A stimulants.

Downers, cramp, weight loss.

You lose your memory.

WHAT OF THE FUTURE:

One of the more striking aspects of the attitudes revealed by these respondents describing their drug misuse is its impermanence. They did not expect it to last, seeing it as a phase in their lives, an enjoyable prelude to 'settling down', starting employment and getting married. These reasons were cited most frequently as the most likely for giving up drugs; also mentioned were the effects on health and the influence of friends and family:

Getting married.

If I met someone who loved me and I loved them.

Having a kid would definitely make me stop.

I suppose if I were to settle down I would have to change.

Getting married and having kids. I want to set a good example.

Health, pregnancy, getting married.

My boyfriend and myself are well into going out at weekends and taking E's or wiz but if we decided to settle down to have kids, there's no way I'd take drugs then.

If I had a child, I'd give up Es, speed and acid but not hash (cannabis) as this is not bad for you.

Health, state of mind.

Seeing one of my friends die.

Getting a good job would make me stop due to safety at work, etc.

I'm getting a job driving buses so I'll have to stop.

When the brain is in demand, e.g. exams.

College commitments and sports ambitions.

If my family found out. It would really hurt and upset them.

My friends, when we decide to call it a day I think we all will.

Friends and social circumstances.

THOSE WHO DO NOT MISUSE DRUGS

While the great majority of the young people interviewed are regular and frequent misusers of drugs, a minority of those who participated in the survey said that they did not do so (13%). Their views are in sharp contrast to those reported above and can be summarized as hostile to drug taking, convinced of its ill effects and determined to resist the social pressures on them to join the majority.

Some people do take drugs, but that doesn't mean that everybody does it and that it's normal.

The majority of people must have been (using drugs) to listen to such music and dance to that crap.

I've always believed drugs to be dangerous, because you often hear of people overdosing. The governments of the world wouldn't be so against them if there wasn't any danger.

I've seen people swallow 3–4 tablets without knowing what they were – if anything is stupid and dangerous, that is. They don't know what is in those tablets.

None of my friends take drugs.

Some of those who had never misused drugs regarded them as extremely dangerous:

> Addictive, destroys the body.

> Can lead to addiction, this in turn can lead to crime to support the habit.

> Dangerous and a lot of people get hooked and die. They are not legal because they are dangerous.

> Drugs can affect the rest of your life, jobs, etc.

> There are definitely risks, overdosing; kills brain cells.

> If you inject you get HIV, hepatitis, etc.

> Addiction, ill health, going crazy (LSD), overdosing.

> Yes, otherwise we wouldn't hear about all the deaths caused in raves with people taking Ecstasy. Once you start it is a slippery slope.

> I value my life too much!

> There is more to life than thinking where your next stash of drugs is coming from.

> Death, prison, being ill, losing your friends and family.

DISCUSSION

This group of recreational drug users is part of a rapidly growing phenomenon which has recently been the subject of research. The literature on these young people is still, however, relatively scarce. They do not appear to have the problems of physically addictive drug misuse nor of injecting. They find it difficult to understand their own drug misuse in terms of clinical problems, but instead see it as relatively unproblematic and similar to alcohol misuse, except that they identify the risks involved. They often consider it as a brief period in their lives which will end when they settle down, get married, have children or take up regular employment.

Only a very small proportion of them injected drugs (8%). These figures are reflected in the qualitative data regarding attitudes to injecting; the great majority thought that this practice was unacceptable within their social group.

The issues of social and cultural context are important as these factors influence the safety of drug misuse. If social consensus were to prohibit injecting and encourage sensible drug misuse, it would have positive implications for both health and HIV prevention. It is possible that, for the majority, the dominant culture may condone amphetamine misuse if it is swallowed or inhaled but not injected. However, this is clearly not the case for all club attenders as injecting still takes place within some social groups. For those recreational drug misusers who do inject and do not attend agencies, it is less likely that they have access to syringe exchanges and more likely that they will share in a nightclub or rave than they would in other contexts. If this is how the misusers see themselves it is likely that oldfashioned help, advice and drug treatment will be seen as irrelevant.

The contrast with the behaviour of problematic drug misusers who attend drug agencies is very marked. In drug agencies in the same area, amphetamine and heroin were the most common problematic drugs and those most usually injected, more than a quarter of amphetamine misusers injecting and a third of heroin misusers. It is also relevant to mention that 53% of agency drug misusers reported injecting at some time and 33% in the past month. (Of those who had injected in the past, 49% had shared and, more significantly, 18% of those who had injected in the last month had shared; see Keene et al., 1993).

It is clear that the rave/nightclub group does differ from drug agency clients; they identify the risks they take and are open to help and advice. The difficulties are targeting the group, defining the group profile, determining the major risks and then offering acceptable forms of help. It is worthy of note also that research in the field of alcohol misuse (Harford, Wechsler and Rohman, 1993; O'Callaghan and Caalan, 1992) indicates that the social context of substance misuse and the reasons for misuse may be of particular relevance. If this is so, it has implications of prevention and health promotion projects for both alcohol and drugs. It may be necessary when targeting particular groups for prevention or harm minimization strategies to consider not only the general characteristics of the target group but also the beliefs and mores of that group:

1. The reasons people give for misuse may be more diagnostic than the amount or frequency of misuse.
2. Such knowledge makes it possible to target this group and it will also affect the content of any campaign. For example, there may be a priority in the group on relaxing and socializing.
3. Incentives and beliefs about drug misuse may act as indicators of behaviour. That is, it is not simply the costs of drug misuse in terms of risk or health which indicate future action, but also beliefs about its positive functions.

Although three-quarters of the respondents report that one of the best sources of information is their own experience and more than half say that friends provided information, it is interesting that more than 40% reported obtaining information from leaflets, suggesting that education and promotional products may well inform a large proportion of young people.

Respondents were often effusive about the positive effects of drugs but were also well aware of the costs of drug misuse. When asked to describe the bad things about drug misuse, the most commonly mentioned disadvantage was the 'comedown' the next day, causing depression, lack of concentration, irritability and feeling ill or tired. Many mentioned the expense and the unknown or bad quality of the drugs themselves. Only two spoke of problems of debt and violence related to their drug misuse. A small number spoke of health problems varying from generally feeling ill the next day to spots and weight loss and one 'the bad effect of speed depleting your body of calcium'. Several people mentioned the similarities with alcohol misuse and commented that although the hangovers from drug misuse were bad, they were not as bad as those to be had from drinking.

The subjective views of the drug users give an impression that they misuse drugs recreationally in a similar way to drinkers and for much the same reasons. The short-term risks are perhaps greater than alcohol and the after-effects often unpleasant. Most respondents were aware of the risks and were also aware that their drug misuse could be made safer if they had access to information and followed simple rules.

The viewpoints of this group of recreational drug misusers give an insight into a particular kind of drug misuse. Chapter 6 will examine the research literature on drug prevention and drug education and consider the appropriateness of interventions aimed at this group.

POSTSCRIPT – FOR RESEARCHERS AS WELL AS OTHER PROFESSIONALS!

It can be seen from the above data that drug misusers view their drug misuse in different ways to many researchers and professionals. This is best illustrated by the responses to a particular research questionnaire designed by the author. Subjects at raves and clubs were asked to fill in questionnaires to assess the respondents' attitudes to changing their own drug misuse before and after a prevention intervention (see Project Pitfall in Chapter 6). The results were the reverse of what might have been expected or hoped for, in that people appeared to be less likely to change their drug misuse after reading project literature. There may be several reasons for this, including the high turnover of nightclub/rave attenders and the incidence of new arrivals.

This result is interesting in itself as it indicates flaws in the content of the questionnaire, which was based on the premise that the drug misuse was problematic. Most respondents did not see their recreational drug misuse as problematic at all; they were therefore confused and did not understand the point or purpose of the questions in this part of the questionnaire.

This issue is discussed later Chapter 6, but it is relevant to point out here that it is not just professional interventions but also researchers and research instruments that can be inappropriate to this group of respondents. When undertaking this research, the author assumed that the subjective understanding of drug misuse in this group would be the same as other groups of drug misusers in which the substance misuse is described in terms of pathology or behavioural disorder. The data reported in Chapter 6 indicate that implicit assumptions in the consequent research design were at fault.

Talking to high-risk drug misusers

3

INTRODUCTION

This chapter will first examine daily and long-term patterns of drug mis-use, followed by reasons for misuse. It will then outline different kinds of risks and problems associated with drug misuse and finally consider what can be done to help.

In the same way as professionals need to distinguish between social drinkers, intoxicated drinkers, uncontrolled binge drinkers and depen-dent drinkers, they will need to distinguish between different categories of drug misuse. As with alcohol problems, the type of problem and therefore also the intervention will vary depending on the type of drug misuse. Unfortunately, there is far less information available about dif-ferent patterns of drug misuse and there are many different types of drugs. Therefore there are dangers in developing simplistic categories which become less than useful as more information is acquired and better conceptual categories and theories are developed. However, in the absence of more sophisticated knowledge it is pragmatic to base def-initions on those we already have from the alcohol field.

This chapter will give the reader a rather different view of drug mis-use from Chapter 2. The respondents here are talking about dangerous drug misuse, where there is a direct risk to their physical health and general well-being. This is not a black and white distinction. There is a continuum from recreational to risky drug misuse, though most do not travel the distance from one to the other, some make the journey very quickly indeed. For example, many users of cannabis, Valium or Ecstasy will be at no greater risk than a social drinker but those who misuse many different drugs in an uncontrolled or chaotic manner will be tak-ing risks more dangerous than those of a binge drinker and those who inject will be taking life-threatening risks.

It is necessary for practitioners to be able to distinguish the type of respondent reported here from those of the previous chapter, as the former are at far greater risk than the recreational misusers and inter-vention becomes a greater priority. The intervention will need to be designed to reduce the risks. It is also necessary for practitioners to be able to distinguish risky drug misusers from those in the following chapter on dependent misusers, as the type of intervention will again be different. For the risky misusers it is necessary to reduce the risks and the potential harm; for dependent misusers it is necessary to do more, by actually dealing with drug dependency itself.

Before going further it is important to highlight the exception that proves this rule. This is the group identified by respondents in the previous chapter as most at risk – young, inexperienced people trying drugs for the first time. It is, of course, possible to take serious risks when misusing drug recreationally on a regular basis, but this is rare. In contrast, the short period when people first start to misuse drugs can be very dangerous as they are particularly vulnerable, largely as a consequence of their ignorance about drugs and about their own individual responses to particular ones. They will not know, for example, what their tolerance levels are or whether they are physio-logically vulnerable to certain drugs. This will be less of a problem if they are part of a responsible social group who will educate them about drugs (such as some of the individuals in Chapter 2) or if they receive timely education from other authorities. If no information is available they may take unnecessary risks and it will take time before a drug misuser learns from experience their own individual responses to particular drugs.

The more experienced misusers identified this vulnerable group of experimenters as most in need of information and advice. It is possible that they require more than this: practitioners may also need to include these risky experimenters with other high-risk misusers and consider more concrete ways of reducing potential harm.

This chapter, then, gives a different picture of drug misusers; the respondents are those who made it clear that their drug misuse was risky, chaotic and uncontrolled. This was often because they misused a wide range or large quantities of drugs and/or injected them, often with few sober periods. They were at risk from overdose and other physical complications but perhaps most significantly, they were particularly at risk from HIV and hepatitis.

These respondents were drawn from different research projects over a period of several years and interviewed in many different environ-ments, ranging from their own homes, cafés, pubs and nightclubs to syringe exchanges, drug agencies and in prison. The respondents were all asked for basic information about themselves and their drug misuse. They were also asked why they misused drugs in a dangerous way, how problems developed and what kind of help they wanted.

Recreational misusers at nightclubs tended to restrict themselves to cannabis, amphetamine, benzodiazepines and alcohol. Those at raves used more halucinogens (LSD and Ecstasy) and only used benzo-diazepines to 'come down' after the event. In contrast, the high-risk respondents misused a wide range of both stimulants and depressants on a more frequent basis, irrespective of the setting or time of day. It was difficult for many of them to give an account of their drug misuse in the past week or predict it for the days to follow; there was less sense of planned or controlled misuse. For some this was not the case at the time of the interview but they remembered recent chaotic periods when their misuse had been uncontrolled. Many had periods of controlled misuse interspersed with periods of uncontrolled or dangerous misuse. It will become clear that during the latter they need help and support.

TYPES AND PATTERNS OF DRUG MISUSE

It is relevant to mention here that particular drugs were associated with more or less harm; for example, Ecstasy, LSD and cocaine were seen as less risky or harmful than amphetamine and heroin. Usually the method of ingestion and the patterns of drug misuse would influence the perceived degree of risk. Polydrug misuse was more frequent among these respondents.

A day in the life ...

I use some drugs as uppers and some to help the comedown.

In the morning I take speed, like. I inject a lot, normally a gram at a time. (I am) late rising with Speed, that is, if I sleep at all. It depends on whether I sleep or not. If I sleep, I get up and have a shower, have a hit then (inject drugs). If I stay awake, I'll wait until I feel a bit psychosed and have one then. Usually within half an hour of waking up or if I'm up all night, about 9 o'clock. It's sort of wild when you're injecting because after the initial euphoria wears off you can't really share anything. If I've got quite a bit, I get a bit greedy and do quite a lot of it but if I've only got a little bit I ration it. I try to. I use Valium in the night, purely to go to sleep. I don't even wait for the buzz, I just let it knock me out like. About 10 o'clock at night. I'd probably be smoking blow (cannabis) all day as well.

Sometimes I have these patches, I've got pretty strong self-control normally, but it's beginning to get out of hand. I had my first withdrawal of amphetamine last week. I've never had that before. I didn't even know you could have withdrawal of it, it's not supposed to be addictive. I had a pretty psychotic head on.

Walking down the street thinking 'Please say something to me, anybody', I like it too much, frightening really.

I am scripted six tablets of dexamphetamine a day but only take five so I am left with seven at the end of the week which I may take but I may not take. At 10.00am in the mornings I usually take two tablets and two Valium. Between 6.00pm and 10.00pm I take another two and one Valium. After 10.00pm I take another Valium to go to sleep. Every day, from 5.00pm onwards I smoke blow. Last week was not a typical week as I was feeling paranoid. Most weeks are up and down but more down than up. The week before was a good week for me and I was not paranoid.

Sunday, a.m., I didn't really have anything. From 8.00pm to 12.30am I smoke a couple of joints throughout the evening. I smoke joints to relax. Saturday, 1.00pm, I had a line of speed, 3–5pm a couple of joints. At 7.00pm a line of speed. At 11.00pm several joints. We had a barbecue in the afternoon and went down the pub with several friends. Friday, 7.00pm a line of speed; 11.00pm another line, a couple of joints for recreational use. Thursday and the rest of the week, a couple of joints for recreational use from 8.00pm onwards. This is a fairly typical week for me, I did the same the week before and most weeks are similar to this. My drug use has got less over a number of years and I am now less likely to go out and look for drugs but I am more likely to stock in case I run out. I am not dependent on drugs. I use drugs for relaxation after a day's work. If drugs were easier to get hold of I would probably use more. I take drugs to make me feel good, I take speed because I can stay awake longer for recreational purposes. There is no real difference in how much I like the drugs since I first started.

My drug use at the moment has gone chaotic. I feel like I am losing grip sometimes. Sometimes I am feeling good and sometimes I feel like I am going back into a rut.

I periodically use speed and when I do I use quite a lot. My drug use has varied a lot since I started taking drugs. I now want dope more and more and I like it more and more than before. However, I do not crave it. It has been so long since I have taken E's and speed that I am frightened to use it again.

I sometimes inject speed three or four times a day depending on how stressed I am.

I normally smoke three or four days out of the week depending on company. I haven't been able to get any dope for about three weeks. My drug use has definitely varied since I started. I used to take speed almost daily and I used to smoke daily for about two years. I have had a couple of bad experiences and this has put me off. If I still lived in London, I would still be using and would probably be using harder drugs. I like E's less than when I first started taking them. With speed and dope I think the more you take the more you like it. I am not dependent on drugs unless I went to a nightclub. For example, I would be bored in a nightclub in London without drugs. I can't imagine not taking hard drugs again. I know I will do it again in my life but don't class myself as dependent in the classical sense of the word. After going through a period of regular use it seems as though that nothing was enjoyable without drugs. I have come close to overdosing on drugs and I have suffered from paranoia and anxiety problems after drug use. I am more likely to do it if I am with people who use drugs but even if I wasn't I would probably do it now and again.

Many respondents were polydrug users and would misuse a range of drugs depending on availability:

Anything I can get, that is available.

I'll use anything to get an effect.

I prefer cocaine, but will use amphetamine or anything else when I haven't got it.

I use heroin but will use other pills if it's not about or if worse comes to worst, I'll drink and use pills.

I use heroin, but the methadone will do if I can't get it.

It is clear that some of the respondents are misusing in safer and/or more controlled ways than others. This illustrates the tremendous range of types of risk. It is also important to emphasize that most of the respondents varied their drug misuse over time; they had periods in the past year and in the past several years when their drug misuse became heavy, uncontrolled or risky in other ways and other times when they regained control. This variation in types and level of risk and fluctuation over time makes it extremely difficult to include such drug misuse in one category, but it is essential to do this in order to identify those most in need of help.

Several themes emerge from the data. Different drugs are misused for different reasons, some drugs may be used as a stimulant and then others to counter the after-effects. Drugs may be mixed to increase the effect and more drugs taken when respondents are already intoxicated and less discriminating. Patterns of misuse vary but on the whole are different to those reported in the previous chapter. Instead of once or twice a week, regular daily misuse is more frequent, with many of the respondents having had periods of daily misuse in the recent past. It is important to point out that the more chaotic misusers were often unable to remember or report previous patterns of misuse. The majority of data describing patterns of misuse are therefore related to the less chaotic users.

REASONS FOR DRUG MISUSE

Respondents were asked what they thought were the good things about drug misuse and the reasons why they misused drugs. While some of the positive responses cited here are similar to those given by recreational drug misusers in the previous chapter, the respondents focused largely on different reasons. Although they emphasized the positive physiological effects of drugs, they were less concerned with hedonistic than with self-medicating functions. Recreational misusers were concerned with the social effects, risky drug misusers were not, for the most part. Indeed, if anything, drugs have an adverse effect on their social life.

HEDONISTIC REASONS

> I will have my speed after and I will be happy then.

> It is what I look forward to.

> I can't enjoy things without.

> Drugs guarantee a good time.

> Good fun.

THE PHYSICAL EFFECTS OF INTOXICATION

> The good thing about drugs is that drugs makes me feel good.

> Speed keeps you going.

> The effects, heightened perception, etc.

> Makes me happy and excited.

> Positive feelings.

USING DRUGS TO COPE

> I just can't cope without them.

> I need drugs just to get through the day.

I couldn't live without drugs, I never could.

When I'm uptight, the first thing I do is go for my tablets.

I couldn't function, I couldn't do anything if I didn't have them. I needed it more and more.

I have always needed them. I should have had them when I was a kid.

It started when my marriage broke down. I can't see the kids. I lost my job. I feel depressed and can't cope at all sometimes. I can't face people. I feel frustrated and want to break things without drugs.

Makes you have fun with friends and not be square and boring.

I got right on top of the situation and I packed in speed virtually. I was just doing it to go out, I packed in E's, all pills and all that, all I was doing was smoking the blow and I met a girl and she went away to work and I was all stressed out about it and she didn't come back and that was the worst time I've ever had. Felt as if I was upside down. I just couldn't handle being straight cos I was having like rages all the time. Just frustration I suppose cos there was nothing I could do to stop her going because I met her too late. People in the world today have so much trouble that it is normal to take drugs. Middle-aged people take benzos, like doctors.

Walking down the street straight and normal is a big difference, I can't do it.

After the detox I was more aware of what was going on around me. Before, I couldn't remember anything. I always used drugs to hide from things. I think I always will.

Aggression-wise, nobody would be able to come near me if I didn't have them (Valium).

If I have had a stressful day. It could relax you.

USING DRUGS TO SELF-MEDICATE FOR SPECIFIC UNDERLYING PROBLEMS

Because of depression. I take downers to take away the come-down and opiates to take away my body pain. I have broken all the bones in my body but I had them fixed because of a number of motorbike crashes and sometimes fighting. Because I can't get doctors I have to rely on tablets because of my emotional and physical state.

You think you can take the drugs but it's like painkillers, it doesn't get rid of the problem it just covers it up and it gets worse then but you don't notice it.

I think I am more anxious and depressed than most people, but it's a circular thing, I get like this when I use too much.

I haven't got worse because of the drugs, it is the other way round, I was always like this.

Without my amphetamines I am too tired and I have no power or energy to survive and must have my Valium to sleep at night, as I am afraid to go to sleep in case of fire in the house.

As the years progressed I became lonely and more often on my own, a recluse. My friends settled down and had kids but I didn't as I had a drink problem and went to jail a lot, as I was stealing to drink. I was mixing pills and drink, but when I was about 21, I realized that this was too dangerous so I decided to use 'calming' drugs such as cannabis. I stopped using pills at this time and only used cannabis, this I think saved my life as I used it as an alternative to alcohol.

My main problem is aggression. I've got a charge for ABH. I was taking temazepam and throwing the kids against the wall. I've been in hospital twice and discharged myself. I can't remember as good as I could. My wording is not very good. I started with solvents at the age of 13.

Reasons for misuse can be split into three categories – hedonistic, coping and self-medicating – although many respondents were confused between the causes and effects of drug misuse.

DRUG-RELATED PROBLEMS: RISKS, UNCONTROLLABILITY AND OTHER DAMAGE

We saw in the last chapter that recreational drug misusers had experienced or were aware of problems but, as we shall see, their responses were much less negative and serious than those reported below. Most of these respondents had experienced some kind of difficulty. The main fears were of contracting HIV or hepatitis and other health hazards. These were followed by uncontrollability and other more general damage to psychological and social life.

RISKS

Most misusers told of bad experiences, such as feeling faint or sick and having panic attacks, while many had more serious continuing paranoia and health problems.

> At one time I let someone inject me with amphetamines and I was taken to hospital as a result and almost died. I thought he was injecting me with street speed but I just don't know what it could have been.

> Had one dirty hit once, but I was lucky.

> I've caught hepatitis B from unsafe sex and I thought I had AIDS. But I never take any risks. I always clean spoon before and after jacking up. I have shared needles but I was lucky. Don't use same needle. I know so many dealers from prison I know who to trust. It's usually the ones that use speed themselves are the ones to watch.

> I take risks when I go collecting. I could be involved with the police. I could be caught dealing. I don't want to mess things up! I am a very sensitive person but my mind gets a bit evil sometimes.

> It depends on what you are taking. With some drug use there are only legal and health risks. If using drugs in a chaotic manner, then there are risks, for example, crime to get money, needle sharing; however, needle exchanges are open so this can be done safely. There is always the risk of AIDS, injecting into an artery, a risk of job loss, a risk of not being able to get visas, the risk of going to prison, the risk of having a prison record, etc. It is relative, it depends on the circumstances. For example, parents, job, etc.

> You always take risks when you take drugs because it is illegal and there are always health risks. It is not risky if you buy off people you know and don't take too much and stop when you have had enough. I have had problems associated with drug use. I have been busted for possession and have had health problems. I always smoke at home and I don't carry drugs with me. I have seen lots of other people with problems because of drugs and I have seen people being busted, people have died who have overdosed, suffered relationship breakdowns and beaten up.

> I think unsafe sex because of drugs is also a very big risk.

> Every time you buy it there is a risk of having bad drugs.

> There is always a risk of buying bad E's in a nightclub but it wouldn't stop me buying them, I would just take half first and see how it felt. If it was bad, I don't think it would kill me.

Yes, overdosing, stress, anxiety, paranoia, psychosis, etc., etc.

Overdosing, people also buy stuff that they don't know what it is mixed with. I was always getting ill.

Infections on the arms and legs where you inject.

Yes, because the quality is not as good as it was and it changes, so you do not know what you are taking.

You don't know how things will affect you, you might go unconscious.

You are sometimes too out of it to be sensible and take more stuff.

Taking precautions

Yes, but there are ways of minimizing the risks.

There are risks of HIV but these can be reduced if you listen to sound advice about safe drug use and of course safe sex. (Take a condom.)

Yes, if you know what you are doing. Never mix your drugs, buy from somebody you trust.

Is the risk worth it?

I think the risk is worth it as long as you know the risks and don't overdo it.

There are loads of risks somebody takes in life, a few more won't make much difference.

I'm aware of the risks but I think it is worth it.

HIV

Many respondents gave reasons for sharing needles and syringes:

I had no money for clean works.

The chemist was closed.

The chemist was too far.

I missed the needle exchange.

I was the only one with works.

We were stoned.

We were pissed.

I was in prison.

It can be seen that the respondents were aware that their drug misuse was risky, many giving graphic details about risks they had taken. But it is less easy to determine if they have an accurate view of the particular risks associated with, for example, misusing drugs while intoxicated and mixing drugs.

The replies to these questions were similar in some ways to those in the previous chapter but 'bad things' are worse and the risks taken more frequent and more serious. 'Addiction' was mentioned more often as a risk.

UNCONTROLLABILITY AND DIFFICULTY IN STOPPING MISUSE

Once you've injected you can't take it any other way. I want it more and more but I hate the drug now, I'm ashamed of doing it. I can't walk around with a T-shirt on (injection sites visible on elbows).

If I pick up any drug, whatever it is – alcohol, drugs, video games, gambling – I don't know when I'll put it down and it will mess up my life in the meantime. Something triggers me off, but I do not know what it is.

Valium – I can never give up but I can give up the rest. It is not just giving up the drugs but it is getting out of the drug circle. I feel like a coward. I am addicted to speed but can give it up. I have always prided myself on my fighting ability but I can't say no. Maybe, if someone wants a kilo of speed, part of me says great, but part of me wants to get on with my life and say no. I feel like a coward. I know from my heart I don't want to deal, I want to get on with my life and go to university. There will be a time though, when I will be able to stand up and say – no, I am not dealing.

Everybody has got a different tolerance and attitude to drugs.

It's getting less. If I smell dope, my head's not in very good shape at all – wouldn't it be nice to have some! I'm still addicted to the drug scene, but I don't have the compulsive urge to use every day.

There is always a temptation to do more than is necessary and some people ruin themselves if they abuse drugs.

Don't know what I'm doing sometimes.

GENERAL DRUG-RELATED DAMAGE

Because of the damage done by my drug use I cannot manage situations which in previous years I would have dealt with.

I have suffered from depression from taking drugs.

Afraid to go to bed in case I go unconscious and then am sick.

The loss of my marriage; wariness of friends. Unpredictability, that type of thing. Paranoia, people are wary of me. Family problems, in loss of communication with my family.

I've been homeless loads of times, slept in streets. But I've never been homeless here. My accommodation is not 100% but I'm not in a position to moan. People are worse off.

My main problem is needing drugs, getting skint and getting into trouble for robbing to buy them.

I have had paranoia, I suffer regularly from paranoia when on drugs but I also enjoy the way my mind goes overactive. I can sit and think about things. I think all drugs can cause paranoia.

I have had problems with taking drugs. When taking a lot of E's I experienced a low at the beginning of the week, after the weekend.The bad things are health problems, legal problems, problems with obtaining them occasionally.

(Since getting in control of my drug use) I've become more the way I used to be, instead of a paranoid, freaked-out wreck. I still get a bit paranoid, but that's to be expected. I feel like my old self again, I've got my confidence back, well some of it. I've put weight back on, I'm not a thief any more.

(When I was taking drugs) I had voices in the head, paranoia, fear. All that's stopped because I'm on methadone and because of the counselling. I was still having panic attacks up to six to eight months ago. They've stopped now. I'm finally getting my life back together.

I sunk all the veins in my arms. I used to do body building but I lost all my weight, all my muscle. I went down to nine stone from 13 stone.

I was doing criminal offences, burglary, to get money for drugs. It would have got me into trouble in the end.

Psychological effects

You don't feel good after.

Coming down the morning after.

Anxiousness.

Paranoia.

Panic attacks.

Depression.

Can't sleep.

Being sick and scared.

Messes up your mind.

Bad trips.

Paranoia.

Tired all the time.

Mood changes all the time.

Health

Spasms and headache.

I had abscesses and swollen-up legs.

You can get infections from using dirty needles.

You mess up veins and get veins blocked with bad drugs.

I've got scabs and sometimes get infected (injection) sites.

I feel ill all the time.

I had thrombosis and my leg swelled up and was extremely painful.

I got jaundice and went yellow.

We all got hepatitis before they gave free needles and things.

I used to use blunt needles and get lots of bad infections, but now I can get hold of clean sharp needles it is much less.

Overdoses.

You don't eat because you don't feel hungry.

Illegality

Getting busted.

Illegal.

Get into trouble with police.

Get record.

Finance, debt, intimidation and violence

I'm afraid all the time. I can't pay back what I owe.

I have to deal because I owe money to make it up.

You just get mixed up in things and owe people favours and then do things you don't want to.

There is a risk of falling into debt and getting threatened.

Falling into debt.

It's too expensive.

It costs too much money.

If something goes wrong you can't call the police like normal people.

Social life

Since I have started using drugs, I am liked less and less.

Loss of friends and loss of your personality.

You lose friends.

When you're taking drugs, your family worries about you. There's a social stigma associated with taking drugs. People look down on you – they turn their noses up.

You get more bad tempered when not stoned.

Zombifies you.

Makes you like a cabbage.

Waking up next to some ugly bastard (the morning after).

Girlfriend doesn't like it. I keep doing dumb things.

DISCUSSION

These two sections have examined the reasons why people misuse drugs and the damage associated with this misuse. Although it is clear that many risky drug users still misuse for pleasure this is complicated by their other reasons. Many say that they no longer get any pleasure at all, that the drug misuse is purely functional.

The misuse of drugs to cope or self-medicate may lead to more serious problems, as the misuser may become psychologically dependent on the drug as a remedy or solution and so misuse increasing amounts on a regular basis. This in turn leads to problems caused by the drug itself and the consequent self-medication of these problems by misusing more drugs. So, for example, respondents may misuse more drugs to compensate for the paranoia caused by initial drug misuse:

> I haven't been drug free for so many years, it just makes me paranoid. I used to be so out of my face.

Not surprisingly, this circular, interactive process of drug misuse as a solution and as a problem causes much confusion about what are the causes and what are the effects of drug misuse:

> I haven't got a clue (what the problem is). I went off the rails – I don't know why. I was crying, suicidal, taking everything, buying off the streets to deal with it. I went to the doctor, but he wouldn't prescribe – he sent me here (drug agency).

It can be seen that while some of these respondents misused drugs for hedonistic purposes, the majority took them as a remedy or a solution for other problems. It should be remembered that all of them had attended agencies of some form for help of some kind, even if this was simply clean needles and syringes. It is therefore possible that there are many risky drug misusers who do not see themselves as having 'underlying' problems, who buy syringes, etc. from pharmacists and therefore see no reason to seek help.*

This chapter is concerned with the views and beliefs of misusers who take drugs in a risky way and attend agencies of some kind for help. Of these, a number saw themselves as in the grip of some kind of compulsion, but most interpreted any problems as precursors of drug misuse and the misuse itself as a coping mechanism or form of self-medication. They were aware of the risks involved but considered these necessary evils. Some were confused about what was the cause of their drug misuse and what was a consequence. Many had periods of extremely unpleasant and chaotic drug misuse during which they took risks they regretted. They were afraid of getting infected with HIV and hepatitis or overdosing. Those who injected had often developed infections from unhygienic equipment and dirty injecting sites. Some got into debt and were afraid of violence as a consequence. Most fell into difficulties at

*For an account of risky drug users who do not use services, see Keene, Willner and James (1996).

times, but felt that on the whole they were in control of their drug misuse and simply needed basic health-orientated help such as clean needles and syringes and health care advice. At difficult times some felt the need for a prescription of substitute drugs to enable them to detach themselves from the street, to stop injecting or to stabilize and reduce their drug misuse.

WHAT CLIENTS THINK SHOULD BE DONE

The main requirements are provision of clean needles and syringes, health care, information about safer drug misuse, substitute drug prescriptions and help with social and psychological problems.

The following section examines the views of respondents about the kinds of help needed and that available for the problems identified above – health problems, risks of HIV and hepatitis, periods of chaotic uncontrolled drug misuse, underlying problems and problems of coping generally. Once the different types of problem have been identified, it can be seen that a different solution may be appropriate for each one. The respondents themselves suggest a range of possible solutions for different problems, from prescriptions to help stabilize drug misuse to basic health care facilities. The first section examines the previous attempts respondents have made themselves to change their drug misuse; the second considers the help they think they need from professionals.

SELF-HELP

Respondents were asked if they had changed their drug misuse in the past and the reasons why they had made these changes. Many had made changes and most were concerned with making their drug misuse safer, protecting themselves from HIV and other health hazards and bringing uncontrolled drug misuse back within their control. Self-help here also refers to service provision such as syringe exchanges where drug misusers can get the information and equipment to enable them to help themselves; people will not stop sharing just because they do not have access to clean needles and syringes.

HIV Prevention

I've stopped injecting, started dabbing.

I've stopped, I would never start injecting again.

I've stopped injecting temazepam.

I don't share needles now.

I use safe and clean works.

I don't share works, needles, syringes, spoons, anything.

I use the needle exchange.

I use clean needles.

Reduce sharing.

Less injecting, smoke or snort instead.

I clean my needles more.

Avoiding addiction or uncontrolled use

Cutting down due to high tolerance levels.

I've cut down.

I do not overdo it now as I know there is always a price.

I've been wanting to stop injecting for a while, but I don't think I want to enough yet. My girlfriend wants to stop too. I'm fed up with being ashamed of it.

I used to inject drugs, stopped two years ago. I was doing too much. I was stoned every day and night.

Change in use of particular drugs

I changed from opiates to amphetamine.

Cut way down off speed.

Ecstasy is duff because of all the shit in them now.

I changed from Ecstasy to amphetamine when Ecstasy was no good any more.

I changed to drugs that I knew were better quality.

Health care

I get help if my health is affected.

If my health goes bad I go to a GP, but I haven't got one at present.

I try to deal with the bad effects from prolonged use and watch for health deterioration.

I look after myself and slow down if I get ill.

I withdraw myself when it gets too much, I cut right down and get back in control.

People have swollen up hands and legs, you should use different veins.

You get lot of trouble with chest infections and lot of flu.

Hepatitis is quite common sometimes, you have got to be careful then.

WHAT PROFESSIONAL HELP DO YOU THINK YOU NEED?

It can be seen that these risky drug misusers were concerned with reducing the risk of HIV by ceasing to share or inject. They attempted to reduce other risks associated with drug misuse by controlling the amount and quality of drugs misused and being careful about hygiene and health issues. What they wanted was the support and resources to help them do this. When asked about GP support, many respondents said that they did not have a GP and were even reluctant to attend accident and emergency departments where they were already known.

In essence, most drug misusers emphasized that they wanted basic consistent health care support to help them to reduce the risks and damage and more intensive support at particular times when their drug misuse or lifestyle got out of hand and became too difficult to manage on their own. Most saw themselves as fairly independent and in control of their lives most of the time, but dependent on the provision of clean needles and syringes and health care support.

Respondents attending drug agencies for harm minimization also described their needs in terms of the prescription of drugs to regain control, help them to cope and self-medicate. They also felt that prescribed drugs were useful to help them stabilize their drug misuse and stop misusing in risky, chaotic or uncontrolled ways. They appreciated access to a psychologist for anxiety and anger control and help for social problems. It should be emphasized that these comments reflect respondents' confusion about drugs functioning as a solution yet also causing further problems. Although some clients did not know what was wrong with them or what kind of help they needed, they did not want help to give up drugs but to reduce the risks.

Reducing the risk of HIV and hepatitis

I need clean works, that's all, nothing else, no counselling, no do-gooding.

I just need clean works and perhaps a script when I get into trouble.

I just need basic things like needles and syringes.

(Showing handful of condoms, from basket in waiting area) I pinched all these – there's no way I'm catching AIDS.

I go down the chemist's to buy needles and syringes or I get them off my friends.

You can get them (needles and syringes) from the chemist, you don't have to go to (drug agency).

Basic health care

It would be good to be able to see a nurse when you need some help, the GPs are no good, they won't ever believe you about anything.

I need something to stop the pain in my legs.

There is a chance of OD-ing (overdosing) and not getting help, because people are scared to call a doctor or an ambulance.

The veins can get blocked and infected, it would be useful to be able to get antibiotics and painkillers without the usual trouble with doctors ... in fact it would be good to have a doctor in the first place!

Abscesses are a problem, but you don't like to go and get something as people will know.

Substitute drugs

I need stability. I need to be stabilized. That's why I went for a methadone programme. I desperately wanted to give up drugs but I couldn't – there were too many about.

With prescribed drugs you know what you are using, on the street it could be mixed with anything.

Practical help

I have never tried to get help for my drug use. If I became dependent and I lost my house and my job or put in prison, then I would get help.

I have never been homeless, I'm not a junkie, but it would be useful to get some practical help with prescriptions even if it's just temazepam, when you need it.

I've always hated probation officers, social workers. Do-gooders who do nothing. I wouldn't listen to someone who hasn't taken drugs cos they don't know anything about it.

I need to move from a one-bedroomed house that's like a shoe-box. I want someone to help me manage my anger. I don't really know what's going on.

The majority of these respondents thought their primary aim was to reduce the risk of HIV and other risks such as infection and overdose. Some respondents wanted to stabilize their drug misuse, by receiving prescribed drugs and/or other methods. The aim of stabilization or maintenance serves different purposes for different types of drug users. It will enable those misusing drugs in an uncontrolled or chaotic way to regain control of their drug misuse and to order their lives so that there is less risk of harm. It gives injecting drug misusers a regular supply of oral drugs to help them stop injecting. It will allow those who misuse drugs to help them cope or to medicate for underlying problems to behave in a safer way.

Prescription drugs to stabilize and maintain safer drug use

I used to take everything – now I stick to diazepam tablets as prescribed. They help me look at things before I act. Sometimes I have problems after drinking.

I've stabilized completely. I don't have to worry about drugs or money or not having enough.

The main thing that's helped me give up drugs is the script. It makes you not want to take drugs. It stabilizes you and makes you your old self again.

Diazepam stops me looking for other tablets and helps me to understand things better and control my temper to an extent.

People use more when they buy on the black market, because they don't know when they'll get more.

I've not been cut down on my methadone script. I'll tell them when I'm ready. It took a lot to give up the other drugs.

To stop me burgling chemists. To keep me out of jail because that's where I would have ended up.

In a way I'm still a junkie because I'm still on the drug. I don't really class myself as a junkie because I see a junkie as a needle user. But I'm still an addict.

Help with psychological problems

Respondents' opinions differed radically when asked about counselling services and social support. Those who did not attend agencies usually felt strongly that this kind of help or therapy would be very unwelcome and made many disparaging comments, whereas those who had received help with anxiety and depression felt more positive.

Relaxation and behaviour control techniques

(The relaxation tape) was working at the beginning. It sent me to sleep. I don't use them any more. They get on my nerves. They don't work for me. I can't be bothered with it now.

They've helped me with my anxiety, using tapes and listening to my problems with an open mind.

I'm less anxious now. If I do feel uptight, wherever I am, I do my breathing exercises. If I'm in the house I listen to the tape and it does work. I thought it was a load of garbage to start with. It took me a year and a half to get into it.

I've done anger management courses. It was a three-way meeting with probation. It was planned for me, with me there. It's so that if I'm ever out, it will teach me to control my temper better. They stimulate situations that might occur. They look at what started the argument, where you could stop. Every argument is either win or lose. They said it could also be win/win where both sides are happy but I can't understand this.

Counselling

She (clinical psychologist) allows you to own the damage done to you. She is very nurturing. She allows you to speak. She doesn't control you in any way. She might just gently bring you back if you go off the subject. She looks at messages from childhood, feelings of abandonment, patterns and how I allowed all these things to happen. It relieves the blaming of myself.

She's giving me faith in myself. I'm learning to re-parent myself. She uses her particular skills to give more choice to the individual. It's nurturing, gentle – very different. She gives me support and invites me to see my own way. I'm looking at childhood issues and core issues ... I'm dealing with the emotional crisis.

It helps me to vent things out when I talk to them. I explain what's going on. They advise me – I've stuck with their advice. They get me to look at things from other people's point of view. They're trying to make me see what the outcomes could be for the other person and myself.

I volunteered to go to rehabilitation once but I couldn't come off temazepam and Valium. I went to a drug agency but nothing really. I have walked away and hurt lots of times. I have asked for help all the time but I don't know what is going on. My wife has left me and I need counselling because of my angry and obsessive nature. I feel like killing her sometimes. I need counselling because sometimes I act on impulse.

Irrelevance of counselling and psychology

I'd stand on my head for an hour each week, if that's what I have to do to get the script.

This (counselling) about your childhood and that, it's all a load of cobblers.

I don't need all this, I just use drugs like you drink, you don't get therapy for drinking, it's just pointless.

The groups are really naff, nobody takes it seriously, what's it for anyway?

I don't want to offend him (syringe exchange worker), after all he's only trying to help, but I can't see the point. (laughs)

I don't mind, I go along with what they want, you have to really.

We have a chat, he gives me the week's works (needles and syringes), he asks if I want any counselling … I want a script but I don't want all the rest of the palaver, so I don't bother with the script.

I couldn't cope with all that muesli-knitting, trendy, do-gooding stuff.

(A friend) gets our works for us each week, he doesn't mind going to (drug agency), but most of us don't go, it's for nutters really.

They don't really do anything (at the drug agency), they just talk and they don't know much anyway.

They're a load of p***ks, what a waste of time.

Help with social problems

> The staff have been good. They wrote a letter for my court case.

> I'm more aware than what I used to be. I'm not so ignorant. I will talk to people. For example, I wouldn't go to the dentist or the hospital – now I will go. I've got more respect for other people's points of view. I'm more polite – I'll hold the door open – little things that have made a difference in me as a person.

> I did try to take the dog for walks but I can't walk far because of my sciatica. I'm doing boot sales on the weekend. I enjoy that. I go round knocking on doors asking people if they've got stuff. It takes my mind off things. I was doing swimming once a week but I haven't done it lately. I can't be bothered on the day. It's too much of a big deal.

> It's just different things to do. If you keep yourself occupied you don't think about drugs.

The benefits of a harm minimization service

> I know I can get help if I need it.

> Before I used to share (needles and syringes) at least sometimes, now I know I've got works at home, I can wait.

> At least the nurse will clean up wounds and things and give me antibiotics.

> I'm more careful now with the injecting and generally, I'm back in control of things.

> My confidence, my self-esteem. I've put on a lot of weight and I feel 100% better.

> I still get a bit paranoid but that's to be expected.

> Before I would do things, make snap decisions. Now I think about what's going to happen ... It works for me nine times out of ten. It delays the violence, but sometimes it comes out in the end.

> I plan things more, I don't run out of works or store drugs so I don't take unnecessary risks.

> It's there and I need it.

CONCLUSION

In talking about their drug misuse, these men and women have demonstrated the meaning and importance of drugs in their lives. They have made a decision to misuse drugs and can manage quite well most of the time, but there are many drug-related health problems, including the risk of HIV and hepatitis, and there are times when drug misuse gets too heavy, dangerous or out of control.

Once these problems have been identified, potential solutions become apparent. First, the most obvious solution is a basic health care service. This is often denied to drug misusers. Whilst this may sometimes be for a good reason, if they are pressuring GPs for drug prescriptions, it is essential to reduce drug-related harm and maintain and support drug misusers. Second, the way to reduce HIV and hepatitis is to ensure that injecting misusers have easy access to clean needles and syringes. Third, a prescription for substitute drugs would help stabilize and prevent further damage, particularly for those periods in drug misusers' lives when they encounter temporary difficulties or for those who have continuing serious problems such as physical dependence. Finally, if drugs are misused to cope with psychological and social problems, psychotherapeutic and cognitive behavioural interventions and social work support may be helpful in some cases.

The majority of respondents wanted help to reduce drug-related harm, but on the whole they did not want to change and as a consequence could not see the relevance of therapeutic interventions. Whilst they saw a clear need for harm minimization for both non-dependent and dependent drug misusers, most of the respondents saw no need at all for treatment for non-dependent drug misuse. At times the therapeutic option appeared counterproductive in that it deterred them from attending harm minimization services, which do not involve treating drug dependency but instead give clients the necessary support to help them reduce drug-related harm.

Talking to dependent drug misusers

4

INTRODUCTION

This chapter will follow a similar format to Chapter 3, first looking at patterns of drug misuse and at how respondents understood their own problems. It will then look at how and why dependence develops and its risks and consequences. Finally, it will focus on the needs of dependent respondents and the possible kinds of help available.

The two previous chapters have used research data to illustrate the meanings that drug misusers ascribe to their behaviour. This chapter also presents a snapshot which is not intended to be an exhaustive and comprehensive portrait of all dependent drug misuse. There may well be other types that are very different from that delineated below. Moreover, the boundary between risky and dependent misuse is not clear and absolute. The dependent misusers described here may also take serious risks and there is therefore a need to identify those who do, as this will affect which intervention is appropriate for them.

People who misuse dependently have different reasons from other misusers. There are several different theories of the aetiology of dependence that are discussed in detail in Chapter 8 but, as in Chapters 2 and 3, the emphasis here is on how these dependent misusers see themselves. The aim is to describe how clients understood their own problems and how they set about dealing with them. It will be seen that some do not see dependence as the worst of their problems and some do not see it as a problem at all, but rather a less than perfect solution to many other difficulties. Many respondents felt that their more serious worries would be resolved by having a long-term prescription for the drug on which they were dependent. They thought that factors associated with illicit drug dependence were more problematic, such as health and legal risks and general elements of the lifestyle. Dependence was seen as aggravating these problems as the need for a continuing large supply of drug increased.

This chapter includes respondents from six different agencies over a period of many years. The agencies included two community drug teams in different areas, a short-term methadone prescribing clinic, two non-statutory drop-in drug services and a Minnesota day centre.

The themes that emerge from the statements below are generally unequivocal and resonate closely with those found among the risky misusers.

- There are different routes to dependency and different consequences.
- The dependent misusers' ideas about dependence are different from those of professionals (particularly in the early stages of the inter-action between them). The respondents tend to emphasize the moral aspects, the effects on others, their lack of will-power and behaviour which is out of control. They are clear that what they call 'addiction' is qualitatively different to social or psychological dependence and that it can be permanent, remaining with the individual even after the drug misuse has stopped, probably for the whole of one's life.
- These respondents have different attitudes to drugs (on the whole) from those in Chapters 2 and 3. They are more likely to see their drug misuse as problematic and are more likely to define themselves as addicted.
- Dependent misusers are very often confused about what is happen-ing to them, particularly about whether the drugs cause their problems or vice versa. They cannot make sense of their dependence, often perhaps because the professionals they encounter are equally uncertain. This confusion is compounded by a lack of knowledge about the drugs they misuse and about the nature of withdrawal. For example, many misusers of benzodiazepines are unaware that the drug can cause anxiety and depression that may be greatly exagger-ated during withdrawal. Respondents who misused opiates spoke of a deadening of their emotional life, apathy and depression, without seemingly considering that these phenomena were drug induced.
- These respondents are all clients of agencies and are therefore likely to have more serious problems than those who do not attend and may also differ from them in other ways. We have no information on these non-attenders but it seems probable that they will be less concerned about their misuse and have far fewer problems arising from or underlying their dependence.
It should be remembered that all the respondents were clients and although the aim in the research was to interview as many as possible prior to contact with staff, it is clear that many of their responses may be influenced by the agency context. Of this group, some were using prescribed drugs and alcohol in conjunction with illicit drugs. As might be expected, those respondents who had attended agencies for

some time tended to understand their problems in the way the staff of that agency did (this will be discussed in more detail in Chapter 8). So clients attending a Twelve Step, abstinence-orientated agency would see themselves as hopelessly addicted, whereas those attending a psychologically oriented agency would see it as a behavioural problem over which they could exert control. There was also a lot of ambivalence among misusers who believed contradictory things about their drug misuse, sometimes seeing it as pathological and sometimes as simply recreational or medicinal.

PATTERNS AND TYPES OF DRUG USE

The respondents in this chapter were attenders at drug agencies, where one could expect to find more heroin users as the agencies were oriented towards prescribing for and treating opiate dependence. The types of drugs preferred are opiates, as these are the most addictive, and not amphetamine as found among the high-risk misusers in the previous chapter. The patterns of misuse are more regular and frequent, most misusing drugs every day. As for the previous chapter, it should be remembered that the least chaotic respondents are those who can remember enough to report their patterns of misuse, therefore there is no record of really chaotic misuse here. However, in contrast to the previous chapter, where the most serious problems were associated with chaotic misuse, more serious dependence is not necessarily associated with high-risk drug misuse. Nevertheless, much dependence was correlated with high-risk misuse at some time.

The more experienced heavily dependent drug misusers were often less at risk than chaotic irregular misusers. It is also important to note that drug misuse varies from month to month and year to year. Most respondents had periods of heavy, uncontrolled dependent misuse, interspersed with periods when they misused less and had their misuse under control. Some felt that help was only necessary in these more difficult periods, which could be avoided altogether if legal prescribed drugs were available.

Every morning I have 14 mls of methadone, which is scripted every day. From 4.00pm onwards, I have a mull (cannabis cigarette) every night of the week. On pay day, I go to town to score some Gee's Linctus which is made up of 40–45 mls of anhydrous morphine. I take it four days a week at mid-day to two o'clock depending on which town I am in. Sometimes I use Codeine Linctus as well but this is hard to get hold of. This is a typical week for me, it is the same most weeks. I usually take all my drugs at home apart from the methadone which I have to take in a chemist.

Yesterday, I took a break from methadone, I only had 10 mls and no cannabis. By this morning, I was really cut up, I couldn't walk – it is one of the worse withdrawals you can have.

This is a typical week for me. I am scripted methadone every day. I smoke £4–5 worth of dope every day, through the day. Yesterday I had 10 mg of Valium at 6.00pm and an injection of physeptone 12.00am (25 ml). Friday I had methadone 11.00am, dope all day. Thursday, Palfium at midday, methadone and dope. On Saturday I had physeptone, methadone and dope. Sunday I had dope all day. The weekend before I had speed. I don't usually 'cos it makes me go thin and I don't like being thin.

I am scripted physeptone – 703ml (which I) take at 11.30am and have joints all day. I have one Valium a day, it varies when I take it. It is the same all week. I have amphetamine about every two weeks but try to avoid it 'cos I don't like the drug. It is more needle fixation.

There have been plenty of times when I really wanted to stop, I have tried, I go on a heavy binge and then stop and then use again in moderation ... I don't drink and I think any drug taking in moderation can be nice. Nowadays I have a script and smoke cannabis.

I am probably at my most stable time now, I was taking heroin and dope in London.

(My drug use) has definitely got worse since I started injecting.

It (drug use) varies. Just as I seem to get on top of it, like, something happens and messes my head up and get back into bashing all the drugs again. Then I gave up Valium completely, I took myself off 'em. I used to sit with 'em by the side of the bed where I could see 'em and I wouldn't touch them. I like to be in control of the situation. I don't like the situation to control me, like, it happens sometimes and eventually gets out of hand.

THE MEANING OF DEPENDENCY AND ADDICTION

Some respondents had a clear idea of 'addiction' as a serious, permanent state, different to more temporary or habitual behaviours or psychological needs, which they described as 'dependencies' (this in contrast to agency staff who either described all forms as 'addiction' or all forms as 'dependency'). Unfortunately, the terms 'addiction' and 'dependency' are often used in a fairly arbitrary way by many professionals; as a consequence it is not possible to be certain about the meaning of the term for each respondent.

Other respondents were very confused about what was wrong with them and could not make any sense of their condition or what was happening to them. This confusion contributed to their suffering. Respondents might have started misusing for one reason, but eventually the drugs got less and less effective and they said they were no longer misusing for the initial reasons (e.g. hedonistic or coping). Yet they could not stop for a mixture of physiological and psychological reasons. It is interesting that while many professionals understand drug misuse as a learned behaviour, the drug misusers themselves have a quite different understanding of it and addiction is seen as something much more than a psychological dependency or habitual behaviour, rather a stronger and more permanent phenomenon.

It should be noted that many respondents say that they misuse in order to cope or to self-medicate for a specific psychological problem. General coping and specific self-medication are split into two distinct categories, but in effect there seems to be a continuum from needing drugs to cope generally with life to specifically self-medicating for a psychological disorder such as depression.

GENERAL

I now maintain my script use because of withdrawal, for example, vomiting, sweating, cramps and because I am paranoid at bed time. I think that scripted drugs are clean drugs but street drugs make me feel dirty, as if I have got the plague. I think that others discriminate against me because of my habit. They call us junkies. They say, oh don't associate with them. I have come to enjoy the drugs less and less all the time and I just want them less and less.

Yes I'm pretty much (dependent). I don't know, 'cos it's more will-power in it. If I can kick Valium, even though I take 'em again now, I'm sure I could kick speed. It's a question of, like, can other people around me handle me when I'm coming out of it, like.

If you stop the script only one source is cut off. Stopping street stuff is going to be hard.

I've not been cut down on my methadone script. I'll tell them when I'm ready. It took a lot to give up all the other drugs when they gave me the script.

Drugs have taken over my life, I am obsessed with them. I also see my drug use and drinking partly self-medicating depression and anxiety. (I have found a way to smooth the edges of life and make it easier).

I got worse after A died, at first it solved problems, and in the end you don't know why you're doing it.

The only person who can stop you is yourself. You only want to stop when you are on drugs ... the pain and paranoia of withdrawals make it too hard.

DEFINITIONS

Addiction

To many of these men and women, the notion of 'addiction' as something very powerful, perhaps even overpowering, had come to dominate their behaviour and their lives. They saw it as usually a permanent state and as providing an all-encompassing explanation for almost everything that occurred.

Once addicted, always addicted.

If you are addicted you need it for daily life. You will organize your life around it. It becomes more important than water and food.

(Addiction is a need for drugs because of) relief for psychic pain. Not just withdrawal because this only lasts a short while.

Most people take drugs because they feel threatened, they are lost souls. Drug addiction is for people with problems. They say, oh I'll have a break today and then want more and more.

Drug addiction is dependence for certain reasons on drugs.

I was totally addicted to drugs. It was causing ill health, mental problems, problems with my family, and a general lowering of my standards.

I'm addicted to diazepam. It's my best friend. I'm crying without them.

I see physical addiction as separate (distinct) from withdrawal symptoms if you stop using.

(I'm physically addicted) because of all the symptoms that come with withdrawal when I try to cut down.

It is like a madness.

Initially tried to substitute pills for alcohol. Result was that I became cross-addicted.

I need it now just to feel normal.

Maybe mental illness goes along with addiction.

My thought processes actually change in the same way each time I go back to using and change again when I stop.

I didn't realize how dependent I was till I tried to stop.

It makes me feel as if I can do things. I don't know where to start with things without my drugs. I was trying to build a shed the other day and I just didn't know where to start. I think I am physically dependent as well because I need energy because of my illness and the paranoia at bed times. I also suffer from withdrawal, for example, sweating, vomiting, stomach cramps.

Physical dependence

Some of the respondents made a distinction between addiction and physical dependence. As the following quotations show, they saw it as precisely defined and different from psychological dependence. Others found the difference between dependence and addiction to be much less clear.

It has affected me physically but not mentally.

I can't stop taking drugs. I'm dependent on the chemicals.

I'm addicted to depressants. I have taken amphetamines, Ecstasy and acid tabs but I don't know whether I'm addicted to them.

I know for a fact if I didn't have them (diazepam tablets) I'm going to fit through the night.

I'm an addict to methadone. I couldn't give it up tomorrow – I'd have to be weaned off it.

Psychological dependence

Psychological addiction makes you feel that you don't have withdrawal but you desperately want to take drugs.

I couldn't cope without them. I just couldn't go out or do anything.

I can't even go outside without my drugs because of my lack of confidence. I think I am psychologically and physically dependent on drugs because of my lack of confidence.

I just need them to get by, for every day.

Willpower and moral explanations

It is not uncommon for drug misusers to cast their behaviour in terms of character and mores. In comparing themselves with others, they ascribe their taking of drugs to a lack of moral strength or weakness of will which does, of course, provide an adequate explanation for their behaviour.

I feel a failure.

I believe a weakness got me into this state.

I feel it is a moral issue. I will have to change my lifestyle and attitudes.

My behaviour has gradually become childish, petulant and inexplicable. I have been shocked and frightened by my behaviour.

It's an addiction, not a sickness. I think a lot of it is greed.

I consider it a weakness, I need self-discipline and self-respect.

I blame myself totally.

I thought I had the will-power but didn't use it, then I suddenly realized the will-power was non-existent.

It is a weakness, something wrong with my personality.

It's up to me really, if I have the will-power, if I am strong enough … I might be too weak.

Uncontrollability

Dependency can, for some at least, lead to a complete breakdown in self-control. There is often a suggestion that this is potentially a risk for all drug misusers, sometimes started by a crisis or some event but also possibly for an inexplicable reason.

I think there is a potential for all of us to become addicted. I used too much until it became uncontrollable.

Something triggers me off, I don't know what.

When you think you are cured, it gets dangerous. Because you think you can handle it but you can't.

I voluntarily detoxed myself. I felt much better. When I came out (of hospital) I thought I had control, but it was soon back to where it was before. I can't see me changing.

I have ruined relationships and lost jobs because of my drugs. I can't control it any more.

Craving

The ideas about addiction or different forms of dependency are sometimes replaced by the notion of craving. It is seen as just as powerful and difficult to deal with.

Thinking about the prescription throughout the day. Mental craving.

I get cravings. I'm thinking about my prescription throughout the day.

Disease of the whole person

A group of respondents believe in the basic premises of the Twelve Step (Alcoholics Anonymous) model, i.e. that addiction itself is a physical and spiritual disease. This category includes those respondents who saw themselves as 'born addicts' or having 'addictive personalities'.

I think that I have had an illness which I am trying to recover from.

I'm a born addict, anything I do, I do to extremes.

I have come to believe that it is in-built, something to do with my make-up.

HOW DID YOU BECOME DEPENDENT/HOW DID YOUR DEPENDENCY DEVELOP?

Again, there is much confusion about what caused the drug misuse and what caused the problems. Drugs solve some problems and cause others. Dependence develops from misusing drugs regularly but it cannot be isolated from the reasons why people misuse drugs. This emphasizes the need to be aware of the underlying psychological and social problems when dealing with dependence itself. This is partly why aftercare and maintenance of change are so important but also this is why dealing with dependence may often include alleviating the underlying problems that led to its development in the first place.

SOCIAL REASONS FOR USING DRUGS

We saw in Chapter 2 that recreational drug misuse when found in raves and nightclubs is a striking manifestation of the importance of peer group influence and the great significance of social interaction. Much drug misuse among those who are dependent is similarly understood as a social activity, affecting status and self-worth. Being and remaining part of a group of friends is to such men and women a necessary condition for taking drugs.

> All my friends were getting interested in pills at the time, so I joined in, I wasn't led into it, it was my own fault.

> The men did the sorting for the women, you just accepted it.

> My husband was doing drugs so I did. I think if he went fishing, I would have gone fishing.

> I used amphetamine and alcohol, I got involved mainly through mixing in particular places ... but I liked those places and those friends, at the time.

> My flatmate offered me heroin to smoke, it seemed nothing, just puff from a bit of foil.

INDIVIDUAL REASONS

A number of the respondents explained why they had continued to misuse or why they had misused more seriously than the remainder of the social group to which they belonged.

> I felt like it was February all of my life; the first time I had heroin, it was like somebody had given me an overcoat.

> First time I took opiates, even four days after taking them I felt lovely but not off my head. So I knew that this was the one I would have problems with.

> When I first started using drugs, I saw them as good because I had a feeling of belonging to my peers. I felt like one of the boys but as time goes on, you are seen as an abuser, a pusher and I feel bad about this. I gave up my script altogether for 11 months and

my life was starting to fall into place but withdrawal made me feel like a 90-year-old man, I felt really ill and I didn't have any energy to do anything, I couldn't even help my wife carry a bag of shopping.

Of our gang, there were five of us who all started smoking dope, then dropping stimulants, but only two of us went on to injecting ... and that was the ruin of my life.

I really like morphine-type drugs, it really puts me on top of the world.

Everybody was injecting heroin, after a while I thought, 'anything for peace' and tried injecting heroin. This was the first time I had injected or used heroin. It was the best and worst thing I have ever done. I should have kept to amphetamine tablets, just bits and bobs.

When you set out to be the man (dealer) ... the following events become part of your life, firstly earning money, becoming big-headed, being ripped off, being arrested, jealousy amongst your friends and animosity from the police.

I was homeless for three years from the age of 13. I think that it is why I started drugs. I started smoking and it led to other things. I couldn't face up to things, I felt like scum, begging on the street. When I started on morphine, all that disappeared. I couldn't give a f**k. I couldn't give a f**k about what the family thought of me, I only cared about where my money was coming from for my next fix. When I tried morphine, it was brilliant and I felt really good for the first time.

COPING

Changes in circumstances or crises which make it difficult for the individual to cope with life are sometimes given as the cause or pre-cursor of drug misuse. From this perspective it is 'normal' or otherwise a rational response to difficulties.

I was depressed and was prescribed pills for years, then I couldn't give up.

My real lapse after 11 months of being clean was because of the death of my father and because of moving back to the town, back to the drug circle.

I take them because I have no friends really. I feel very lonely.

I can't cope with nightmares and depression if I don't take them.

I've had bad experiences in my life and drugs help me forget.

If I had a row or a problem I'd use pills to just obliviate myself out of life. In two years I started to lose control.

They help with anxiety.

When I'm uptight the first thing I go for is my tablets.

Every now and then when you come across a problem, the drugs are a shoulder to lean on.

I use when I'm under stress to calm me down.

I used a lot of antidepressants when my first marriage broke up. This was 10 years ago. After that I used pills to help me cope, eventually using pills to cope with everyday life and looking after my son.

CONFUSED

I was depressed, but now I don't know if the drugs came first or not. I don't know.

I needed to talk and felt a lot about the loss and grief of my husband dying. But I have no idea why I was prescribed the tablets.

SELF-MEDICATION FOR UNDERLYING PROBLEMS

A common explanation for drug dependence is that it is a form of self-medication for a variety of problems, in much the same way as an antibiotic is used to deal with an infection. To those who hold such a view, drug misuse is a rational and often necessary means to an end. It is also possible that some respondents may in effect be self-medicating for withdrawals without realizing it.

Some people manage to settle (drug free), but some become mentally disorientated and confused, they don't know how to keep a relationship going. There are problems if you haven't been taught to cope with life, then there is a weakness which you cover with drugs or drink. These are my problems.

When I left school I became interested in other drugs, opiates, I didn't touch anything for a while because I was working for a newspaper, but I started drinking and eventually turned to heroin as a cure for alcoholism.

When I was 17–18 years old I had a nervous breakdown and was given prescriptions by my doctor.

I think I was always very anxious and a bit depressed, but it gets worse with the drugs.

So yes, I am getting fed up with it, in a way I haven't even started my life yet because I am still young and it is already going downhill. As time has gone by, it has become normal to pick up my script but I do agree, it is not normal. For me, it is just like drinking a cup of tea. I see it as, like, a normal thing to me. I do it every day, it has become a normal everyday thing but it is wrong.

I think my problems are panic attacks and anxiety; I don't know how to get rid of this without pills ... but pills are not really the answer.

It's the fear you need to get rid of. Drink is pleasant if you can control it. I couldn't, so I used safer drugs, to make me happier and calmer.

STOPPING AND STARTING

As we have noted in earlier chapters, drug misuse may follow a variety of paths or careers, even for the same individual. It is not surprising, therefore, that these men and women frequently mentioned starting to misuse drugs and their (sometimes repeated) efforts to stop.

I was in a mood and didn't get my spirit back. One day I had a religious experience and opened my mind and it kept me going and made me feel good. Before, I had been feeling bad for so long, that I didn't realize that I was feeling bad.

My husband died and two years later my fianceé died. One thing led to another and I had a relapse. I'm not strong enough and didn't understand how difficult it was (to stop). I didn't need to inject because I could afford to smoke it or do what I wanted.

Every time I gave up, it was because I was bored with drugs. Also the ties. Had to stay in London to score. Not keen on methadone, I'd rather be on heroin. I am bored with drugs at the moment and want to do other things. I've done them before and I know I can do them but not in my state.

I couldn't say why or how I stopped, I just decided to.

I don't know why I did it (stopped) that time and not before and I don't know why I started again.

Drugs caused problems with my parents, but they tried to get me a job and made sure that I was trying to get a job. Found a job and stopped taking and started doing sport.

Since I have stopped using my old personality and my sense of humour are coming out again.

I have managed so far to stay off, but I could not pin down why or what made me stop, apart from I wanted to.

WHAT ARE THE GOOD THINGS ABOUT DRUG USE?

When asked about why they misused drugs, almost all the remarks fell in the 'helps you cope' or 'addicted'; categories, rather than the reasons previously cited for recreational and risky drug misuse. Reasons for mis-using were largely negative and there were few positive responses. But on the whole these men and women felt that this was not a relevant question. People did not think they misused for rational reasons but because they 'needed to', not because they enjoyed it any more. There were fewer clear or rational reasons and more tautological replies or concepts, such as 'hooked', compulsive, can't help it, some kind of ill-ness or disease. This is a different perspective from the recreational and risky misusers.

NO RATIONALE

There is nothing good any more, I am just addicted to methadone.

I don't know why I use, I can't stop.

I'm not hooked, there is no reason why.

A MISTAKEN RATIONALE

I confused 'not feeling' with 'feeling good'.

I thought I could control it but I can't.

CONFUSED

I really don't know. I'm out of control, can't help myself.

I just can't understand it, different people are telling me different things.

THE RISKS AND CONSEQUENCES OF DEPENDENT DRUG MISUSE

These people recognized that there were many ill effects of drug misuse but when discussing its consequences, there are two factors which compound the problem. As mentioned earlier, people were often unable to distinguish between causes and consequences. They were also unable to distinguish between the effects of drug misuse and the effects of withdrawing from drugs, as both were an integral part of their pattern of drug misuse. Most people's drug supply was not stable enough to guarantee a consistent amount of drug in the bloodstream and the common experience was a continual series of ups and downs as the drugs took effect and wore off.

The most commonly mentioned consequences of dependent drug misuse were depression (almost all respondents mentioned being depressed at some time), followed by apathy, lack of energy and mood swings. Inability to feel emotion while misusing and inability to cope with feeling when stopping were common problems.

GENERAL RISK AND CONSEQUENCES

When asked about risks, those associated with injecting, such as HIV, were most commonly mentioned (as in the previous chapter), but a larger number of respondents mentioned the dangers of overdose, together with risks to long-term physical and mental health. Risks to social life and relationships were also common; many people mentioned destroying relationships, losing jobs and accommodation and, more generally, being rejected by the normal social world.

> I see tremendous risks associated with drug taking, for example becoming dependent, feelings of no self-worth, damages the health, family problems, relationship problems. There is also a risk with sharing needles, risks of unknown dealers, overdoses because you don't really know the percentage of the street drugs and injecting causes clots. I have taken a lot of risks, I have shared needles and I have used the same needle on myself up to seven times. I tried to manage the risks by sticking to my script and staying away from street drugs. I also stick to one dealer but I never used to. You can also cut down the risks by not letting someone else hit (inject) you because you just don't know what's in it.

> I see drug taking as risky because it causes family problems but in my case, my family problems were the cause of the drug

taking. I think that if you take drugs you have either got or you will have problems, for example, family problems, financial problems. The problems will have a snowball effect and will get worse when you start taking drugs and this will lead to other problems.

I was drinking on methadone once and I almost killed myself on my bicycle riding home. Methadone and alcohol is a really bad mix, you black out and don't know what you are doing. I could have overdosed easily but I didn't give a f**k. One time, I took two bottles of Gee's Linctus as well as my methadone and I could well have overdosed. It was like being on smack. I was too afraid to go to bed in case I spewed up. I s**t myself!

I'm lying down all the time. I'm desperate as I'm always feeling ill, I can't think straight. I'm having suicidal thoughts.

I feel I am going mad.

My memory is going and I'm always feeling ill.

EFFECTS ON OTHERS

Some of the respondents referred to the ill effects of their drug misuse on their family and friends. As we noted above, the social dimensions of drug taking are often a common theme in the statements.

It affects me badly and causes too much trouble to other people

I didn't care about anyone more than drugs.

It affects my husband and son now because I'm stupefied by my addiction.

You can't avoid having negative effects if on drugs, because it is a negative thing (even if it is not stealing, etc.).

HIV AND HEPATITIS

The fact that all these respondents are agency clients undoubtedly has a bearing on their awareness of the risks of HIV and hepatitis. It is not mentioned as often as one might expect, perhaps because it is now firmly part of the taken-for-granted perspective of drug misusers.

There's not a serious risk of AIDS for me because I am careful about things like that. I have shared with (ex-boyfriend) but it was safe as he was clean and I knew I was.

In the town, I was mugging, begging, etc. to get money. One day, I had to share a needle because I had no money and the needle exchange was shut. I was sharing with a friend and I caught hepatitis B off him. I collapsed and was put in hospital for two months, then I went through 'cold turkey'.

We were washing them (syringes) in between.

Used to share needles but not any more.

I keep needles clean, make sure of that.

INJECTING UNKNOWN OR CONTAMINATED SUBSTANCES

I rarely had bad hits and when I do, it's on speed 'cos of what's in the speed and not 'cos of messy equipment.

Have had bad hits, terrible headaches but I got over it.

SOCIAL AND LEGAL RISKS

All the respondents who are dependent, just like those discussed in the other chapters, are aware that their behaviour has social consequences and may well be illegal. Many of them have experience of the courts and the police.

I have to drink my methadone in the chemist. It is the new policy. I was drinking my script one time and my aunt walked in and caught me, now it has gone straight through my family and I haven't seen my mother for weeks.

Because I am a registered drug user, I couldn't go to certain countries because I would be classed as a smuggler. If I went for a job it f**ks you up. Because I am moving, I have been looking for a chemist with a back room for me to take my methadone so no one can see me.

Parents don't know that I'm back on them. Would cause problems if they did.

I lost my first wife through my drug use.

Heroin was too expensive and hard to get hold of and often cut with other things. It was also only possible to obtain through prison contacts on their release.

My main problem is with police and courts. I can't go on because of the way they treat you in police cells.

OVERDOSE

Don't really see any risks as long as I'm careful. I might snuff it if I take a big dose but I'm careful.

I've known a lot of people who died on it and it's really upsetting. left me feeling really bad about it, but it didn't stop me.

If this guy hadn't seen me collapsing, I would have been dead now.

WHAT DO YOU NEED TO HELP YOU DEAL WITH YOUR DRUG DEPENDENCY?

All the respondents were asked what their needs were and what kind of help they wanted. They identified a wide range of different needs arising from the previously identified problems and gave some idea of the kind of services they found useful. Not all needs were directly related to dependence, nor were all services requested concerned with overcoming this problem. Some respondents felt that the answer to dependence was a regular script for substitute drugs, but most agreed that solving related problems would help them deal with dependence itself.

NEED TO DEAL WITH THE DRUG DEPENDENCE

I've had psychiatric help and group psychotherapy which did me some good, but they were concentrating on problems, not drug use.

I like NA because they offer intensive treatment which is very good.

I'm not going to let go of my own responsibility for myself. The battle is still mine. But I need help.

I want to stop, I can't do it on my own.

NEED TO BE ABSTINENT

I think abstinence is the only way.

If I stopped I'd change automatically.

I think complete abstinence is silly.

You need to stop dependence to any drug.

NEED TO LEARN TO COPE GENERALLY WITHOUT DRUGS

You have to learn to cope with emotional problems without running away to drugs.

Have to stop psychological need for heroin and divert energies given to drugs to practical problems.

NEED TO DEAL WITH UNDERLYING SOCIAL AND PSYCHOLOGICAL PROBLEMS

I had to get rid of panic attacks and anxiety, not addiction, and a lot of pills are not the answer.

Abstinence is not the answer because drugs are not the problem, they are only a symptom.

NEED TO CHANGE YOURSELF

I think you need to leave yourself open for people to criticize you or you will not change or get better.

I have to change my personality to be more assertive and adult about things.

I will need to change my mental obsession with pills as a coping mechanism.

NEED TO CHANGE YOUR LIFESTYLE

I think I will have to become less solitary and more gregarious.

I will have to change my lifestyle and way of thinking.

The main problem is time on your hands and not knowing what to do.

WHAT KIND OF HELP OR TREATMENT WOULD BE USEFUL?

A common problem with consumer surveys is that customers will often state that they want the kind of help with which they are familiar. It was difficult for those who did not have experience of a range of services to choose, but the following data give some indication of the kind of help respondents felt they needed, together with their opinions of the usefulness of the actual help on offer.

PROFESSIONAL UNDERSTANDING OF THE PROBLEM

Never underestimate the indescribable difficulty of giving up heroin.

The government's heroin problem has got nothing to do with my heroin problem.

She's not going to know what the problem is if she doesn't listen to me. I don't know what the problem is myself, mind.

NEED FOR A PRESCRIPTION

Withdrawal

The difficulties of withdrawal were commonly mentioned as an important justification for the issue of a prescription.

I am sticking to my script now and I want to wean myself off with the doctor's help and with counselling.

The script should be reduced at the individual's own pace.

I'm against methadone, because withdrawal is much worse and takes months. I think they should prescribe heroin.

I was detoxed but not given enough tranquillizers to stop the craving. You should have a script and then have counselling to decide if you want to give up completely or to control it.

You should give a quicker reduction at first (e.g. from 50 mls of methadone down to 15 mls within two to three months), then a much slower reduction while you sort things out … and prescribed sleeping tablets are essential as script reduces and afterwards.

Maintenance

If I am provided with a methadone supply, I find I have no problems except possibly money.

I need to have enough methadone so I can cut out the heroin completely.

I think methadone is the only way for me, I had a script for five years and I was all right then.

I have tried to get help in the past. I wanted a script to get off street drugs. I went to my doctor for help but my doctor was unsympathetic and told me that it was up to me if I wanted to buy street drugs.

Inpatient withdrawal (detoxification)

Many respondents emphasized the need for better inpatient detoxification services and aftercare.

Detox facilities are really bad and degrading.

Accommodation after detox is necessary, as many people are homeless and go back to drugs.

Detox should be in general hospitals, not in mental hospitals.

DRUG AGENCIES AND TREATMENT PROGRAMMES

Drop-in agencies and day centres

Need for a day centre where one could meet and discuss problems with people with similar problems.

A day centre which is open until 10.00pm and at weekends.

Some kind of day centre with an emphasis on self-help, both educationally and socially.

More people to listen without judging you.

Follow-up and support after treatment to avoid relapse.

Individual counselling

People should be given more individual counselling, help and more attention, also some sort of communication with our GPs.

We need one-to-one type therapy instead of so-called peer groups.

I feel staff here give clients time without the need to use heroin and without all the allied problems.

I haven't been asked to do anything – give up drugs or cut down on anything – I've just been counselled and helped and that way I've given them up myself. They've never pushed me into anything. They've just helped by supporting me. They've never gone against me. If I ever need to see them any time I just make a phone call and help is there.

They listen and give thingy (feedback) back to me. They feed back in such a way that I come to terms with it and think about it a lot. They ask how I'm keeping, how I'm feeling, what I've been doing in my spare time, anything I need help with. They just listen and give me feedback.

They advise you to talk about problems that you bottle up ... It helps me to know that there's someone who'll listen without being judgemental – who knows what you're talking about. It helps on the mental side.

She sits there and listens to me. That's the good thing about her. She doesn't butt in, she doesn't preach, she just seems to know what to say to me. She says 'Why do you feel like that?' ... . I don't know where I'd be without her – the bottom of the river I expect. If it wasn't for this place, I think I'd be in prison or homeless or broke. She helps me to keep a cool head – to keep a lid on it. She gives me advice but by the time I'm out the door it's forgotten. It just feels better talking to her.

Group counselling

I appreciate the unique value of having to 'share ' with others your predicament.

You can be helped by talking to people with the same problems.

I think more of the groupwork type of programme is necessary.

I would cut down the drugs more quickly and have more group-work as a support.

They told me to open my gob and introduce myself. They drilled 'Trust, Risk and Share' into me. It's saying how you feel and asking other people the same thing. They do it because it works – well, it did in my case.

You share things you don't share with your partner or family. You get feedback from the group or just listen. They help you along, they help you to cope and find solutions you can't find yourself. You talk out of the problem.

It's to express feelings. To share problems. To get feedback off the group. It's made me look at myself in depth. You can't see yourself.

(They give me) a shove in the right direction, a kick up the arse. Just guidance to show me there's a better life. To work – eventually. There's always people worse off than yourself. Confidence. Basically a new way of thinking. Everything I did revolved round drugs.

I needed a focal point – somewhere where I had to be every week. to know other people were feeling like I felt.

I find that methadone with a weekly group meeting does not offer satisfactory support for someone coming off heroin.

Don't have groups, it is easy to make contacts in them.

It's bad for you. You come out gasping for it. You can't lie to them, mind.

Cognitive behavioural techniques for relaxation, anger and anxiety management and self-control

Before I would do things, make snap decisions. If someone is saying something now, I think about what's going to happen. Combined with the tablets, it's making me feel slower.

It's got better but I haven't changed. The violence has stopped. I've replaced the doors and I haven't broken any more. I'm eating a bit more humble pie.

I used to have a lot of trouble with anger. Now I feel I can step back from it.

Residential places

I think the best thing is residential rehabilitation. Relaxed and responsible, long-term (one to 12 months).

General advantages of treatment

I see (the drug agency) as a sanctuary, I feel like I am safe from everything in here and nothing can hurt me.

You need structure and support, a place to go.

Disadvantages of treatment

I expected the so-called treatment to improve the quality of my life, in fact the reverse has happened.

I thought that by coming for help, I would again become a free-thinking, capable and questioning member of society and lose the 'junkie' label, whereas in fact I'm now weighed down by blood labels. I go to drugs agencies but only for drugs, they are all pr**ks.

SUPPORT AFTER TREATMENT

General

Follow-up support should continue for a long time after treatment and staff should consider and plan this carefully before treatment is ended. I think that relationships are the most important thing, then health, then security ... you need all this if you are going to stay off.

Place people in a detox unit for a given period and give them support on a day-to-day basis from a counsellor when they leave the unit.

You need support with follow-up to avoid relapse.

You need the group process again after treatment, to discuss any emotional feelings, psychological feelings, any serious problems we're experiencing. I have found certain parts of it helpful. I realize other people are feeling like I'm feeling, with problems of how to live again.

Special accommodation is vital as many people are homeless.

I think of it as a detox from treatment. You can't take treatment into the real world. I think aftercare should be for life.

I need support on a day-to-day basis from a counsellor after in-patient detox.

Relapse prevention techniques

It's a role-play of situations you're going to find yourself in – working through different options. It helped me to see there's more to life than just drugs. It's made me say no. It's made me avoid them situations. It's made me paranoid as well!

You need more help with staying stopped, it is easy enough to come off, but really hard staying off.

Anybody can stop, the problem is to do with starting again, because things happen and people come round and offer you things ... and you think, well, why not really.

I can stay off for quite a long time, then something bad happens and I go back to it.

You don't get any help afterwards, you finish and they say 'Right, off you go' ...

Lots of people have been here before, they get better and leave and then can't cope and they come back.

You need to know you can come back when things go wrong, that somebody will be there.

Help with life without drugs in the future

Need help dealing with a life free of drugs.

Need help to sort out what to do with my life in the future.

It is a long-term thing.

Help with social isolation and lack of social support when you give up the old drug-using social life.

Help people to feel they are not cut off or isolated with their problem.

Either I keep coming here and talking about things or I hang around with them (drug users) – they're a bunch of nutters.

Alternative activities

They've arranged an activity course for me – that's one way they've helped me. It's all sorts of things – archery, assault courses, general activity centre. It's to keep me occupied, to stop me getting bored.

You need something to do really, something to take up the time.

Housing

I'm homeless at moment but hope to get a flat in the next two weeks.

I am living in a shelter for the homeless at the moment.

Support from local GP

Need support from local GP.

DISCUSSION

It will have become clear from the way in which dependent misusers talk about themselves that the meanings they give to their behaviour and circumstances are complex, change through time and are not easily categorized. Their statements should alert professionals to the differences between their own perceptions of dependence and those of their clients. It will also be clear that these men and women see themselves as clients, that is, requiring help and support, particularly in the form of counselling informed by a sympathetic understanding of their problems.

Just as there are many different individuals, so there are many different meanings; these need to be understood because they can often become incorporated into the respondent's self-image and consequently also become a vital part of their social relationships. The quotations have shown vividly how different understandings of dependence can influence the client's self-image and so also their beliefs about their own problems and capacity for change. This in turn influences what they feel they need and has important implications for the kind of support that professionals provide.

Institutional drug misuse   5

INTRODUCTION

The earlier chapters in this book distinguish between recreational, risky and dependent drug misuse to give some order to a complex and ever changing phenomenon. As we have seen, recreational misuse may sometimes be risky and risky drug misuse encompasses recreational and dependent misuse. Up to this point, little mention has been made of the broad social environment in which the drug misuse occurs. It may be at home, in a rave or nightclub or in the street and the setting influences the consequences of drug misuse but does not necessarily determine it.

In this chapter we consider the limiting case, that is, those situations where the environment has a direct and overwhelming effect on drug misuse. These occur in 'total institutions', first characterized by Erving Goffman as 'a place ... where a large number of life-situated individuals, cut off from the wider society for an appreciable period of time, together lead an enforced, formally administered round of life' (Goffman, 1968, p. 11). Examples are prisons, mental hospitals, army barracks and monasteries.

The example chosen is one where a majority of the inmates have substantial experience of the misuse of illegal drugs – a local prison. The topic of drug misuse in prison has been widely aired in the media and is seen as a matter of significant public concern. As with drug misuse more generally, common perceptions are often quite different from reality, tending towards the view that all drugs are addictive and that drug addiction has become a serious problem in all prisons. It is also widely assumed that prisoners take more drugs than they would do in the community.

Policy and practice in the criminal justice system in Britain, as elsewhere, centres on the issue of control of those in custody. Such control obviously implies the prohibition and prevention of all illegal activities as well as behaviour which threatens the establishment and maintenance of order. The enduring problem for those responsible for the prison service is that drug misuse, seen from the perspective of society at large, is essentially a public health problem but prisons are not primarily concerned with health care.

This chapter therefore sets the context for an examination of the role of the prison and its control regime in determining the patterns of drug misuse as seen through the eyes of inmates and ex-prisoners. Chapter 9 will examine the views of staff, explore the research literature in this field and consider the implications for policy and practice.

A STUDY OF INMATES' DRUG MISUSE

The study was undertaken in 1993 and examined drug misuse before, during and after prison in a sample of men who were in custody (134) and a further 119 who were ex-prisoners on probation. All the respondents filled in self-completion questionnaires, which were supplemented by a series of interviews with inmates, prison staff and ex-inmates. All the inmates were given questionnaires and just over a half completed them. Probation officers were asked to give question-naires to all their clients who had custodial sentences in the past five years: they estimated a response rate of about 70%. All the ex-prisoners lived in the catchment area of the prison, but the two groups in the study do not necessarily come from the same population. In fact, well over four-fifths of those on probation had attended the same prison as the inmates. It is important to note that the prisoners reported only on drug misuse before custody and those on probation on drug misuse during and after custody. For a full account of the methodology used in the study, see Keene (1996).

In the prison, about four-fifths of the inmates are on remand and roughly one-third are young offenders. The study cannot therefore be employed to generalize about those prisons with older, more stable populations.*

A BRIEF QUANTITATIVE REVIEW

It has already been noted that the two groups of respondents come from similar but not identical populations. Moreover, the nature of the research design makes it impossible to draw conclusions about the effect of imprisonment on long-term drug misuse since, although custodial and postcustodial drug misuse can be compared, precustodial misuse is

---

*This research was conducted before compulsory urine testing in prisons. There is much anecdotal evidence to suggest that the consequences of this testing have been extremely detrimental, as prisoners change from using cannabis (which is traceable for 4–5 weeks) to heroin and other opiates which do not remain in the urine for more than a few days. It would not be possible to repeat this type of research in the current prison environment, where drug use has 'gone underground' in effect.

taken from a different group. Nevertheless, there were very few statistically significant differences between the two sets of responses in spite of a wide range of different answers. This may be taken as a consistency of response between individuals which is unlikely to occur by chance. The results of the study also accord with those reported in recent research in Britain, as we shall see in Chapter 9.

The two most striking findings of the quantitative analysis are, first, the drug profile of more than half of the young men in this study resembles closely that of male drug misusers attending raves and nightclubs (described in Chapter 2) and differs in much the same way from that of drug agency clients: second, nearly a quarter of the prisoners had been clients of drug agencies when in the community and have patterns of misuse that reflect this fact. There is also a small minority of inmates who do not attend agencies but whose behaviour and perspective define them as problematic drug misusers. (Less than a quarter had never used drugs.)

Just under three-quarters (74%) of the prison group and just over four-fifths of the ex-prisoners (82%) said they misused drugs in the community; three-quarters of the latter group reported that they took drugs in prison. These proportions are very much higher than those given in most recent research reports but it must be remembered that the range and type of drugs included in this study are much greater.

The age structure of both groups of young men is much the same. Both had relatively few under 18 – less than one in 10 – and just over a fifth (22%) over 30. The proportion of young offenders in the prison group was high at 45% compared with the ex-prisoners group where 39% were under 21. As with the young people who attend raves and nightclubs, age and drug misuse are related (but not in exactly the same way since those who are in custody or on probation are somewhat older). Drug misuse was greater among the younger respondents. Thus nine out of 10 of the men under 21 reported misusing drugs, while only two-thirds of those who were 21 and older did so.

As to type of drugs used in the community before and after prison, cannabis headed the list with 74% of all drug misusers in the study, followed by Valium (59%), LSD (49%), amphetamine (47%) and Ecstasy (31%). The 'hard' drugs – cocaine (24%), heroin (15%) and crack (12%) – were misused much less. Not surprisingly, the levels of drug misuse reported for the period in custody were generally lower than in the community – Valium (55%), LSD (39%), amphetamine (27%), Ecstasy (20%), cocaine (16%) and heroin (10%). The exception was cannabis, where the rate for prisoners was 75%, slightly higher than in the outside world. The rates of injecting (14%) and sharing needles (9%) were about half those found in the outside world, where 28% had injected drugs.

About a quarter of the young men in this study used the services of a drug agency. Their drug misusing behaviour is, as already noted above, quite different from their drug misusing peers who did not. The agency clients who reported misusing heroin, cocaine, crack and methadone constituted about half the number or more of misusers in each case, whereas those who had used amphetamine, Valium, Ecstasy, LSD and cannabis were one-third or less of all misusers. There are therefore sound reasons for supposing that the agency misusers are problematic in their behaviour, while most of the others misuse drugs in an essentially recreational way. The same pattern is evident when one looks at injecting and sharing needles. Two-thirds of those who shared needles and over half of those who injected (53%) were agency clients.

Of those ex-prisoners talking about their drug misuse in prison, one-fifth who reported using heroin did not attend an agency in the community and the proportions for those misusing other drugs were cocaine (53%) and methadone (43%). Over a quarter of those injecting (29%) and one of the 10 men who shared needles belonged to this small group. They are neither recreational nor problematic misusers. The meaning of these figures is considered below. Perhaps the most significant finding was that 21% of the respondents reported using a needle and syringe exchange or buying needles from a pharmacist when not in prison.

QUALITATIVE DATA: THE DRUG MISUSERS SPEAK

In general, men who misuse drugs in the community before they are sent to prison and who resume their misuse after discharge are in no doubt that drugs are important in ameliorating the custodial regime. It appears as if they see it in much the same way as self-medication for a chronic and troublesome illness.

THE RELIEF OF TENSION

Where men live close together in confined circumstances and a controlled regime, the possibility of tension arising and leading to conflict and violence cannot be disregarded.

Drugs make your life in prison a lot easier as everybody is calm and not so violent.

I think downers (tranquillizers/night sedation) should be allowed in prison because it relieves pressure and tension between cell-mates.

Although I have never needed treatment in prison, there are boys who need it. I have been in cells with boys on 100 mls of Valium on the out and need it. Then when they are put in here they start to withdraw and then you cut them down again so all the time they are withdrawing ... which makes them nasty and edgy. Then officers want to know why they are in a mood all the time, when they are bringing it on themselves.

Turn a blind eye to cannabis because it calms a lot of the boys down.

People should be allowed to smoke cannabis in prison. It helps people to overcome aggression and violence.

It helps calm the prisoner down. It enables him to face prison life.

A MORE HUMANE REGIME?

A majority of these young men believed that misusing drugs ameliorated the harsher effects of imprisonment. Their view was that drugs had a humane purpose seen in terms of the inmates' welfare.

Most people in prison who have never used certain drugs often start to take them to feel a bit better whilst serving their sentences.

It is harder to stop taking drugs whilst in prison as when you are banged up it passes the time. And outside most people take drugs as there is nothing much else to do.

I think all prisoners should have a little medication to help with the trauma of prison life.

I think that if I never had drugs in prison I could never have served my sentence.

PROBLEMATIC DRUG MISUSE

The great majority of the respondents had considerable knowledge of drug misuse and were aware that a small proportion of inmates had serious problems which required treatment. As we shall see later, this is an issue of policy and practice which is by no means straightforward. Those men whose drug misuse is problematic are not a homogeneous group, hence their further division into dependent, self-medicating and risky categories. In Chapter 1, the complex and changing relationship between these three categories was explored: here we see it reflected in the way inmates talk about their misuse of drugs.

Dependent

Someone with specialized knowledge of drug misuse should be on the staff to safely wean a user off drugs ready for the move to another prison or to return a 'non-junkie' back to society. Not just physically weaned, but mentally weaned into being able to cope without them. The mental care is probably more important than the physical for permanent effect.

Being an addict to temazepam, I find it very hard to cope with other inmates' attitudes due to my loss of the drug. I think that if this prison did have facilities to cope with drug users then the inmates would feel that they haven't been cut off from help from the outside.

There is not enough treatment used for coming off the actual drug, (reduction) of drugs is too fast.

If people are addicted or heavy users they should be helped, not punished.

The help I did get in prison was totally useless. The doctor here prescribes you the minimal amount of any drug and it is good for nothing. Probably many of us in here have been prescribed drugs on the outside for quite some time. I have been taking drugs for nearly five years and the doctor expects me to come off them in as many weeks. So I think something needs to be done about this soon, perhaps it will stop some of the drugs being smuggled in and help people to come off properly instead of waiting for their next hit when they get out.

Medicating

I'm getting diazepam from the prison doctor, he is cutting me down all the time. The reason for taking diazepam is my anxiety state and nervous disability, but my head is going and I am getting very depressed. I need the diazepam to calm me down, but the doctor doesn't see this problem.

A bit more help with the doctors would be good as you get no attention to what you say about what is wrong, you just get ignored.

They did not give me my proper medication in prison.

The doctor prescribing drugs in prison should check with the client's doctor to see what he is prescribed and why.

Risky

Even though I used needles I didn't share and I was on heavy prescribed doses of legal heroin substitute. Now (in prison) I am finding it difficult not to share and get no script.

> If people are addicts they'll use these (an accurate drawing of a syringe and needle is inserted) anyway, so why not supply new ones to avoid the chance of infection of any kind.

> The syringes and needles were very old and because of the multiple usage going on, the HIV risk was crazy.

> Too much multiple usage. There was one syringe and needle for the whole landing.

> Needles when available were really old and used by God knows who and how many people.

> Of the 12 of us in the hospital wing all the others were HIV. I might have got it if I used their works.

WITHDRAWAL DANGERS

The need for appropriate recognition of serious problems of drug misuse and the requisite response goes hand in hand with comments on the dangers of withdrawal symptoms and issues of social control posed in a closed environment.

> When I came in, they gave me 30 ml of Valium. I was on these for nearly a month. I stopped them after three weeks and it is making me violent ... it is cracking me up.

> I think the prison doctor should think a bit more about people who come through reception with withdrawal symptoms. I came in last time with heavy withdrawals and the doctor didn't give me much Valium to help me through and he took them away too soon.

> I was on drugs on the out and I was suffering from withdrawals and the doctor would not prescribe me anything.

PRISON AS REHABILITATION

There were some drug-misusing respondents who took a different view and saw custody in more positive terms. They were, however, a small minority.

> Drugs are evil things that ruin your life. I can happily say that while inside I have given up drugs and smoking for two months and it's the best I've ever felt. I'll try my best to stay off them in the future.

Put it this way, I took lots of drugs on the outside, but I won't take any drugs in here as I want help from them in here to stop me re-offending.

THOSE WHO DO NOT MISUSE DRUGS

The quarter of the respondents who stated that they did not use drugs resembled in many ways their non-using peers who attended raves and nightclubs. Their perspective demonstrated the same hostility to drug taking and emphasized the need for strict policies of control.

I hate drugs and why if someone gets sent to prison for a crime should they be allowed to carry on doing the same crime? I have a problem too and it's not drugs so why can't I get a questionnaire on housebreaking? Don't let them have any drugs and you're half way to getting rid of the problem.

Drug use is epidemic in prisons over the last five years, it should be stopped.

DISCUSSION

Research has not demonstrated a clear causal relationship between criminality and drug misuse (Hammersley, Forsyth and Lavelle, 1990), but the close association between crime and drugs is wider than the incidence of drug-related offences and the prison environment brings drug misusers and non-drug users closely together. It is more pertinent that a significant proportion of all young men misuse drugs and therefore the rates reported for inmates do not indicate a problem characteristic of this environment as such but rather of this population of young men.

The analysis has shown that the broad distinction between recreational and problematic drug misuse still holds in prison although the context sharpens and to some extent transforms the meaning of the activity. Recreational drug misusers do not see themselves as addicts or dependent but view drug misuse as a rational response to custody in broadly the same way as their peers attending raves account for their behaviour. However, as we shall see in Chapter 9, for many recreational drug misusers their behaviour causes problems not encountered in the community. The problematic misusers evidently regard prison as exacerbating the difficulties they encounter in life and their circumstances as compounding them. Risky and dependent drug misuse becomes more problematic because the agencies and services in the community which exist to support their clientele are no longer available, so the possibility of contracting HIV is always present. Moreover, it is not unusual for inmates to be unclear about their problems and therefore confused about what they need. When a respondent says he 'needs' drugs, it is unclear whether this is because he is addicted, because he is withdrawing too quickly or because he requires them as medication for problems. He may well be confused himself.

It is the nature of total institutions that behaviour is redefined for those who are admitted to them. 'Normality' for the long-stay mental patient, the resident in an old people's home or the prison inmate can only be understood in terms of the philosophy and rationale which informs the particular institution. Prisons, then, redefine drug misuse so that it becomes qualitatively different from the outside world. In talking of their drug misuse, inmates are aware of the illegal nature of the activity. The costs and consequences of discovery by staff and are therefore unlikely to reveal much, if anything, of such matters as injecting or sharing needles. The curtailment of their freedom is reflected in the way they express their views to outsiders. The consequences for dealing with drug misuse are explored further in Chapter 9.

PART TWO
Research and Practice:
Talking to Professionals

INTRODUCTION

Part One has dealt with the experiential knowledge and understanding of drug misusers themselves. The second part of the book will examine the knowledge and understanding of different professional groups, from the scientific researchers studying physiology and behaviour to the clinical knowledge of different types of practitioner.

The views of these various professionals will be influenced by their different knowledge bases and the types of drug misuser they meet (in the same way as the drugs misusers' knowledge is influenced by their own experience and that of their peers).

The following chapters have been arranged into three categories: prevention and education, harm minimization and treatment. The fourth chapter is concerned with the control of drug misusers in institutional environments. It can be seen that there is a match between each of the chapters in Part One and its counterpart in Part Two.

Recreational drug misuse – Prevention and/or education
Risky drug misuse   – Harm minimization
Dependence   – Treatment

This simple framework allows professionals to classify the types of problems and match these to the most appropriate form of intervention. It is clear that one category of problems does not exclude others, as many misusers will have problems from more than one category, but it serves as a rough practical guide for practitioners, not only to the kinds of drug misuse likely to be problematic but also to the different kinds of problems associated with different types of drug misuse.

Drug prevention and education

6

INTRODUCTION

The data in Chapter 2 gave insight into and understanding of the views of some young people about drug misuse. This chapter will analyse the research literature and examine contemporary practice methods which the reader can then place in that context, specifically taking into account the perspectives of those who are not yet misusing, those early experimenters and those who misuse recreationally on a regular basis.

This chapter outlines the rationale for drug education and for prevention campaigns in particular and examines where they are appropriate and why. It then describes the different methods of drug prevention and education, ranging from basic literature to community projects. Literature concerning the effectiveness of drug prevention in different contexts with different target groups is examined to enable professionals to use the approach most effectively. The chapter ends with guidelines for those developing their own prevention and education programmes and a case study.

Drug education and drug prevention policy and practice are based on sets of assumptions and initial premises that are more often than not unstated and even unexamined. Where that is the case and where these fundamental notions are grounded in inaccurate data and oversimplistic, the end result may well be ineffective and wasteful of resources.We have seen in the early chapters of this book that there are good reasons for believing that the common understandings of drug misuse are not well founded.

WHAT IS IT AND WHAT IS IT FOR?

Drug prevention means different things to different people. It can mean stopping people starting to misuse drugs or stopping people misusing drugs. It can simply mean educating people about drugs or it can involve 'demand reduction' (Dorn and Murji, 1992). Or it can mean pre-venting the harm caused by drugs (often called secondary prevention) and teaching people healthy lifestyles (Davies and Coggans, 1991). Drug prevention should not, however, be confused with HIV prevention or prevention of other drug-related harm, as the aims and priorities of each are different.

Secondary prevention will be dealt with in the following chapter under the heading 'harm minimization'. This chapter will cover primary drug prevention and drug education generally.

It is often assumed that the aim of drug education is drug prevention. In local and national policy documents the distinction between the apparently straightforward aims of drug prevention and the broader remit of drug education is often not clarified. It is perhaps implicit within many policies that drug education and drug prevention are the same thing, that if you educate people about drugs then this will prevent them using them. This is clearly not the case for the respondents reported in the second chapter; most had extensive knowledge of drugs and a substantial proportion had acquired it through health education and drugs prevention literature. Few saw the rational and sensible response to this information as not starting to misuse drugs or stopping existing drug use.

This is not to say that either the policies or the drug misusers are right or wrong but simply that if there are two radically different inter-pretations. It is as well to consider whether the protagonists are really considering the same phenomenon and consequently question whether we are educating the appropriate people about the relevant things.

Specific prevention approaches to public and school education have caused much controversy regarding when and whether they ought to be used and if they are effective. Swisher (1993) identified the following areas of controversy:

> Evidence that prevention makes a difference; difficulty in agreeing how to demonstrate the effectiveness of prevention strategies; con-fusion regarding the differences between treatment, intervention and prevention efforts; and concern about the ultimate purpose of prevention for the target population.

For the purposes of this chapter, the term 'drug education' will be used to incorporate drug prevention with other educational approaches; the term 'prevention' will be used only for programmes designed specifically to prevent drug misuse. There can also be confusion about the content of prevention programmes (as many include alcohol and cigarettes); it should therefore be pointed out that most studies included here deal with illicit drugs. It will also be necessary (as for the following two chapters) to refer occasionally to research dealing with a mixture of drugs or specifically to alcohol, cigarettes and prescribed drugs. This is the case because studies often include more than one substance, but also because research on legal drugs is more extensive and comprehensive.

UNDERSTANDING THE PROBLEM

It would appear that the general public have an exaggerated view of the dangers of drugs. This may be indicative of the effectiveness of media and school prevention strategies in providing a successful form of prevention (Leitner, Shapland and Wiles, 1993). However, it is possible that this type of prevention is only useful if aimed at people who have little or no contact with drugs or drug misusers (it may also influence the general public to vote for general drug policies based on inaccurate premises).

The viewpoint of young people starting to misuse drugs and misusing drugs recreationally (elaborated in Chapter 2) has given us an understanding of drug misuse which is somewhat at odds with received wisdom. In that chapter it was demonstrated that non-drug users have an oversimplistic view of the dangers of all illegal drugs, but that when they start to misuse drugs and/or meet people who are knowledgeable about them, their reasons for misusing or not using particular drugs change as a consequence. It follows that the reasons for not misusing drugs if you never come across them are completely different from the reasons for not misusing drugs if everyone else does. The latter are perhaps closer to the rationale for stopping (e.g. relationships, work and issues of responsibility).

Where drug misuse is more common than not among young people, it may be more sensible to find out why people do not misuse drugs rather than why they do and base general prevention programmes on this information. If drugs are acceptable within a social group, the reasons for not misusing may be similar to those for alcohol. It is therefore necessary to find out what your target group already knows and what they are likely to find out, before designing an education package. It would be unrealistic to give non-drug users information about safer drug misuse, such as safer injecting sites, and by the same token, unrealistic to tell cannabis misusers that all drugs are equally dangerous or recreational cocaine misusers that all drugs are inevitably addictive.

In order to understand the problem it is necessary to examine studies of the initiation of drug misuse. This work is mainly limited to school children, though there is some work on young adults. The results give us information about the social context of first drug misuse and sequences of behaviour leading to initial drug experimentation.

GATEWAY DRUGS

Forty years ago research conducted with adolescents and young adults indicated that young people who misuse drugs pass through a sequence of stages, from cigarette misuse to alcohol misuse to cannabis misuse and finally to hard drugs (Guttman, 1950). The goal of developing effective education and prevention programmes can depend on the identification of reliable predictors of onset. Therefore if Guttman's work can be generalized to all other populations, it would be possible to identify which children were more likely to misuse drugs and so target education at vulnerable groups, 'since (if) adolescents do not progress on average to a given stage without having experienced all prior stages, the population at risk for any given stage is that which is at the preceding stage' (Kandel, 1982).

Guttman developed a scaling technique focusing on the sequential and hierarchical ordering of drug misuse. This led to the notion of 'gateway drugs', that is, the hierarchical progression from smoking cigarettes to drinking alcohol to misusing marijuana to misusing other drugs. This research has recently been replicated by Andrews and colleagues (1991). A similar study indicates a relationship between drug misuse and 'delinquent' behaviour; for example, early onset of cigarette and alcohol misuse appears correlated with cannabis misuse (Yu and Williford, 1992). A longitudinal study of self-reported drug misuse (and urine analysis) identified prior 'delinquent' behaviour as a predictor of cannabis and cocaine misuse over time (Dembo, 1991). Kandel and Yamaguchi (1993) found that first drug misuse at an early age is a strong predictor of further progression. It has also been suggested that increasing frequencies of alcohol and cigarette misuse may also be markers for more serious patterns of substance misuse (Bailey, 1992).

Most studies indicate a sequential progression from legal to illegal drugs but there are exceptions. Blaze-Temple and Kai Lo (1992) have illustrated that this does not necessarily apply in all cases. They interviewed a group of people who misused dance drugs and found that the majority had used tobacco and slightly less had used alcohol, but that 29% had never misused marijuana. More than a quarter of their sample did not therefore progress through the 'hierarchy of drugs'. They concluded that it may be possible that young people misusing dance drugs have not progressed through a hierarchy of gateways.

It is important to note that, even if this sequential progression is valid, it does not imply causation (Clayton and Ritter, 1985). Despite the lack of causal evidence, research has led to the generally accepted belief in some circles that a predictor is a causal factor, that is, that misuse of one drug causes misuse of another. Some researchers are clearer about the limitations of their results than others; for example, Mensch and Kandel

(1988) demonstrated a relationship between dropping out of high school in America and early drug misuse, but stressed that they were less clear whether prior drug misuse increased the propensity to drop out or dropping out led to drug misuse.

It is now thought probable that independent predisposing factors explain all drug experimentation rather than one drug leading to another. There is no doubt that it is risky to see drug misuse indicators as explanatory or view them in isolation from differences in vulnerability at different ages and within different family and social contexts. The existence of predictors or markers of later drug misuse – early onset of cigarette and alcohol use – may be useful in targeting education programmes.

Although evidence is scattered at present, there is some indication that social influences are strongest in the period of initiation into drug misuse rather than increased misuse (Steinberg, Fletcher and Darling, 1994). As a consequence, the proposal that we should develop an integrative theoretical framework for the development of prevention strategies and an accompanying content which are matched to the developmental stages of drug misuse seems a reasonable way forward (Werch and DiClemente, 1994). DiClemente's own work on stages of change in the treatment field has radically changed practice in many drug agencies (see Prochaska and Diclemente (1986) and Chapters 8 and 10 below). However, this process approach to treatment has not yet been developed in prevention.

Scheider, Newcombe and Skager (1994) stress the importance of considering age-related developmental phenomena in any drug programme. This is particularly pertinent for Britain, where the developmental or maturational approach to education generally is more pronounced than in America, and it may therefore be appropriate to develop this approach within the latest (1995) initiative on drugs education in British schools.

Unfortunately little work has been carried out differentiating between individuals in terms of cultural factors or gender. The studies quoted in this chapter range from America to Australia and Britain. It is difficult to make comparisons, as different studies highlight different variables. A study in Trinidad and Tobago by Singh and Mustapha (1994) demonstrates high correlations between drug misuse and less religious involvement and less commitment to traditional values. Catelano et al. (1993) have found higher rates of tobacco and alcohol initiation among European-American children than African-American children; although the same risk factors predict the variety of substances used, there were differences in the level of exposure to risk factors. There has, however, been an emphasis on sociocultural factors in some alcohol education programmes (Globetti, 1988). Cheung (1993) considers the limitations and utility of a range of different approaches to ethnicity. Raffoul and Haney (1989) consider the issues for older people in the light of ethnic differences.

There are few studies examining gender, but these have identified clearer differences; for example, Kandel and Yamaguchi (1993) found that alcohol was more important among males and cigarettes among females in the progression to illicit drugs. Hammer and Vaglum (1990) found that unemployment was linked to alcohol misuse for men but there was a negative relationship for women.

Finally, it is difficult to find correlations with personality (Hammersley, Lavelle and Forsyth, 1992), although there are some data to indicate that people reporting a history of child sexual and physical abuse and mental health problems may be more likely to misuse drugs (Shearer, 1990). It has also been suggested that those adolescents who are apparently invulnerable to drug, alcohol and nicotine misuse are more healthy, physically and mentally (Marston et al., 1988).

In order to understand the influence of educational approaches on beliefs, attitudes and behaviours, it is necessary to consider briefly the theories underlying cognitive and behavioural psychology.

THE SOLUTION: EDUCATION AND PREVENTION IN THEORY, POLICY AND PRACTICE

THEORETICAL BACKGROUND TO HEALTH EDUCATION AND DRUG PREVENTION

Practitioners in health education and drug prevention often tend to ignore a great deal of the theory and research in their own field, particularly that concerning the influence of attitudes on behaviour (Kirsch, 1983). However, when the research is analysed, this may seem to some a rather sensible decision on their part.

Much of prevention and health education research is psychological, based on social learning theory, cognitive theories and methods derived from these approaches (see Kim, McLeod and Shantis, 1989). This work on health behaviour is summarized below. It should be noted that drug misuse and health are not strongly related (Hammersley, Lavelle and Forsyth, 1992) and that young people, particularly students, are less likely to respond to health messages about substance misuse than adults (Duran and Brooklyn, 1988).

Although psychological models can be useful, it is important to understand their limitations. The two main models of health behaviour derived from this psychological base are the theory of reasoned action and the health belief model. These models explain how cognitions are linked to health behaviour but they do not account for the very small variance in outcome studies and serve no real predictive function (Silver and Wortman, 1980; Turk, Rudy and Salovey, 1986). There is little evidence to show that cognitions actually influence behaviour change and some psychologists have themselves come to doubt the usefulness of these theories.

One of the early models derived from this base was the health belief model (Becker, 1974), which emphasizes the relevance of individual beliefs to the consequences of an action (for example, smoking), likelihood of getting a disease and the perceived costs and inconvenience. Leventhal and Nerenz (1985) reject both social learning theory and the health belief model because both assume that people react to events such as illness in particular ways (e.g. in terms of perceived seriousness and vulnerability). In other words, that ordinary people function in the same way as scientists and statisticians using abstract and general notions. Leventhal and Nerenz suggest instead that people think and react to situations in a specific rather than a generic way, in concrete rather than abstract terms; they do not react to the world as statisticians do. An abstract or generalized intention does not therefore necessarily lead to a concrete or specific behaviour.

Cognitive behavioural theories derived from social learning theory also ignore the contemporary importance of advertising and the media and, perhaps more importantly, they underestimate the importance of social factors (Cartwright, 1979). For example, cigarettes perform a 'social' function for the young and provide a means of social barter and bonding. Young people are often aware of the hazards of smoking, yet they perceive that benefits in social terms outweigh the perceived potential problems. As far as smoking is concerned, then, ensuring the establishment of non-smoking as a social norm appears to be crucial to the prevention of smoking in the longer term (Royal College of Physicians, 1992).

The theory of reasoned action is a slightly more complex approach also derived from social learning theory and used in health promotion and prevention work, but placing more emphasis on the socioenvironment. The key components are intention (the beliefs and evaluation of consequences) and the socioenvironmental component (the individual's perception of how peers or friends react to that person's smoking in general). The theory stresses the importance of an individual's beliefs about the personal consequences of a particular course of action (Sutton, 1992). These theories lead to an analysis of what is seen as rational behaviour in terms of the individual beliefs and social/cultural context of subjects.

Psychology has tended to focus on outcome studies of particular interventions, as a scientific methodology is designed for this purpose. However, a recent model of change processes has highlighted the limitations of this approach. This model of attitude and behaviour change is the 'stages of change' model of Prochaska and DiClemente (1983). The authors developed a model of change from their work with smokers, incorporating the notion of the sequential process of change. They then modified it to develop a practical motivational counselling method for work with drug and alcohol misusers. The method is based on a model of individual change consisting of four stages: pre-contemplation; contemplation; action, and maintenance of change (see also Chapters 8 and 10 below). This process model of change has yet to be applied to the area of drug prevention, but it is possible that preventive interventions will be more effective at particular stages, for example when an individual is contemplating or intending to experiment with a drug.

What, then, are the practical implications of psychological research and cognitive theories of health beliefs? Drug education and prevention programmes usually function by giving information and/or changing attitudes, assuming that this will influence future drug-using behaviour. Psychological researchers have created theories but cannot demonstrate a direct link between cognition and behaviour. Cognitions have been shown to be strong predictors of behavioural intentions (Fishbein and Ajzen, 1985) but not of actual behaviour. There have been few attempts to examine the factors leading from intentions to behaviour. As might be expected, these results are reflected in the evaluations of drug prevention programmes designed to affect behaviour; while knowledge can be shown to alter attitudes and beliefs, this does not appear to have a direct effect on actions.

The limitations in the research in this field are partly due to the inherent constraints within the psychological discipline itself. Because its subject matter is restricted to individuals and small groups, psychology tends to ignore the complexity of social context and because its methods are largely scientific outcome studies, the relevance of process in individual change is disregarded. Despite these considerations, the theories of health behaviour help to interpret the complex interactions of cognition and behaviour. Because research cannot demonstrate that changes in cognition will produce corresponding changes in behaviour, it does not necessarily follow that they are not related. It is equally possible that cognitions are related to behaviour but that other factors are also influential. Unfortunately there is much less research about such influences. The focus of drug prevention therefore remains firmly within the remit of health educators and teachers, many people arguing that changes in knowledge and attitudes will lead to changes later in life and that changes in other health behaviours can be attributed to education (Gonzalez, 1988).

POLICY AND PRACTICE IN DIFFERENT COUNTRIES

As might be expected, there is in effect little agreement about what the aims of drug education should be or which professions should be involved in drug education or prevention. This is significant as different professions have different priorities and will therefore give different messages. So the police might be expected to stress drug prevention, law and order and perhaps the moral aspects of personal responsibility, whereas a health worker may be more likely to teach harm minimization and emphasize health risks associated with particular methods of drug misuse.

The focus of drug education in Britain, Australia and America has also been different, as have their policies towards harm minimization and treatment. America has stressed the role of the police and the specific remit of drug prevention. Although there is a range of programmes in America, from the peer-led SPARK programme in New York to the Person Education-Development Education (PEDE) programme from Minneapolis (see Kim, McLeod and Shantis, 1989, for list of US programmes), it is the police-led programme DARE that is more widespread, having been adopted by approximately 50% of local school districts nationwide (Ennett *et al.*, 1994). This programme is largely didactic, 'with officer as expert and makes frequent use of lectures and question and answer sessions' (Ennett *et al.*, 1994). While there is some evidence that police involvement has improved police relations with school children and the community as a whole, there is less evidence for other benefits of police involvement in drug education in schools (Williams and Keene, 1995).

In contrast, Australia stresses that drug education should be part of a comprehensive health education programme. There is no reference to police participation in the extensive Australian review of projects (James and Carruthers, 1991). Instead, there is a strong emphasis that drug education belongs in a comprehensive health education programme (NCADA Task Force Evaluation, 1988). This review concludes that:

> Research conducted in Australia by Irwin *et al.* (1990), Homel *et al.* (1981), Thompson (1988), Garrard and Northfield (1987), Garrard (1990) and Ballard (1988) indicates that teachers, when properly supported and resourced, can develop relevant and effective health/drug education programmes for little extra cost to the system.

Dickerson (1991) recommends that teachers should be trained, together with other generic helping professions, to present primary prevention.

In Britain, the Advisory Council on the Misuse of Drugs (1993) has emphasized the health aspects of drug misuse and the need for schools to be more involved in drug education, reflecting the emphasis of the National Curriculum Council (1990) (see Williams and Keene, 1995). Traditionally police have had a limited role in drug education in schools. However, several pilots of the American DARE project have recently been tried here (Whelan and Moody, 1994; Keene and Williams, 1996). Although the involvement of police in the United States is widespread and fairly well integrated with the content and methods of US school curricula, this is not the case in Britain.

The Advisory Council on the Misuse of Drugs (1984) carried out a review of research on drug education in schools. They concluded that it is possible to increase knowledge about drugs and their effects, but attempts to influence attitudes towards drug taking show mixed results, with more negative than positive findings, and that studies looking at projects aiming to reduce drug misuse mostly show no change in level of misuse. The British National Curriculum Council (see Williams and Keene, 1995) also state that information alone is unlikely to promote healthy behaviour.

It should be noted, however, that evaluations of all prevention programmes tend to rely on short-term outcome studies of samples of non-drug-using school children. If there were any long-term effects on drug-misusing behaviour, these would have been difficult to identify. The authors who have reviewed research (Goodstadt, 1974; Berberian *et al.*, 1976; Schapps *et al.*, 1981; Dorn and Murji, 1992; Williams and Keene, 1995) have come to similar conclusions – that it is impossible to say whether these programmes have been effective or not.

While most drug education and prevention is aimed at young people in schools or involves the distribution of health education leaflets, in recent years drug prevention in Britain has been developed outside the usual school-based work. There have been several education campaigns based on dance drugs attempting to reach young people who attend clubs and raves. The Lifeline project in Manchester produced what is probably the earliest and best known literature with the 'Peanut Pete' campaign. This involved the design of a cartoon leaflet aimed at recreational drug misusers – at the time a newly identified group. Lifeline identified them as 'A group dedicated to lively enjoyment ... requiring a different treatment and educational approach from opiate addicts' (Gilman, 1992). The author described the ways in which this agency attempted to attract these misusers to its service:

> The task of attracting these newer users fell into four stages. Firstly we had to get a clear idea of who they are – an 'identikit' of our target group. Secondly we had to decide how to advertise our service ... Thirdly we had to develop our service response to their needs ... fourthly we had to arrange monitoring of our contact (p. 18).

A similar project was developed in Brighton by the drug agency DAIS. It also began to identify a different type of drug misuser from the older heroin addicts. Clients in the drug agency started to present with problems of stimulant and hallucinogen misuse. This spurred the drug worker to develop health education strategies to deal with these problems outside the agency (Fraser, 1991). The agency developed an educational risk reduction campaign in an attempt to reach the '... happy consumers ... There could be at least a thousand and perhaps as many as two thousand at any one time' (p. 13).

There is a consensus between all three countries that young people are effectively targeted through schools, although there is a growing awareness of the need for community-based programmes and the 'social influences' approach (Kim, McLeod and Shantis, 1989).

> School based education is a necessary but not sufficient ingredient in preventing drug abuse. Schools provide an opportunity for imparting accurate knowledge, systematically examining values and practising decision making and learning to cope with social forces within a controlled setting. Schools by themselves, however, cannot counter the range of powerful forces that operate outside the walls of the classroom and school. Therefore, schools based programs required the support of complementary home based and community based programs. (Goodstadt, 1987)

Although it seems sensible, when considering a new programme, not to decide which profession should carry it out before determining what needs to be done and the best methods of doing it, the reality is often that the most enthusiastic individuals from any profession take the lead. As we have noted above, the focus of the professional is likely to vary. Teachers would probably take a more sophisticated approach to the teaching itself, emphasizing learning as a developmental process, and the methods would be likely to differ from those of police and health workers. For example, in British schools interactional skills-based learning within a developmental framework of personal and social skills is common practice.

Perhaps the role of the educator in determining the content becomes even clearer if the newer approach involving peer-led education is examined. Here, school children or young people are recruited to educate others. This seems initially to be effective, for the reasons that Jessor, Collins and Jessor (1972), Davies and Coggans (1991) and Conrad, Flay and Hill (1992) suggest, but there is a need for high-status individuals to be tutors and facilitators. They may not have the same beliefs or priorities as the organizers.

PRACTICE: METHODS, DRUG EDUCATION AND PREVENTION

There are various models underlying drug prevention and education methods. Botvin (1990), Davies and Coggans (1991), Dorn and Murji (1992) and Williams and Keene (1995) identify the following:

- transfer of knowledge or cognitive methods;
- the life and social skills approach;
- social and community approaches;
- a mixture of knowledge and skills, with back-up in the family and the community.

Dorn and Murji sum up the aims of some of these methods as follows:

- providing information to individual decision makers;
- seeking to remedy supposed deficits of moral values or living skills;
- bolstering peer resistance strategies in the context of anti-drug norms;
- providing alternatives to drug misuse through youth and community participation.

Transfer of knowledge or cognitive methods

The credibility of the teacher is important here. We know that peer influence is greater than that of adults as children get older (Jessor, Collins and Jessor, 1972), which may explain why peer-led programmes appear to be more effective. Eiser, Eiser and Pritchard (1988) carried out a study of the effects of videos on young people and found that the entertaining video had a greater effect on attitudes than no video or a didactic programme. Davies and Coggans (1991) suggest that adult credibility is rapidly undermined by delivering inaccurate or prejudicial information. As they point out, teaching young people to:

> 'just say no' under the guise of life skills ... is potentially problematic in that many young people may realize that they are not really being helped to enhance their decision-making skills as such; rather, they are being told what to do (p. 56).

The life and social skills approach

This category includes general life skills programmes and projects designed to teach specific drug resistance skills. The former work on improving coping skills, self-respect and self-confidence generally; the latter emphasize assertiveness and communication skills and specifically resisting peer pressure.

These strategies are to some extent based on an understanding of experimental drug misusers as inadequate, unskilled and pressurized. They also emphasize the need for young people to develop alternative ways to achieve the social status acquired by experimental drug misuse.

Although there is some evidence that low self-esteem and drug dependency are correlated, this does not tell us if experimentation and self-esteem are related nor does it tell us which led to what. The resistance training was specifically designed to counter pressure to misuse, which was thought to be one of the main reasons that young people would start misusing drugs (Ringwalt, Ennett and Holt, 1991). For this reason, much drug prevention education is targeted at younger school children, before they have reached the likely age of experimentation.

Although evaluations show little positive change in terms of drug-misusing behaviour, social skills and other skills-based, or affective, programmes have been shown to have a greater effect on knowledge and attitudes than information-only approaches (Rosenbaum *et al.*, 1994).

Social and community approaches

These comprise approaches which use social environments, whether family or community networks, and those aimed at providing alternatives to drug misuse through youth and community participation. Ives (1988) argues that many young people have not misused drugs and have a negative attitude towards them and therefore that there should be more involvement of these young people in prevention schemes, as they are most likely to be listened to by others.

A mixture of knowledge and skills

Eggert and Herting (1991) have shown that peer social support is a useful means of counteracting probable antecedents of drug misuse, such as prior drug misuse, family disorganization and poor school attendance. Klitzner, Gruenewald and Bamberger (1990) found that parent groups had an effect on parental control of children's social activities but not on their drug and alcohol misuse. As with school programmes, they also found that those parents who attended parent groups were not those with children at risk of misusing drugs.

DOES IT WORK?

The answer to this question is directly affected by the amount and nature of research which has been undertaken. There is a clear need for longitudinal studies of prevention and a corresponding one for studies of individual differences. There is also a paucity of detailed ethnographic studies to allow thorough understanding of the target group and of studies of particular subgroups to identify distinguishing features of at-risk populations and cultural beliefs systems which maintain at-risk status. The few longitudinal studies indicate that once the lessons have stopped, the programmes' effects also stop (Ellickson, Bell and McGuigan, 1993).

Drug prevention programmes have largely been aimed at children and until recently have been dominated by schools-based education following an information and scare tactic approach. The first flickers of doubt about the utility of this approach came with the work of De Haes and Schurman (1975), who compared different forms of this type of drug education only to find that none worked; that is, no method stopped pupils starting to misuse drugs. If anything, the fear-arousal and infor-mation strategies simply increase curiosity and experimentation with soft drugs. Reviews of the literature report that the impact of these programmes is minimal (Goodstadt 1974, 1987; Schapps et al., 1981; Dorn and Murji, 1992). There is a consensus among researchers that, while drug information programmes do increase knowledge about drugs, they do little to change attitudes and behaviour.

These results have not led to a focus on different target groups but on different methods of teaching children in schools, resulting in the development of skills-based interactive learning programmes and often including peer education and the involvement of family and commu-nity. Research carried out with school children suggests that social competency and resistance skills-building programmes are effective to some extent in prevention of drug misuse and smoking. Alberts et al. (1992), in an American study, have shown that peer pressure was applied in up to 70% of cases, usually after drugs were refused. Results from this type of local survey are indicative of significant geographical and local variations.

These conclusions are based on the notion that children misuse drugs because they do not have the self-confidence or social skills to say no or alternatively, that they misuse them because they are unskilled and unsocial. 'The programmes (and research reports) are based on the premise that deficiencies in psycho-social skills in youth contribute to vulnerability and substance misuse' (Gross and McCaul, 1992). These assumed premises of the programme managers and researchers are not proven, nor shared with drug misusers themselves.

The general influence of family and friends may be more relevant. In a study of the influence of family and peer group on adolescents' drug misuse, Lopez, Miron Redondo and Leungo (1989) demonstrated that the parents' drug and alcohol misuse and their relationship with the adolescent influenced the latter's drug misuse. Orford and Velleman (1990) and Maltzman and Schweiger (1991) indicate that parents' increased drug and alcohol misuse leads to children more likely to start alcohol misuse earlier and more frequently and to misuse other drugs in their teens. (However, there are possible genetic and biological effects which complicate these data.) There is some evidence that influence of family and friends can reduce the amount of abuse of prescription drugs amongst adults (Kail and Litwak, 1989).

The family and peer group can therefore have both a good and bad effect on drug misuse, either preventing or modifying problems or contributing to the early onset of alcohol and drug misuse. Hsu (1993) suggests that the family should therefore be taught to help identify and prevent drug misuse. Hoffman (1993), Sarvela and McClendon (1988) and Swadi and Zeitlin (1988) indicate that although family factors are important, it is peers that have the strongest effect on drug misuse. Steinberg, Fletcher and Darling (1994) examined the influences of parental monitoring and peer influence on 6000 adolescents over time. They found that, although parental monitoring was associated with less drug misuse, the strongest influence was the drug misuse of peers. These factors are therefore relevant both as useful indicators of possible future drug misuse and as a possible positive influence through peer group and family interventions.

There is less information regarding the success of media and general public health information campaigns but the effects seem similar, in that prevention-orientated information increases knowledge and may change attitudes or even intentions, but even the best seems to have little effect on behaviour (Makkai, Moore and McAllister 1991). Information about controlling drinking or stopping smoking aimed at adults, for example through GP surgeries, has more success (see Botvin, 1990, and Royal College of Physicians, 1992).

There is less research available regarding drug prevention work with adults. It has become clear from studies on the prevention of smoking that changing adult behaviour is the precursor of changes in the behaviour of children. Reduction in smoking prevalence seems to be a result of people giving up smoking rather than fewer people starting. None of the British child-orientated interventions have had much effect on teenage smoking, because attempts to change the behaviour of children directly have largely failed if the behaviour of their parents is unchanged (Royal College of Physicians, 1992). In America the same conclusions were reached by the US Department of Health and Human Services and described in a report by the Surgeon General (1989). This concludes that the general fall in young peoples' smoking in the United States had followed and not led the major decline in adult smoking since 1960. Despite this knowledge, '... there have been few educational programmes specifically targeted at young people outside the school environment' (Nutbeam, 1988). It is unclear why the focus of prevention campaigns should remain on children in school.

The report from the Royal College of Physicians also emphasized that, while children are influenced primarily by their parents' behaviour, influences on young adults are different: 'Although the family has the first impact on the child, as he or she grows older the influence of friends becomes extremely strong ... Many studies have shown that best friends smoking is one of the most important factors related to the uptake of smoking' (p. 60). The effect of peer influence is also stressed in other studies of adolescents and young adults (see Conrad, Flay and Hill, 1992, and Sutton, 1992).

The research therefore shows that the influences on children are not the same as those on adults. This is compounded by the likelihood that children do not function in the same way as adults: 'It seems likely that children's attitudes towards smoking are intrinsically different from those of adults – comparably more intuitive, less rational, less stable and probably even less closely related to behaviour' (Goddard, 1990). Therefore it would clearly be unwise to generalize from research concerning children to young people or adults.

The conversations in Chapter 2 of this book may help us understand why. The way drug misusers see themselves and their drug misuse itself seems very different from the views of young children, drug educators, the media and general public. First, the respondents talked a great deal about the good things associated with drug misuse and saw their decisions to start and continue using as rational and sensible. They did not recall being pressured or bullied into taking drugs, nor did they see themselves as lacking in social 'resistance skills'. Instead they had acquired a great deal of information about the effects of drugs and the dangers of unsafe drug misuse from their friends and any self-respect may have been derived from what they saw as 'sensible' drug misuse.

Second, many people who misused drugs did not see themselves as 'drug abusers' or 'drug addicts', but simply people who occasionally misused recreational drugs socially as if they were drinking alcohol. This is borne out by the work of Coggans et al. (1990), Newcombe (1991) and Solowij, Hall and Lee (1992), all of whom point out that drug misusers see their own misuse of drugs as different from that of junkies or addicts. If the target group do not see themselves as the appropriate audience, but instead as radically different from the stereotypical drug misusers portrayed in education and media, then their lack of response is more understandable.

The implications of these two different pictures of drug misuse throw some light on why drug education projects based on the traditional stereotypes of the drug misuser may be less than effective.

CONCLUSIONS: TARGETING

Research shows, then, that it may be worth targeting those who are apparently at higher risk of drug misuse (those who start smoking and drinking earlier than average), even though the 'gateway theories' of drug misuse should be treated with caution. It may also be necessary to target in terms of geography and locality, i.e. in areas where drug misuse is known to be greater. However, there are several points to be borne in mind. First, the drug-abusing person should not be seen out of their socioeconomic context (Mason, Lusk and Gintzler, 1992). As can be seen from the studies of nightclub attenders and school children, many youngsters now misuse illicit drugs; they cannot therefore be categorized as part of a small minority of delinquents.

Secondly, drug misuse among school children likely to drop out of school is very different from the national norms for drug misuse. This means that the types of risks each group is exposed to are different, which has implications for the content of the information directed at each group.

Eggert and Herting (1993) demonstrate that although 'drug access' and 'drug misuse control' are similar for both high-risk and low-risk groups, frequency of drug misuse itself and adverse consequences of misuse are higher in the high-risk group:

- **High risk**: frequent misuse, adverse consequences, easy access, poor control;
- **Low risk**: infrequent misuse, few adverse consequences, easy access, poor control.

They stress that different populations of youth will require varying strategies because of different patterns of involvement. These authors also suggest elsewhere that these populations may also have significantly different patterns of change over time (Eggert et al., 1994). These issues have important implications in terms of targeting appropriate interventions at different groups of people and underline perhaps the most important aspect of drug prevention or education – that the target group must be thoroughly understood in terms of knowledge, attitudes and risk behaviours within the context of their social environment.

It is also important to note that schools-based interventions may fail to reach some at-risk adolescents who may be irregular attenders. An example of a more appropriate intervention for this group might be outreach targeted at vulnerable groups outside the school environment.

It can be seen from the earlier part of this book that young people are often well informed about drug misuse and they often talk of engaging in drug misuse as part of a lifestyle common to the high-risk environment in which they live. Knowing that certain behaviours carried a high risk was not enough to change these behaviours. It therefore seems clear that drug education strategies for adolescents in high-risk environments must address the issues posed by these environments, when determining content and method of intervention.

Lamarine (1993) argues for a new approach to drug education programmes which would direct intensive interventions at the minority of youth who are particularly susceptible to drug problems. McLaughlin *et al.* (1993) found that teachers could be trained to identify students who were at risk from substance misuse. Hawkins, Catelano and Miller (1992) also recommend a risk-focused approach, emphasizing that it requires accurate information to allow the identification of risk factors and also to enable identification and matching to the most appropriate methods and content. This focus is advocated by several researchers, notably Scheider, Newcombe and Skager (1994), who stress the importance of content of drug education and emphasize that prevention is only one type of education and that risk reduction may be more appropriate for some populations.

PRACTICAL GUIDELINES FOR THOSE INVOLVED IN DRUG PREVENTION

The two important factors are **message** and **target group**.

Much hinges on whether providers are concerned predominantly with messages of drug prevention or harm minimization. It is in the discussion of these concepts that the aims and priorities of providers can be clarified. If the aims and objectives of any intervention are made explicit and the target audience clearly defined, the skills and priorities of relevant professionals can be matched to the programme and audience. It also becomes easier for different professions to collaborate in providing education both in schools and in the community.

For example, in Britain policy guidelines recommend a multi-agency approach in schools, involving teachers, health workers, drug specialists and police. Policy makers and providers will need to be aware of the different priorities, content and methods of each profession and conversely of the audiences, drug-related knowledge, attitudes and experience and learning abilities at different ages. With regard to schools, first the aims are various:

> In any one class of primary and secondary school pupils, we can anticipate that there will be pupils for whom the emphasis should be on prevention, whereas for others harm minimization should be a priority (Williams and Keene, 1995)

Second, to achieve curriculum objectives for drug education:

> the most up-to-date and effective teaching methods must be employed ... teachers and others need to have access to interactive approaches and should not be reliant on didactic and out-dated methods. The production of pupil manuals and audio-visual aids is only a part of the design of appropriate learning experiences for pupils. (Williams and Keene, 1995, p. 11)

It is necessary to find the most appropriate medium for the type of message to be conveyed. There is much emphasis in the literature on the need to locate drug education in a comprehensive course of health education. It is included in the statutory core and foundation subjects and in the non-statutory crosscurricular theme of health education. Many schools have incorporated drug education in programmes of personal and social education or personal, health and social education or guidance education.

Third, programmes need to assess the nature of the local drug cultures and the variations of drug experience and risk, according to pupils' ages and stages of social development. They also need to match the skills, knowledge and priorities of the providers to the needs of the target group. 'Drug education programmes must be locally sensitive and learner sensitive if they are to be credible and effective' (Williams and Keene, 1995).

Future research and practice in drug education should perhaps focus on high-risk individuals, within particular social groups and geographical localities. Those most at risk may be less likely to respond to traditional prevention methods and messages rather than to peer programmes and community-based back-up programmes offering a range of information tailored to the knowledge, attitudes and behaviours of those targeted.

OPTIONS

Professional involvement

It has been suggested that teachers should be drug educators in conjunction with the police, as they have experience and expertise and could take this role with little training or resource implications.

Action: In-service training for teachers.

Teaching methods

It is also suggested that didactic methods be replaced with general interactional and skills-based methods which have been found to be more effective. These methods should be integrated with other core curriculum and educational process structures.

Action: Develop drug education as an integral part of life/social skills packs.

Content and target audience

Overall, it is suggested that drug prevention content should be integrated with other life and social skills programmes and not taught as drugs-specific education. Specific at-risk groups should be targeted for specific health-related drugs education.

Action: Develop exploratory ethnographic research and pilot project, targeting at-risk groups in schools.

TASKS

Therefore there are three basic tasks:

1. Clarify aims and objectives. Drug prevention, general drug education, harm minimization.
2. Clarify target audience. Non-drug users, inexperienced drug misusers, experienced drug misusers. School children, young adults, general population. Research your audience and adapt content to locality, knowledge, age, etc.
3. Match appropriate professionals to aims and audience. Police priorities are community relations, law and order. Health workers' priorities are drug-related harm and health care. Teachers' priorities are child development and education. All have different kinds of knowledge about drugs and different kinds of skills in communication.

PLANNING AND DESIGNING YOUR OWN DRUG EDUCATION PROJECT – QUESTIONS

Questions you should ask yourself:

What are the aims?

Prevention of drug misuse; stopping children starting or stopping existing misuse. Harm minimization; safer drug misuse, fewer health risks.

What are my specific objectives?

- To give information.
- To change attitudes.
- To change behaviour.

What is the extent of drug misuse in my target group now and what else do I need to know about the target group?

- Level of knowledge and experience.
- Attitudes.

- Social context.
- Alternatives.
- Costs and benefits of drug misuse.
- Geographical local differences.
- Individual differences (within different age groups).

Although the consensus of academic opinion seems to be that targeting education at specific groups or individuals is the answer, there is very little concrete guidance as to how to do this. Psychosocial measurement scales have been developed to assess indicators of likely drug misuse (e.g. Henley and Winters, 1989), but there is little evidence of their effectiveness.

Who should receive it?

Everyone or targeted at at-risk children only? Groups of children at different ages can vary and individuals within groups can vary.

Who should provide it?

General teacher, specialist teacher, visiting speaker (health or police) or drug agency worker (how will this affect the message?).

In what context should I provide it?

- In school – curricular or extracurricular (separate or integral part of social and life skills programme).
- One-off or several over time.

What methods should I use (which package)?

What content should I use (which package)?

How should I evaluate it?

Answers to these questions will require a small pilot study of your target audience. Do not decide on your message before you have information on your target audience, as you will need to modify the content of any programme or package to suit the knowledge and experience of your target group. For example, as can be seen from the respondents in Chapter 2, there will be little point in using a drug prevention pack for those children or adults who already know more about drugs than you do.

TARGET GROUPS – ANSWERS

A total population or a high-risk sample?

If you wish to target a whole population you will need to take account of individual differences within it. If you wish to target those at greater risk from drug misuse you will need to identify them in some way. At present we do not have accurate predictors of who will have drug problems, though we do have pointers as to who is more likely to misuse illicit drugs.

Prevention: stopping people starting

Rosenbaum and colleagues (1994) suggested that drug prevention should be targeted at the age group most likely not to have started experimenting with drugs. This makes sense if the aim is to stop people starting rather than to persuade people to give up. The ACMD and Health Education Authority (1991) in Britain recommend resistance training prior to the transition period from primary to secondary education (10–14 years). This is similar to the American target group.

General drugs education: reducing the harm and stopping people using

If you want to target children or young people already misusing drugs, you may wish the emphasis to be less drugs prevention and more drugs education generally, including information about the health risks attached to particular drugs and methods of misuse. You may find it practical to tailor different types of programme at a total population (e.g. prevention) and a high-risk sample (e.g. harm minimization).

Programme providers

You may wish to provide the programme yourself or to involve others. Check on the relative skills and priorities. For example, teachers have teaching skills and knowledge of the target audience, health workers have knowledge of health issues, specialist agencies have knowledge about problematic drug misuse, police have knowledge about the legal aspects of drug misuse.

Programme content

One of the drugs education packages could be used (see p. 121 for a list of training and educational packs) but check that the information is relevant to your area and to the target audience. Local differences can be substantial and vary with time. Curriculum issues are relevant if your programme is to be school based. It may be sensible for it to be located in the National Curriculum structure with its emphasis on continuity of development. Drug prevention in schools should be adapted to fit existing philosophies and structures within schools and developed as an integral part of general interactionist skills-based learning in the PSE teaching (i.e. part of the process of developmental, maturational learning). There is no reason why there should be a specifically large input of drug prevention as research shows that it has little effect. There has been very little research on targeted harm minimization with young people; this may well be more effective.

Programme evaluation

It is difficult to evaluate prevention and education projects because short-term outcome gives little indication of long-term changes in attitudes and behaviour, nor any detail about the processes involved. You cannot place your own evaluation within the context of previous research as the deficits highlighted earlier restrict the conclusions that can be generalized to your own work. Because usefulness of outcome studies is limited, it also makes sense to carry out a longitudinal study of your project to describe what happens during targeted intervention over time.

It would be useful to develop research looking at the practicality and efficacy of targeting those at risk with tailored interventions. The criteria for targeting are at present unclear but this could be done in terms of knowledge and attitudes, social environment, crime, other problems. In schools, small groups of at-risk children could be identified for extra teaching or specialist input.

CONCLUSION

Despite the knowledge that social and family relations are important to the initiation and maintenance of drug misuse, it is important to stress that there is not enough evidence to give us any clear idea of how resistance to drugs can be increased in children or adults. Nor is it apparent which social or life skills would be useful in preventing people experimenting with drugs. Issues of morality and social responsibility are in all probability of great importance, but these are not within the remit of contemporary researchers, who are not therefore equipped to offer specific guidance in these areas. We do know that it is possible to carry out both prevention and harm minimization, by giving general skills-based education to all children and targeting at-risk children for harm minimization. We also know that prevention is only really effective for non-users and experimenters; once people become recreational misusers, they need education and public health safety measures (a social form of harm minimization).

It is important to note that general prevention programmes are not the only alternative. While they serve an educative purpose, it may also be effective to target smaller groups of drug misusers or high-risk pupils with harm minimization information and equipment to enable them to misuse drugs in a safer way.

A CASE STUDY. POSSIBLE PITFALLS

Project 'Pitfall' (not its real name) was an initiative designed to prevent drug misuse among young people attending nightclubs and raves. The project lasted for approximately six months and involved a team of workers attending nightclubs and raves on a regular basis to hand out literature about the dangers of drug misuse and gadgets promoting drug prevention. It was targeted at the population described in Chapter 2. It highlights many of the problems inherent in this type of programme and offers many learning experiences.

The project began without a clear vision of what its message would be or exactly what its target population was. The project staff made contact with young people in a variety of nightclubs and raves and distributed a substantial quantity of pamphlets, postcards and gadgets, e.g. key rings and T-shirts. They offered advice and arranged for some of the young people to contact other agencies. The initial lack of direction created difficulties which are discussed below.

The author carried out a study of the process of setting up and developing the project. The outcome study was quantitative, while the process study used qualitative methods – document analysis, non-participant observation and interviews.

PROJECT PITFALL: RESULTS

The project staff had conflicting aims and methods: some wanted to achieve drug prevention through public education; some wanted to achieve harm minimization through outreach work.

The prevention aims of the project will be considered here as a guide to avoiding pitfalls in prevention work. Although respondents generally said the project was well organized and the information attractive and well presented, there is little evidence of attitude or behaviour change as a consequence and few said that they found it useful for themselves.

If the attitudes and opinions of the young people in Chapter 2 are recalled, we begin to have a clearer understanding of why prevention might not work with this type of population. If they already had a realistic understanding of the risks, as the majority did, the prevention literature was unlikely to influence them. It is plausible that even a badly tailored project would fit some of the target group and it did; new young attenders at raves were less knowledgeable and perhaps more in need of advice and information and there is evidence that Project Pitfall changed attitudes among this small proportion of its overall target group. Results indicate that younger subjects contemplated the advantages and disadvantages of drugs and considered changing their behaviour more frequently than older subjects. The main target group themselves recognized the usefulness of the project for certain types of person: 'I think it's a really good idea, especially for the kids who don't know what they are doing. They should also do something about the people who drink a lot' and 'I'm willing to say it could be a problem in some cases'. This highlights the necessity of targeting any message at an appropriate group.

PROJECT PITFALL: PROBLEMS

The project team included professionals concerned with prevention and also with harm minimization. The respondents' attitudes to their own drug misuse were different from those of the prevention project staff, including the researcher; many did not understand the point or purpose of the project or the research questions in the questionnaire. In other words, the beliefs of the target group and their reasons for using were completely at odds with those of prevention strategies and the complementary research questionnaire.

It was implicit in the prevention part of the project literature and is implicit in many substance misuse questionnaires that once substance misuse is identified as risky or problematic, respondents will then be likely to want to reduce the risk. This is clearly not the case for many of the respondents in this study.

Over 200 respondents were questioned about Project Pitfall. On the whole they were well aware of the dangers of drug misuse: more than 80% of the respondents were well informed about the risks of drug misuse prior to the intervention. It can be seen that if these young people already had a realistic understanding of the risks, the prevention literature would then be unlikely to influence this further. As might be expected, there was consequently little attitude or behaviour change in the group as a whole.

PROJECT PITFALL: LESSONS

Start out as you mean to go on

Decide what your aims are and what messages you want to convey and then decide which methods and what medium you will use.

Project Pitfall provides a good example of how not to develop an education or prevention project. It provides an apt illustration of the problems arising from muddling up the prevention approach with a secondary prevention or harm minimization approach and provides a general insight into the difficulties of choosing appropriate methods.

These two philosophies traditionally have different priorities and different methods; the following methods from each were used in the project.

| Philosophy: | Drug prevention | Harm minimization education |
| Method: | Public education | Individual outreach work |

In this project the aims, methods and medium were not clearly defined at the start.

Common sense would suggest that you work out exactly what it is you want to communicate right at the beginning. It was only after Project Pitfall had started in earnest that people began to suspect that something was wrong. Differences between staff began to surface concerning what was actually going to be communicated to whom. All had initially taken it for granted that everybody was trying to do the same thing and everyone knew what everybody else was doing. Too late, people realized that this was not the case and that the views of the project workers themselves were in conflict. Among staff, comments included:

It gradually became clear that most voluntary and statutory organizations connected with the project had no idea of the relevance or effectiveness of drug prevention.

I don't think that primary prevention by itself works, especially with young people. It needs to be backed up by non-judgemental and accurate information on drugs with respect for the young people's rights as adults to make their own decisions.

If they want to take drugs, it is their choice and I am there to provide advice and information on the legal status of drugs and the ways of preventing harm and health problems through drug misuse.

These views became clear in one worker's response to the evaluation questionnaire which carried the implicit message that drug misuse was problematic and needed to be prevented:

The research questionnaire was assuming things that weren't acceptable to the attenders ... i.e. that drugs were something that you would have good reasons for wanting to stop. I don't feel that people took it seriously, the questions were not relevant and I found it difficult to explain to people what it was about. People were asking 'What is the point?'. I could only say that it was to see if people were ready to change. I myself didn't see the need to do that part, I would tell people they didn't have to do it.

Problems with the definition and communication of the message itself are not easily resolved as differing beliefs are hard to reconcile; it is difficult to integrate primary and secondary intervention (harm minimization) messages.

The methods

Initially, the target group itself was not clearly defined. For prevention purposes, the wider the target group the better, but for harm minimization, outreach may be the preferable option as for some groups it is more effective to be more focused. Therefore the conflict concerning the content of the message caused inevitable disagreements as to the means of communicating it. Those project workers concerned with harm minimization felt that the project should have been tailored to a small, clearly defined group of younger drug misusers at raves who were at risk. In contrast, the prevention workers felt the message should be aimed at a far broader group of young people, particularly non-drug users:

In the beginning I knew the clubs we should be targeting from my own experience (personal and professional) of attending different venues. But there was pressure by the prevention staff to extend and target a wider audience.

One worker identified clear differences between separate groups of illicit drug misusers and suggested that each group be targeted in different venues, in different ways:

Whereas drugs such as Ecstasy, LSD and amphetamine are used mainly in the rave culture ... it is mainly amphetamine in the nightclubs. There was a greater need for harm reduction in raves than other clubs because even though they misuse drugs in all clubs, they misuse more in raves and the age of attenders at raves is lower. This type of regular outreach is not so appropriate for other wiser target groups. There is no need to do it on such a regular basis in the non-rave clubs. However, in the raves it is necessary on a regular basis because there were a lot of new people appearing every week and a lot of these were youngsters with very little knowledge about drugs ... Even with these people it takes time to get to know them, but it was easier to contact new people in the raves through the existing contacts.

It can be seen from the workers' views that there had been an inherent conflict of interest early on in the project but this was not obvious at the outset. Instead, each partner had taken for granted their own philosophical assumptions about aims and priorities and had concentrated on the practical issues, emphasizing commonalities and focusing on the development of a concrete product. As the project progressed it is possible that the continuing tacit nature of disagreements caused unnecessary confusion and hindered their resolution.

The medium

The main lesson to learn from this project is always to begin by deciding on the aims, clarifying the message and designing appropriate methods to achieve your ends. However, even if these crucial matters are dealt with effectively and imaginatively, it is still possible that the enterprise as a whole will fail because of ignorance about the vital significance of the medium. Non-statutory and statutory health and welfare services often lack awareness of the importance of promotion and this is compounded by a lack of resources and/or expertise.

Despite its inability to resolve issues of message and method, Project Pitfall did manage to avoid the third pitfall – underestimating the importance of the medium. The project produced a polished, high-profile image and, as a consequence, despite a muddled message and methods, gave a good impression to the general public. It had this in common with many similar prevention projects. As with drug education in schools, the byproduct is improved community relations and political popularity, whether this is through police officers in schools or central government initiatives.

Many of the difficulties in the development of local non-commercial prevention projects are found in the area of promotion. Therefore the ways Project Pitfall dealt with these issues are of direct use to anyone setting up a similar project. These problems were resolved by the involvement of advertising and media specialists and having the resources necessary to develop a cohesive profession campaign. These experts felt that most voluntary and statutory organizations have no idea of the importance of effective communication: 'The crux of it is that social care agencies are providing a service that they don't feel they need to advertise, so they are decades behind commercial companies'. They therefore have no idea how to obtain the type of products they need for their work: 'They had no experience of being clients to media or an advertising organization, that is, no idea of what they could expect and what would be expected of them, because they have no idea of the processes involved'.

So, contrary to its name, Project Pitfall effectively identified possible dangers in this area early on and effectively avoided every one. Social care practitioners are not trained in public communication, they tend to work on a one-to-one level. They are not trained to use advertising agencies or market research, while communication and advertising agencies themselves are not used to the needs of such client groups.

A member of staff explained that previous projects had been undermined by a lack of promotional expertise. 'Basically it came down to the lack of publicity. They had simply stuck a handful of very badly designed photocopied posters in the market a week before the start date.' He then compared this to commercial work: 'For example, look at smoking advertising, it's professional! Then look at health prevention promotionals ... they are not the same standard, nor the same levels of market research. However, 'Nicorette' is advertised as professionally as cigarettes because they are selling it'.

Project Pitfall therefore had a planned publicity campaign which involved local newspapers and radio, together with the compilation of a list of agencies at which to target information. As a consequence, the response rates to the Project Pitfall events were good and the project profile remained high. A phoneline was inundated with enquiries following promotional events appearing on radio and television and in local newspapers.

This understanding of the need for effective professional presentation and proactive publicity resulted in a high-profile project with professional publicity material. It is hoped the reader will be able to achieve this without experiencing the earlier pitfalls.

MATERIALS FOR DRUG EDUCATION AND TRAINING FOR PROFESSIONAL FACILITATORS

- *Drugs: Responding to the Challenge. Facilitator's Manual.* (B. Howe and L. White)

Health Education Authority
78 New Oxford Street
London W1CA 1AH

This gives a very general overview of drugs and includes many games to influence beliefs and attitudes. This pack was first published by the Health Education Authority in 1987 and is due to be updated in 1996. It includes perspectives of drug use and misuse, stigma and stereotyping, defining drug problems, getting the local picture, educational strategies, skills for education, fieldwork problems and course assessment.

- *Taking Drugs Seriously. A Manual of Harm Reduction on Drugs* (J. Cohen, I. Clements and J. Kay)

  Healthwise Helpline Ltd
  4th Floor, 10/12 James Street
  Liverpool L2 7PQ

This pack was first published by Healthwise Helpline in 1992. It gives more specific practical advice and focuses on health and harm minimization, rather than drug prevention. It includes facts on drug misuse, personal drug misuse, harm reduction, the law, giving and receiving help (skills) and community action.

- *Don't Panic: Responding to Incidents of Young People's Drug Misuse.* (J. Cohen and J. Kay)

  Published by Helpwise Helpline Ltd (as above)

This includes patterns of drug misuse, attitudes, signs and symptoms, first aid and medical services, talking to young people and contacting parents.

- Police and Education Joint Training Packs for Drug Prevention Education in Schools

These are based loosely on the American DARE programme

1. RIDE: Resistance in Drug Education (Metropolitan Police, London)
   This includes lesson plans for personal safety, drug misusing society, risks and consequences, dealing with emergencies, peer pressure, stress, self-esteem, assertiveness and personal support.
2. Getting It Right: Helping Keep Our Children Safe (Hampshire Constabulary and Education Department
   This includes lesson plans for 'golden rules', skills development, understanding children's learning, session planning, evaluation, being effective as a visitor and making agreement with schools.
3. DARE, Mansfield
   Initiated by the North Nottingham Health Promotion and Nottingham Drug Prevention teams, this is the most closely related to the American DARE programme and includes similar lesson plans.

- United States

The most widespread drug prevention programme here is the police-run DARE. The guidelines are to be found in 'Officers' Guide to D.A.R.E., Los Angeles Unified School District, 1994.

The lessons include personal safety, drug use and misuse, consequences, resisting pressure to misuse drugs, self-esteem, assertiveness, stress, media, decision making and risk taking, alternatives to drug misuse, role modelling, forming a support system, pressure from gangs and taking a stand.

- Keeping Ourselves Safe (New Zealand Police and Department of Education)

This is a more general package including drugs and child abuse. It is available from the New Zealand Police National Headquarters, Wellington.

Harm minimization

7

INTRODUCTION

This chapter is divided into five parts. The first, an introduction, examines problems of definition; the second analyses the historical phases of the development of harm minimization; the third considers the contemporary response and practice; the fourth presents a case study; and the fifth gives practical guidelines.

WHAT IS IT AND WHAT IS IT FOR?

For the purposes of clarity, the term 'harm minimization' will be used to refer to strategies for reducing the harm associated with drug misuse; it will not be used to refer to the treatment of drug dependence.

Harm minimization is concerned with helping drug misusers limit the physical and social damage caused by drug misuse. The aims of treatment are, in contrast, concerned with helping them to stop misusing drugs or deal with the drug dependence itself.

The methods of harm minimization range from basic health care and drug prescribing to issuing clean needles, but they do not involve therapeutic change. Whilst some behaviour change may be necessary, particularly during periods of uncontrolled or chaotic drug misuse, this will be kept distinct from psychotherapeutic treatment for drug dependence. The term 'drug treatment' will be used to refer only to methods for stopping, reducing and/or controlling drug dependence. These will be examined in the following chapter.

Both the aims and the methods are therefore different – harm minimization uses health and social-related interventions to reduce drug-related harm and treatment uses therapeutic interventions to control or stop drug dependence. In practice, however, the terms 'harm minimization' and 'treatment' are often used interchangeably. This is partly because the term 'harm minimization' is itself ill defined and partly because the concept of 'treatment' has become anything, from simply talking to people or helping with problems to treating the phenomenon itself.

Harm minimization and treatment may both be accompanied by attempts to deal with a range of other client problems, from housing to anxiety and depression; multidisciplinary teams often employ psychologists and social workers to deal with specific psychological and social problems respectively. Generic workers may choose to help with these problems themselves or refer to the relevant specialists.

It is necessary to clarify these three categories of 'harm minimization', 'treatment' and 'help with other problems' for practical purposes to ensure that the appropriate aims and methods are applied to particular problems. It is also important to add a qualifier. In reality, it is as difficult to categorize interventions here as it was to categorize people and problems in Part One. In the same way that some people fell into more than one category of drug misuse, agencies often fall into more than one category of intervention. That is, they provide both harm minimization and treatment as part of the same package, reducing harm and treating dependence at the same time.

This adds to the confusion in the field about what kinds of help are available for what kinds of drug misusers. While some methods can be used for both harm minimization and treatment, such as relapse prevention techniques and prescription of substitute drugs, it is important to remember that the purpose of each method is different. It is also, of course, not unusual for clients to be both risky and dependent misusers and it is important to emphasize that agencies and generic workers can deal with both types of problems at the same time, by offering both harm minimization and drug dependency treatment.

The aim of this chapter is to clarify the basic harm minimization provision necessary for risky drug misusers, independent of any kind of drug treatment. It gives the rationale for the harm minimization approach and examines when it is useful and why. It explains in detail the range of distinct methods that can be used independently of each other within this approach and gives basic intervention plans for professionals.

Despite a lack of comprehensive texts on the subject or a clear exposition of the theory and practice of harm minimization, it is possible to give an overview of both what it is and what it is for. Harm minimization stands between drug prevention and drug treatment as an alternative approach to drug misuse. Unlike drug prevention and drug treatment, there are few texts on harm minimization and no specific theoretical base. It is instead a pragmatic approach to reducing health and social damage among a large population of risky drug misusers, including a small proportion who occasionally lose control of their drug use and start to misuse chaotically. Generic professionals are likely to become involved with many risky drug misusers who simply need basic health care interventions and to know where to find clean needles and syringes, but some chaotic drug misusers can also benefit from basic cognitive behavioural interventions for changing behaviour patterns and preventing relapse back into old risky behaviour patterns (see practical guidelines in Part Three).

The basic premises of harm minimization are the same as the psychological foundations underlying drug prevention and health promotion discussed in the previous chapter, in that clients are treated as rational beings who will make sensible decisions if given the information and support necessary to do so. The following models of health psychology offer reasonable interpretations of human attitude and behaviour change that can be applied to harm minimization practice: the health belief model (Rosenstock, 1966; Becker and Maiman, 1975), the theory of reasoned action (Fishbein and Ajzen, 1975) and the self-regulation model of illness (Leventhal and Cameron, 1987). These models conceptualize the individual as actively solving problems and see their behaviour as the result of a reasoned attempt to achieve a sensible goal. They emphasize the importance of the individual's cognitions or subjective understanding of the health implications of their behaviour and the efficacy of their own actions in avoiding harm.

The basic principle underlying harm minimization is not to change individuals themselves or make fundamental changes in their lives, but simply to give each person the information and the means necessary to enable them to make sensible decisions and behave in safer ways. The aims of giving information and practical help are to ensure that the situation does not deteriorate and if possible to enable the client to make small steps to improve their quality of life. Agencies and professionals using this approach will be far more concerned with a client's exposure to HIV, hepatitis C and their health and social care needs than the degree to which they are drug dependent. This approach is easier to understand if placed in the context of other social and health care service provision in Britain. For example, the emphasis on psychodynamic change in social work became redundant some time ago and the majority of social work practice is seen as practical help and maintenance rather than diagnosis and therapeutic change (Davies, 1985).

It is not realistic to try to make radical interventions in social work or health care with the majority of clients/patients. Instead of changing the client or curing the illness and getting rid of the problem completely, it may be more pragmatic to reduce environmental risks, lessen vulnerability and make life more comfortable (e.g. stairlift to prevent accidents, health checks to identify risks, etc.). With recurring mental illness in the elderly, for example, instead of trying to achieve a permanent cure, professionals look at ways of improving social and individual circumstances. With child care, instead of trying to achieve perfect parenting or removing the child from home, social workers focus on maintenance and prevention services, offering support and help to families when necessary.

In common-sense terms, the harm minimization approach to health and social problems associated with drug misuse is the same as the professional's approach to any other social or health problem and drug misusers are no different from any other client group. In this sense, it can be seen that professionals should do what they are trained to do and have experience doing. The reason they do not is because emotions, morals and history get in the way.

UNDERSTANDING THE PROBLEM

The viewpoints of young people who misuse drugs in a dangerous or risky manner outlined in Chapter 3 have given us an understanding of drug misuse that is very different from that of young people misusing soft drugs on a recreational basis. Although it is clear that no drug misuse can be entirely safe and that it can be risky to misuse any drug even on an occasional basis, it is also fairly obvious that the majority of occasional or recreational drug misusers do not come to much harm. A smaller minority of drug misusers are taking greater risks, either by misusing many different drugs in a chaotic or uncontrolled way or by injecting illegal drugs. The type of drug misused, the patterns and the methods of misuse all influence the amount of risk involved.

This chapter is concerned with those drug misusers who take more risks than average and, as a consequence, need help to guard against possible harm. The harmful effects of drugs outlined by the respondents in Chapter 3 were not necessarily concerned with dependence but with a range of different types of drug-related harm. The kinds of help discussed in this chapter reflect the wide range of problems experienced by drug misusers and can range from protection against HIV and hepatitis to health care for abscesses and infections and housing for the homeless.

The development of harm minimization in Britain has been, to a large extent, the consequence of a pragmatic response to HIV prevention. Many services were developed through central funding for HIV prevention. This has determined the type of service available, which tends to focus on drug injectors and the necessity to provide clean, accessible needles and syringes (Raistrick, 1994). The underlying philosophy of damage limitation was already present in many non-statutory drug agencies offering social and medical support to drug misusers, which has since been extended to include HIV prevention with other forms of harm minimization.

Harm minimization is a response to individual health risks, but also to the wider public health risks of HIV and the health care costs engendered by risky health behaviours. As Stimson (1990) has pointed out, the period from 1986 to 1989 has been a time of 'crisis and transformation' in British drug services, resulting in a change from the individual pathology model of drug misuse to a 'public health' model. This is best illustrated by the shift from trying to prevent or eradicate drug misuse to stopping injectors sharing needles and other damage limitation measures. Drug policy in many countries is a delicate balance between the social issues and the public health costs of drug-related harm. Erickson (1990) points out that social policy must find a means of reducing the health and safety costs to the lowest possible level. He states that the challenge for the 1990s is to implement and test a comprehensive public health model for effective control of the demand for illicit drugs. However, this challenge is not peculiar to the 1990s and harm minimization is not exclusively a response to HIV. It will therefore be interesting to see if any lessons can be learned from the past.

HARM MINIMIZATION BEFORE THE CONCEPT OF 'ADDICTION'

Attitudes to drug problems before the concept of 'addiction' can be best illustrated through controversies surrounding opiate use at the turn of the century. An article by the Surgeon General, Sir William Moore, sums up the general attitude to opium use in the latter part of the 19th century in 'Opium: Its Use and Abuse' quoted in Parssinen (1983):

> There is no organic disease traceable to the use of opium. Functional disorders, more or less, may be induced by the excessive use of opium. But the same may be said of other causes of deranged health: gluttony, tea, tobacco, bad air, mental anxiety, etc.

Parssinen (p. 89) notes that the argument that 'Some persons can use moderate amounts of intoxicants, even opium, without damaging themselves' was frequently made in Britain in the 1890s. Opium was considered a much less serious problem than alcohol at that time. This is still the case today: the seriousness of the problem of alcohol abuse is often compared unfavourably with the less significant problem of drug abuse (see *Alcohol: Our Favourite Drug*, Report of the Royal College of Psychiatrists, 1986).

While at present the government is publicly seen to endorse the American 'War on Drugs", in the late 19th century it openly supported opium use in order to maintain its lucrative opiate trade in India. The Parliamentary Blue Book, presented by a committee commissioned by Gladstone and the Liberal government of 1893, stated that there was little evidence that opium eating led to the physical or moral decay of users. To the contrary, they found that opium was used intelligently as a medicine and in moderation as a stimulant and that to deprive the natives would cause great suffering.

This report fuelled the continuing controversy between the medical profession who asserted that opiate use was a medical problem and those who argued that with proper controls drug use was fairly safe. Similarly today there is a conflict between those who understand dependence as a medical problem and those who see it as a public health issue and within the remit of social and health care professionals. Before 1850, 'The compulsive use of opium was not considered to be an issue' (Harding, 1988, p. 53), but by the end of the century a new theory of dependence had emerged which emphasized the physiological and therefore medical components, together with Quaker philosophy and beliefs about a 'concept of a moral faculty of the soul that could be pathologized' (Harding, 1988, p. 37).

The medical and moral disease concepts of dependence will be discussed in the following chapter but it is important to point out here that both the medical and the 'disease' model of Narcotics Anonymous were developed from ideas from the late 19th century. For example, Levinstein, in his work Morbid Craving for Morphia (1878), defined the disease as 'The uncontrollable desire of a person to use morphia as a stimulant and a tonic, and the diseased state of the system caused by the injudicious use of the said remedy'. Parssinen (1983) states that Levinstein was 'insistent upon the somatic origins of the drug addiction because only by doing so could he convince his medical colleagues to take it seriously as a disease'. Dr Kerr, a prominent medical figure, helped popularize the notion of addictive disease as the result of a structural alteration which can be pathologically verified and he modified Levinstein's conception of the disease of addiction by stressing that it had a psychological aetiology and by emphasizing the difficulty of reversing a patient's opium habit once it was fully established. These ideas can be seen reflected in 20th century literature on addiction, particularly in the Alcoholics and Narcotics Anonymous theory of spiritual disease, developed in the early part of the century.

Not only did ideas of addiction as a medical or physiological phenomenon develop quite recently, but these notions were mainly limited to those people with whom doctors had contact, those who injected opiates (morphia). Much of the medical literature of the time focuses on therapeutic morphia addicts and there is very little interest shown in the extent of general opiate use in the community. Medical interest in opiate use increased as morphia began to be prescribed in an injectable form soon after the hypodermic syringe was produced in the second half of the 19th century. 'Medical men were, for the most part, uninterested in the threat to public morals presented by opium smokers, and even the threat to public health presented by the occasional dispensing of opium' (Parssinen, 1983).

Modern medical practitioners and researchers have also, in common with their predecessors, been concerned with only a small proportion of opiate misusers, those most likely to present for medical help for physical dependence. As might be expected, they have little to say about the majority of drug misusers, who are not in contact with services.

It will be seen in the following chapter that both the medical and disease perspectives can be useful for the treatment of dependence, but they may be less than useful for dealing with non-dependent drug misuse and its associated problems. It is interesting therefore to look also at ideas about drug misuse that existed before the medical models. In the 19th century opium was commonly sold as a medical treatment for pain and various illnesses (there was no restriction on its sale until the 1920s). The restrictions on opiate use were nothing to do with the medical profession but were instead left to the developing pharmaceutical profession and to public health campaigners.

The Poisons and Pharmacy Acts of 1868 and 1908 regulated the sale of opiates in order to reduce 'poisonings' (accidental deaths by overdose and suicides) and ensure that anyone buying opiates was known to the pharmacist. Parssinen (1983) adds that these two Acts, by labelling opiates as poison, produced a 26% and 20% decline respectively in the mortality rate.

So it seems we have come full circle, as in modern literature the prevention of the spread of HIV tends to be seen as a public health issue and is dealt with within that paradigm (Stimson *et al.*, 1988; Stimson, 1989). Non-medical drug services are developing theories and methods of working that have little in common with the theories, aims or methods of the disease and abstinence perspectives. Focusing on 'harm reduction' and 'controlled drug use', clients are seen as responsible citizens, behaving in what can be regarded as a rational way and quite capable of modifying their behaviour in their own interests and for the public good. In a similar way, pharmacists, as part of a general trend to become more involved in health promotion, are involved once again with drug misusers, selling or exchanging free syringes and offering advice and help.

It can be seen that the medical and disease approaches have developed as a reaction to drug dependence rather than drug misuse as a whole. The following chapter will therefore deal with these perspectives. This chapter is concerned, in contrast, with other risks associated with drug misuse in general.

As indicated in Chapter 1, sociological research has shown that medical consultants had limited contact with drug misusers and therefore tended to see it as pathological whereas sociologists saw it as a widespread phenomenon involving many more drug misusers and different types of use. This chapter will consider the perspectives of modern practitioners who have come into contact with a range of different types of drug misuser before and after HIV.

PRE-HIV HARM MINIMIZATION (1970s AND EARLY 1980s)

Prior to the 1970s a small group of doctors prescribed injectable and non-injectable opiates as a therapeutic treatment for drug dependence, as had doctors at the turn of the century. However, medical policies towards drug misusers changed in the 1970s and long-term prescribing of injectable drugs became a thing of the past. In its place drug treatment centres grew up providing short-term withdrawal programmes for those who were prepared to give up opiates. In addition a new group of non-medical workers began to tackle drug problems. These were the non-statutory organizations, staffed by people who were often not professionally trained and who viewed drug misuse in a different way from the medical profession. They saw it as related to other social problems and not necessarily the cause of these problems at all, simply part of a wider social malaise. They therefore tried to deal with these drug-related problems rather than the drug misuse itself. Perhaps because they were doing social work, they were not caught up with the medical model of individual pathology, but saw individuals instead as having social and health problems, in common with the newly developing social work ethic.

These workers were often targeting young people and responding to generic youth problems such as homelessness and unemployment. This form of harm minimization was not public health oriented but based on the individual's social and health needs. The newly developing non-statutory agencies were largely non-professional and in the 1970s and 1980s this implied lack of status, influence and funding. One implication was that the beliefs, values and priorities of these agencies differed from those of the medical treatment centres. Clients were not referred by other professionals, nor diagnosed and treated; in contrast they were often seen as at least as knowledgeable and 'expert' as the staff and therefore as capable of defining their own problems and determining their own needs. Self-referral as opposed to professional referral was a common form of entry to these 'low threshold' or easy access agencies and clients were helped or enabled to make their own decisions.

This philosophy of client participation and enablement did not increase the professional standing of the agencies as it became difficult to define clearly the expertise of the staff and the theories underlying their work. Staff ranged from nurses and social workers to unqualified workers and ex-users. Experience became an important component of this approach, together with practical common sense. It is perhaps because harm minimization has developed from this type of agency, rather than from coherent theoretical models, that it has evolved a pragmatic approach to a range of problems associated with drug misuse.

DEALING WITH DRUG PROBLEMS

This task can be achieved by: helping clients to obtain scripts; acting as client advocates to persuade doctors to prescribe drugs; helping clients to obtain syringes; providing lists of shops/pharmacists where syringes could be bought; festival support services; dealing with problems of intoxication, excessive misuse and overdose.

DEALING WITH HEALTH PROBLEMS

These mainly cover infections and viruses related to injection (hepatitis) and general health related to drug misuse.

DEALING WITH SOCIAL PROBLEMS

These include housing, welfare rights, employment, child care and court work.

DEALING WITH CHAOTIC BEHAVIOUR AND LOSS OF CONTROL

This involves stabilization, prescription drugs, self-management and behaviour change techniques.

These agencies also serve a political function as advocates of broader political change.

MODERN HARM MINIMIZATION: THE RESPONSE TO HIV

HIV TRANSMISSION AND PREVENTION

Research has highlighted the correlation between HIV positivity and the following: needle sharing; the frequency of injecting with used needles; sharing needles with strangers; the frequency of injection; length of drug misuse; past imprisonment; the number of sexual partners injecting drugs; and travel to HIV epicentres (Raymond 1988; Van Den Hoek, Van Haastrecht and Continho, 1989; Des Jarlais and Friedman, 1988; Schoenbaum, Hartel and Selwyn, 1989; Siegal, Baumgartner and Carson, 1991). Drug injectors, as with the rest of the population, were also exposed to

HIV infection through unprotected sexual intercourse as most people who inject drugs are sexually active (Donoghoe, Stimson and Dolan, 1989). The high fertility of female drug injectors, as evidenced by high rates of pregnancy, also indicated the risk both of infection through sex and the risk of perinatal transmission (Brown *et al.*, 1989; Dolan *et al.*, 1991).

In the late 1980s much central funding was allocated to develop syringe exchanges and harm minimization strategies, in an effort to reduce the spread of HIV. Although the response was similar in the Netherlands and Australia, other countries were less clear in their approach (Strang and Stimson, 1990; Gould, 1993). Although there are dissenters (Raistrick, 1994), there is much evidence from Britain to suggest that this programme was effective (Stimson, 1988, 1990, Keene *et al.*, 1993). There was an enormous reduction in needle and syringe sharing in the early years of the HIV epidemic, accompanied by an equally significant change in sexual practices among homosexual men (Bloor, 1995). There are different ways of understanding the relationship between drug misuse and sexual activity (Stall and Leigh, 1994). Research shows that consistent condom use among the young is rare but that the likelihood of condom use does not decrease after drug or alcohol use. It is simply that the likelihood of sex increases (Strunin and Hingson, 1992). There appears to be little empirical certainty about the causal relationship between the pharmacological effects of drugs on sexual activity and the social, situational and cultural factors which may influence sex and safer sex (Rhodes and Stimson, 1994).

Research has shown that, if drug misusers receive information about HIV they modify their risk behaviours, initiating changes in their needle sharing and, to a lesser extent, their sexual behaviour. There are still large differences between risk recognition and risk reduction (Stimson *et al.*, 1988; Huebert and James, 1992; Frischer *et al.*, 1993). There is evidence to show that drug injectors have knowledge about the transmission of HIV and are concerned about the risk (Lungley, 1988; Stimson *et al.*, 1988; Des Jarlais, Friedman and Casriel, 1990). They adopt various measures to prevent HIV transmission: using new needles, cleaning needles, reducing needle sharing, marking personal syringes; partner selectivity – assessing the known or supposed HIV antibody status of the prospective partner; hiding syringes for use in emergency; not sharing when blood is visible in the syringe (Burt and Stimson, 1990); and using the syringe first. Syringe sharing has become more selective, with less indiscriminate sharing (i.e. with strangers), but with possibly fewer changes in rates of sharing with close friends or sexual partners. The syringe sharing that remains is largely between sexual partners, close friends, relatives and small social groups (Bloor, 1995). Burt and Stimson (1990) report that many people who inject drugs no longer view sharing syringes as normal behaviour.

McKegany and Barnard (1992) list six possible influences on continued sharing behaviour: access to clean equipment, the need to inject, assessment of limited risk and inadvertent sharing. Added to these are social influences: the nature of the relationship between sharers and the local cultural norms which stress obligations to neighbours or friends. Bloor *et al.* (1994) have shown that much injecting and sharing takes place within small groups of intimates who are unlikely to share outside these groups. This explains to some extent the limited spread of HIV in some areas (such as the locus of their research in Glasgow). Once again the social patterns of misuse become a useful predictor of likely problems. For example, this social information points to the particular dangers of HIV transmission in prisons and among the homeless, where there is more connection with strangers When sharing does take place in the community, it is more likely to happen away from home (Darke *et al.*, 1994). Factors associated with sharing and other risk taking include criminality (Darke *et al.*, 1994), less time in treatment or not using a pharmacist as a primary source of syringes (Saxon, Calsyn and Jackson, 1994). It is also significant that prostitutes are much more likely to have misused drugs than the general population (Kuhns, Heide and Silverman, 1992).

HIV PREVENTION METHODS

A wide variety of measures was introduced in Britain to try to prevent or limit the spread of HIV infection through drug injecting. These measures were aimed at encouraging drug injectors to change their behaviour to reduce their risk of infection from HIV or of transmitting it to others. These strategies are based on the belief that the immediate public health task is to work with the people who inject drugs or are at risk of doing so in the future. The priority is to help reduce HIV risk, whether drug misusers are willing to stop injecting drugs or not.

HIV prevention measures are aimed at the general population and at specific target groups such as homosexuals and drug misusers. The methods include:

- information and clean equipment;
- national and local informational campaigns;
- advice;
- information and counselling about HIV risk behaviours and protective strategies;
- the provision of needles and syringes;
- providing information about syringe decontamination and providing supplies of suitable decontaminants;
- encouraging agencies to provide a harm minimization service and improve access to drug treatment;

- adapting existing treatments (such as methadone prescribing) to HIV prevention;
- developing outreach to hard-to-reach populations;
- encouraging changes in local drug-misusing communities (Strang and Stimson, 1990).

A new interest has been taken in those drug misusers who do not attend drug services. One of the key issues for HIV prevention is the need to make contact with more drug injectors, whether through making treatment available on demand (the recommendation of the US Presidential Commission on AIDS) or by making treatment and help more accessible (as recommended by the UK Advisory Council on the Misuse of Drugs (1988). Many people who misuse and inject drugs may not see their drug misuse as a problem and are therefore unlikely to attend specialist drug services. This is particularly the case for polydrug misusers or amphetamine misusers, where treatment involving a prescription is unusual. This problem has been tackled elsewhere by changes in the services themselves as they become more 'user friendly' and offer practical help and by establishing prescribing services and syringe exchanges. Various outreach and community initiatives have been tried as adjuncts to an office-based service. Some help people into service contact, others promote and facilitate change directly in the community where drugs are misused to encourage safer drug misuse and sexual behaviour. Outreach and community organization for drug injectors have been developed most extensively in the US; these projects often promote change in the drug-misusing community using 'indigenous' workers who are current or former members of target groups (Keene *et al.*, 1993).

HIV PREVENTION AND DRUG TREATMENT

In Britain, drug services were tremendously influenced by the government's response to HIV, a response which was largely unpolitical and pragmatic. The advent of HIV, in effect, actively changed the focus and priorities of many drug agencies and reinforced and made respectable the earlier non-statutory harm minimization agencies, through an influx of central government funding.

The governmental Advisory Council on the Misuse of Drugs declared early on that 'The spread of HIV is a greater danger to individual and public health than drug misuse' (ACMD, 1988) and 'AIDS presents a far greater threat to the individual and to society than drug use' (1989). It went on to recommend that 'Prevention of the spread of HIV infection amongst injecting drug users must be a priority for all drugs workers'. This was extremely influential in limiting the spread of HIV among those drug misusers who came in contact with drug workers themselves.

Drug agencies were encouraged to change from treating drug dependence to providing individual health care and a public health service. Programme staff found themselves faced with the need to reduce sharing of equipment and change the sexual behaviour of their clients. They needed to encourage drug misusers to accept and use clean needles, syringes and condoms. They could provide these free of charge to their own clients but were also required to take a public health approach to the whole drug-injecting population rather than the small proportion of problematic IDUs they had seen previously.

If workers were already concerned with preventing drug-related harm, preventing HIV infection often became an integral part of their everyday work, whereas if the focus was on preventing drug dependence and encouraging abstinence it was less easy for workers to be seen to condone drug misuse.

These differences are reflected in international variations in HIV prevention strategies. For example, in the United States where abstinence-based treatment programmes are common, there are few harm minimization services such as syringe exchanges (though methodone treatment is available for those who wish to modify their drug misuse). In Britain and other European countries harm minimization and HIV prevention strategies have been more readily accepted as they fit in with broader policies on controlling and minimizing the harm from drug misuse.

Agencies in Britain dealt with this problem in various ways, sometimes incorporating harm minimization objectives into existing treatment regimes, sometimes using harm minimization provision as a low threshhold entry point into treatment. These developments served as a means of rationalizing old services in the context of the new harm minimization objectives. In areas where drug agencies were unable to integrate public health-orientated harm minimization into their service provision, circumstances forced the development of community services based in pharmacies or health care centres; though slower and more difficult to develop, these often proved more useful in the long term. These developments are examined in the case study at the end of this chapter.

Existing drug treatment philosophies and policies therefore influenced the development of HIV prevention services (Keene and Stimson 1996). This process also functions in reverse in that the syringe exchange philosophy reinforces and emphasizes the harm minimization message within the host agencies themselves. Thus in addition to providing a direct service for clients, HIV prevention services such as syringe exchange had a more general impact on services and service philosophy. In America the policy response is the inverse of the British approach; there is instead a responsibility put on HIV services to deal with drug misuse (Selwyn, 1991).

In Britain the changing priorities necessitated by HIV prevention resulted in provision of injecting equipment, health care facilities and prescription drugs. Drug agencies became more 'user friendly' and therefore more accessible to a wider range of drug misusers. However, the resulting amalgamation of harm minimization with drug treatment caused a growing confusion about which was which and which was best for whom. The targeting of services was determined by the services provided. Although the new agencies reached more drug misusers by providing a package of services which included both harm minimization and HIV prevention, it became impossible in some areas to obtain a harm minimization service without drug dependency treatment (Keene, Parry-Langdon and Stimson, 1991).

HARM MINIMIZATION AFTER THE ADVENT OF HIV

It is clear, then, that harm minimization has been greatly influenced by HIV prevention, in terms of theory, policy and practice. However, it should not be forgotten that the aims of harm minimization are far more wide ranging than the prevention of HIV and the populations targeted far broader than simply those at risk through injecting drugs.

The earlier sections covering harm minimization before the advent of HIV demonstrate how different policies and practice were developed in order to limit the damage done by drug misuse in different target groups, whether this entailed labelling opiate sales to prevent overdose among the working class at the turn of the century (the equivalent of a government health warning on cigarette packets) or offering basic health and social care to destitute young people in the second half of the 20th century.

Although there is no clear 'theory of harm minimization' in the sense that the approach prescribes anything that is effective in reducing the harm associated with drug misuse, the aims and target groups of this approach can be clarified.

The wider target group include all those drug misusers who take health risks and a smaller group who lose control of their drug misuse and start to misuse chaotically for brief periods in their lives. Generic professionals are likely to become involved with both groups. Life-saving interventions include health care, needles and syringes, stabilization prescriptions and basic cognitive behavioural interventions for changing behaviour patterns and preventing relapse back into old risky behaviour patterns (see practical guidelines in Part Three).

THE NEW RECREATIONAL DRUG MISUSERS

Recreational drug use has been dealt with in the previous chapter on prevention, where it was argued that education and harm minimization may be a more appropriate response to these misusers. It was pointed out there that recreational drug misusers are more likely to be at risk in the early experimental period, but it has become clear in this chapter that some recreational misusers continue to misuse more and more drugs and their drug misuse becomes more risky. This group of high-risk stimulant misusers has been increasingly presenting at traditional drug agencies, indicating changes in the general trends of drug misuse. Some of these drug misusers may present when their drug misuse becomes chaotic or uncontrolled. Newcombe (1992) reiterates the conclusions of the previous chapter when he suggests that agencies respond to this group by retraining their staff, providing on-site health education for recreational misusers and developing harm minimization outreach for those particularly at risk. Gilman (1991) and McDermott et al. (1992) also focus on distributing literature, emphasizing that most recreational drug misusers do not have problems, though some may develop them in the future.

SPECIAL GROUPS

Harm minimization in the 1990s has highlighted the problems of drug-related harm in particular groups. High-risk drug misuse is more prevalent among some groups than others and, perhaps more importantly, some groups are less likely to attend agencies or receive harm minimization support. Professionals therefore need to be aware of these issues.

Mothers and children

This group is mentioned here as there is much controversy concerning the problems associated with drug-misusing mothers, whether during pregnancy, early babyhood or later childhood. It is generally argued that a punitive approach discourages mothers from attending agencies and instead that they will 'go underground' through fear of incarceration and/or loss of their children (Poland et al., 1993). Funkhouser et al. (1993) have demonstrated that drug misuse and socioeconomic status are significantly correlated with less use of prenatal care facilities. It should be mentioned that fathers seldom feature in the literature on drug misuse, as do children, except by virtue of their absence: there is some indication that they are perhaps more likely to have left the mother to cope on her own (Bekir et al., 1993).

The mothers' fear of attending services is to a large extent warranted. Drug-misusing mothers are more likely than non-users to be separated from their babies at birth, usually through health concerns for mother and/or infant (Mundal *et al.*, 1991). It has also been suggested that, although infants of drug-misusing mothers are no different on non-verbal development scales, they may have specific difficulties in early language development (Van Baar, 1990). It is of course difficult to determine if this is a result of maternal drug misuse or a result of other variables associated with it, such as poverty, homelessness and depriva-tion. Even if these factors are considered in combination, the effects seem small. A study examining the effects of addiction, poverty and serious developmental delay found that these only had an effect in four out of the eight dyads studied (Brinker, Baxter and Butler, 1994). Instead, the authors found that mothers naturally increase their sensitivity over time in response to the infants' involvement in the interaction.

It should be remembered that if mothers are leading a certain lifestyle, governed, for example, by prostitution or the need to obtain expensive drugs, this may result in a need for alternative child care facilities for certain periods of the day, as for working mothers. Society has responded to the problem of maternal drug misuse in different ways: by removing the children from drug-misusing mothers; criminal prosecu-tion for neglect or for health reasons; giving the mother harm minimization help or drug treatment (Chavkin, 1990).

It is clear that a balance is needed between the rights and needs of both mothers and babies. It would be irresponsible to ignore the possible risks of neglect and unrealistic to expect mothers to come for help if they are at risk of losing their children. In many areas this has brought about a more therapeutic and less punitive approach to women drug misusers. But the low numbers of women attending drug agencies and health clinics indi-cate that there is much progress still to be made. (See Part Three for guide-lines on working with pregnant women and drug-using parents.)

Although women have been focused on mainly in the context of children, it is important to highlight other issues related to gender differences; for example, the often male-dominated drug culture may involve the male partner procuring and controlling the woman's misuse of drugs. Women often only arrive at services when their partner goes to prison. Taylor (1994) gives a detailed account of the lives of women drug misusers from their own perspective.

The homeless

It has been mentioned that particular groups and lifestyles may be correlated with drug misuse. For example, homeless adolescents are more vulnerable to HIV risk from drug injecting and sexual activities

(Rosenthal, Moore and Buzwell, 1994). Blankertz *et al.* (1990) have also identified distinct subgroups of homeless and mentally ill drug mis-users. It should be emphasized that it is often the lifestyle or behaviour that present the health risks rather than the drug misuse itself. For exam-ple, homeless adolescents and post adolescents are more likely to mis-use drugs (George, Shanks and Westlake, 1991; Warheit and Biafore, 1991) but they are also more likely to have high levels of depression, social dysfunctions, low educational achievement, few job skills and lim-ited interpersonal coping networks (Warhiet and Biafore, 1991). Although drug and alcohol use are correlated with homelessness, drug misuse appears as only one of a range of debilitating factors and is not necessarily causing or even exacerbating other problems. Berg and Hopwood (1991) surveyed the homeless and reported that most were unhappy with drug services, feeling that they did not fulfil their main needs. McCarty *et al.* (1991) reiterate this point and emphasize that one of the essential adjuncts to work with this group is drug-free accommodation.

Drug misuse may influence the ability of the homeless to compete for scarce resources; in the same way that mothers are less likely to use health facilities, fearing loss of their children, homeless people may fear criminal proceedings for illegal drug misuse. The plight of homeless women with children is obviously compounded by both factors (Robertson, 1991).

Ethnic minorities

By the same token racial and ethnic differences may influence drug misuse. When both background and lifestyle factors are controlled, many of the racial and ethnic differences in drug misuse are consider-ably reduced. Instead, lifestyle factors such as educational and religious values and time spent with peer groups are more likely to predict drug misuse (Wallace and Backman, 1991). More important, perhaps, it has been assumed, for example in the UK, that white males are the main misusers of drugs because the proportion of white males using drug services is far greater than that of females or ethnic minorities. It is far more likely that there are reasons why neither of these groups uses services. An examination of these reasons would be useful.

Older drug misusers

It should be pointed out that, while there is a clearly identified popula-tion of older heroin addicts attending treatment agencies, most sociological studies of drug misusers not attending agencies have identified only younger people. This may be because such groups are often accessed through probation services or other agencies where younger clients are in the majority. Comprehensive population surveys of drug misuse do not identify older drug misusers (Leitner, Shapland and Wiles, 1993): the age group most commonly reported is in the range 16–25 years (see also Mishara and McKim, 1993). It is therefore as well to be aware that this area is underexplored. Dembo, Williams and Getrev (1991) emphasize the dangers of drug misuse among older people, particularly benzodiazepine misuse. They suggest that drug agencies would do well to broaden their age range and consider early intervention for older people with drug problems, particularly in hospital settings.

People in institutions

The extent of risky drug misuse and the limitations of harm minimization services in prison are discussed in Chapters 5 and 9. Drug misuse is also a growing concern amongst children in care (Meyers and Moss, 1992).

Learning difficulties

Little work has been carried out with people with learning disabilities, though work by Elmquist, Morgan and Bolds (1992) indicates that there is little difference between the drug misuse of this group and others.

Mental health

Associations between drug misuse and mental illness vary from Barbee, Clark and Crapanzano (1989), who identified 50% of schizophrenics as having drug and/or alcohol problems, to Woogh (1990), who identified a link between major functional disorders and drug misuse.

THE SOLUTIONS: HARM MINIMIZATION INTERVENTIONS AND PRACTICAL GUIDELINES

The confusion between treatment of dependence and reducing health-related risks is widespread. It is important when planning interventions to be clear about aims and priorities, as the aims of treatment of dependence and harm minimizations may conflict. The solutions to harm minimization problems range from help with accommodation to increased hygiene and safety in the environment, to basic health care and prescribed drugs:

1. health care,
2. substitute prescribed drugs,
3. social and psychological support,

4. outreach,
5. syringe exchange.

HEALTH CARE

The health problems described below are associated with drug misuse. It is useful for professionals to be able to recognize them and ensure that the client receives some form of health care. Unfortunately drug misusers as a group have done much to alienate health care professionals and it is therefore often difficult for individuals to obtain help. These difficulties should not be underestimated and it is not unusual for a client to have no GP. In the long term it is necessary to inform the Family Practitioner Committee but in the short term it may be advisable to go with the client to the local hospital accident and emergency department.

General health problems

There is a risk of overdose, adverse reactions and misuse of dangerous unknown substances with non-injected drug misuse.

General physical effects of drug misuse can be divided into two categories – depressant and stimulant effects. Stimulants (drugs which stimulate the central nervous system) and depressants (drugs which depress the CNS), as might be expected, have very different effects. It is also the case that the withdrawal effects of a drug are commonly the opposite to the actual effect of the drug. This gives a very rough and ready picture of drug misuse effects.

Although it would obviously be best to stop sharing to prevent HIV infection and to stop injecting to prevent other infections, other methods of drug ingestion, while far safer, are not entirely risk free. A recurring trend appears to be the mixing of a range of drugs including alcohol which greatly increases the risks associated with any one substance (McDermott, Matthews and Bennett, 1992).

Problems related to method of ingestion

The second largest category of health problems are those other than HIV that occur as a consequence of injecting drugs. These include hepatitis (particularly hepatitis C), skin infections and the detrimental effects of foreign bodies in the arteries and veins, such as thrombosis. Although non-injecting drug misuse is almost invariably safer, there are also various risks associated with other forms of ingestion, from swallowing a pill in a nightclub to heavy inhaling of amphetamine or cocaine. They range from a high incidence of asthma to more serious bronchial and heart problems.

Hepatitis B and C are far more common among injecting drug misusers than any other group and they are the most commonly transmitted viruses. Hepatitis A is probably transmitted through other routes when hygiene is poor, but hepatitis B is passed through unsterile equipment. It is said that more than half of injecting drug misusers have had hepatitis B (Farrell, 1991). Farrell states that approximately one in 10 of them will become a carrier and it is they who may develop active chronic hepatitis and liver failure. Most people with hepatitis B have only a mild illness, whereas hepatitis C is more worrying. This virus has only recently been named: previously called non-A, non-B, little was known of its seriousness. Among drug misusers with hepatitis C the virus can lead to chronic active hepatitis and chronic liver damage in up to half those affected.

Abscesses and/or skin ulcers are common in injecting drug misusers. Abscesses need not be infected but can simply be the result of injecting ground-up tablets and accidentally missing the vein. If unsterile injecting equipment is used an infected abscess may be the result. These can be treated with antibiotics unless they are serious enough to need incision and drainage. Some infections can cause cellulitis and if this inflammation extends along the vein it may result in thrombophlebitis. Septicaemia can also result as a consequence of a localized infection caused by unsterile injecting equipment and may result in cardiac lesions (Farrell, 1991; Banks and Waller, 1988).

Thrombosis develops from a sinus at a regular injection site and obstruction of the lymphatic drainage system (resulting in puffy hands) may be the consequences of long-term injecting. That is when drug misusers have either overused one site or moved on to more dangerous veins as they have used up the superficial veins. Arterial occlusion can be the result of one injection of particulate matter (resulting in a swollen pale limb) (Farrell, 1991). Perhaps more serious is the gangrene that may occur if a misuser injects into an artery by mistake. This is why it is dangerous to inject into veins that are situated close to arteries, such as the femoral vein (Banks and Waller, 1988).

Overdose

It is often not clear to either professionals or misusers whether overdoses are deliberate or accidental. These categories are not really of much use in understanding this phenomenon and it may be easier to appreciate that individuals who are suffering or depressed may be less concerned about overdose and so less cautious in their drug misuse. It is also likely that people who are living chaotic lifestyles or are too intoxicated to determine the amounts they are using may overdose without realizing it. Ignorance can lead to overdose, lack of knowledge about either the strength of the substance used or individual levels of tolerance.

Overdose results in coma with slow, shallow breathing. The unconscious person should be placed in the recovery position, with the airway open. Opiate overdose can be reversed by the administration of the opiate antagonist naloxone, though this should be delayed until the arrival of an ambulance unless breathing becomes very shallow and the lips cyanosed.

There is also a serious risk of overdose with barbiturates (where there is a very small margin of safety between feeling the effects of the drug and overdosing) and an equivalent risk of fitting when withdrawing (Farrell, 1991; Banks and Waller, 1988).

Respiratory problems

Depressant drugs affect the respiratory system. Large quantities or mixtures of depressant drugs (for example, methadone and alcohol) can cause death through respiratory failure. These drugs also have a depressant effect on the cough reflex, allowing secretions to gather in the bronchial tree, causing bronchitis and pneumonia (Banks and Waller, 1988). A possible complication of inhaling any drug (including stimulants) is asthma (and accentuation of the problem in previous asthma sufferers). If cocaine vapours are inhaled it can lead to lung damage and other respiratory complications. Solvent misuse can result in asphyxia or inhalation of vomit (Farrell, 1991).

Debilitating short- and long-term effects

Hallucinogens, such as LSD and the hallucinogenic mushrooms such as psilocybin, are not physically addictive but can lead to stomach and bowel disorders. Most importantly, intoxication with these drugs can lead to distorted perceptions ('bad trips') and hallucinations, resulting in serious accidents.

The most significant effect of immediate or short-term misuse of depressants is the slowing down of reaction time and disorientation of psychomotor function. Cognition and memory are also affected. This has obvious implications for people engaged in difficult tasks, whether at work or not, such as driving or operating machinery. Any depressant drug misuse is therefore likely to have a debilitating effect on performance and increase the risk of accidents. Much commonplace drug misuse in small quantities, such as benzodiazepines or cannabis, can have serious effects, if dangerous tasks are undertaken, but there is little information available. The stronger depressants such as the opiates may also have serious effects but even here, continued misuse of steady amounts of opiate are unlikely to cause much trouble unless the misuser is falling asleep.

It is those misusers who are unaware of the effects of their drug misuse and those who are chaotic, using different types and quantities of drugs in a haphazard manner, that are most at risk. The most significant effect of long-term misuse is depression; it is also likely that depression will occur in the postwithdrawal period for drugs such as opiates (see Chapter 10 and Appendix for details of drug effects).

Stimulants

Stimulants, as might be expected, speed up the body's functioning by increasing the levels of adrenaline and noradrenaline. Therefore they are often used to increase concentration and prevent the detrimental effects of tiredness, for example for long-distance driving or night work. Whereas depressants lead to depression as a consequence of misuse, stimulants can help prevent depression (and were used as antidepressants in the 1970s) but depression is a consequence of cessation of misuse. There has been much debate concerning whether stimulant drugs are 'dependence inducing' or 'addictive'.

However, stimulant misuse can result in serious health costs. Continual misuse has a detrimental effect on overall health, as misusers get little sleep and often do not eat well. There is not a clearcut physical withdrawal syndrome like that associated with opiate withdrawal, but withdrawal is often unpleasant and leads to increased tiredness, lethargy and depression (see Chapter 10 and Appendix for details of effects).

PRESCRIBING ALTERNATIVE ORAL AND/OR LEGAL SUBSTITUTE DRUGS

As described earlier, historically, illicit drug misuse was controlled by the judicious prescription of substitute drugs by doctors. The advent of HIV and serious strains of hepatitis have led to the revival of this old-fashioned medical prescribing, but for different reasons. Rather than control the social availability and dependence itself, the aims are to control the spread of infections, maintain health and reduce crime. Crime and drug misuse will be discussed in Chapter 9. This section will consider the effectiveness of substitute prescribing on personal and public health.

Methadone prescribing for harm minimization

The question here is not whether methadone maintenance programmes are a useful 'treatment' for heroin addiction, but whether they decrease drug-related harm. Do they reduce the risk of HIV and hepatitis transmission among injecting drug misusers? Do they reduce general health risks? Williams *et al.* (1992) found that subjects in continuous treatment report less needle sharing and fewer needle-sharing partners than those not in treatment. It is, of course, possible that those who shared less would be more inclined to attend treatment in the first place. Longshore *et al.* (1993) and Chalmers (1990) looked in more detail at changes during a programme and found less sharing of syringes. Caplehorn *et al.* (1993) report that the more methadone that was prescribed, the less the chance that programme participants would misuse heroin as well. They found that the likelihood that someone would misuse heroin was reduced by 2% for every one ml increase of methadone.

Methadone maintenance over long periods of up to 18 years has been shown to be relatively safe in comparison with long-term heroin misuse, with no methadone-specific effects not visible in heroin misuse (Novick *et al.*, 1993). A univariate analysis of variables associated with HIV infection carried out by Serpelloni *et al.* (1994) indicates that long-term methadone treatment reduces the chance of HIV infection. This is in all probability because of reduced injecting/ sharing behaviour but may be due to other factors. It is possible that methadone may affect the immune function, but a recent review by McLachlan *et al.* (1993) suggests that this is not the case and consequently they recommend that methadone is safe to prescribe for people with HIV.

One difficulty with methadone prescribing for harm minimization is in determining who should and should not receive it. Traditionally methadone has been prescribed for those dependent on heroin, rather than occasional misusers. However, with the increasing interest in HIV and harm minimization, methadone prescribing is becoming more flexible. Bell, Digiusto and Byth (1992) suggest that the traditional policy actually lengthens the period of illicit drug misuse for individuals by putting off treatment. This may be the case for some, but it is unclear how many occasional heroin misusers stop spontaneously after short temporary periods of misuse.

The risk of HIV, hepatitis and other health problems can therefore be reduced by the use of the oldfashioned method of methadone maintenance among heroin addicts. But what about recreational heroin misusers and misusers of other drugs such as amphetamine and cocaine? There is much less research on the prescribing of drugs to non-dependent users or on the use of stimulant substitutes such as dexamphetamine, which has only become common in the past year or two (see Chapters 8 and 11 for methadone prescribing for treatment).

Amphetamine prescribing for harm minimization

Methadone can be prescribed in order to stabilize and maintain opiate users, to help prevent injecting, sharing and other dangerous and illegal practices. Amphetamine can be prescribed for the same reasons; to convert misusers to a legal, oral drug, to stabilize lifestyle and maintain contact in order to encourage use of other services.

As with methadone prescribing, it is difficult to determine which amphetamine misusers should be prescribed a substitute drug and which should not. Useful guidelines were developed by the Petersford Community Drug Team (reported by Peter Ford in the SCODA newsletter, Institute for the Study of Drug Dependence, December 1992–January 1993):

- The client must show evidence of regular frequent amphetamine misuse (at least every 72 hours), extending over a period of six months or more (this may be corroborated by their use and knowledge of effects, street terms and prices)
- The use of amphetamine should be problematic and abstinence presently not an option.
- Psychological and health problems would not be exacerbated (e.g. blood pressure and paranoia).

The Petersford CDT dispense dexamphetamine in 5 mg tablets from specific community pharmacists. The maximum amount dispensed is 50 mg dexamphetamine a day. Clients are given high or low threshold contracts (50 mg or 30 mg), depending on extent of previous use. The contracts include the conditions that clients keep regular appointments and do not misuse other drugs except cannabis. The team deal mainly with chaotic and injecting drug misusers and have seen evidence of increased stability and use of other services.

Despite the increasing acceptance of harm minimization as a rationale for prescribing substitute drugs, there are many medical practitioners who still perceive methadone prescribing only as a treatment for dependence and so cannot view amphetamine substitutes in the same light because amphetamine is not a physically 'addictive' drug. It is therefore often easier for clients to procure a heroin substitute such as methadone than an amphetamine or cocaine substitute because amphetamines or cocaine are less easily justified within a treatment framework.

This ambivalence in prescribing practice can lead to the wrong drug being prescribed for the wrong reasons, as when stimulant or polydrug misusers are prescribed opiates for harm minimization purposes, simply because these drugs are traditionally prescribed in drug treatment.

It is therefore perhaps more important in this area than anywhere else to be clear about the purpose of prescribing; is a script prescribed in order to reduce drug-related harm or as a treatment for dependence? If the former, then the most appropriate prescription is a legal substitute for the illicit drug of abuse, whether this drug is physically addictive or not; if the latter, prescribing is limited to physically addictive substances for physically addicted patients.

A drug may of course be prescribed for both reasons, in which case there will be two objectives (harm minimization and treatment). The problem here is that the aims of the script become confused. If one aim is discounted (e.g. treatment ends), this is then not a reason to discontinue prescribing (for harm minimization purposes).

Urine testing

A common method for monitoring treatment compliance has traditionally been urine testing. However, the utility of urine testing when pursuing harm minimization purposes is less clear. This is discussed in greater detail in the prison context (Chapter 9). Urine testing in the context of treatment prescribing may be useful to enforce contracts and the same may be true for harm minimization prescriptions; however, prescribing to reduce drug-related harm is more flexible and there need not be the same emphasis on hard and fast rules of 'treatment compliance'. It should be stressed that the least harmful drug, cannabis, remains in the urine for 4–5 weeks, whereas more harmful drugs such as heroin are not detectable after several days. Urine testing may, therefore, encourage clients to switch to more dangerous drugs (as happens in custodial institutions).

Practical prescribing guidelines are given in Part Three.

SOCIAL AND PSYCHOLOGICAL PROBLEMS ASSOCIATED WITH DRUG-RELATED HARM

Contemporary drug agency-based harm minimization often includes reducing social and psychological problems associated with drug-related harm. Many agencies employ social workers and psychologists to work with the relevant clients. Generic professionals may wish to deal with these problems themselves, refer clients or work together with social workers or psychologists if they identify problems in these areas.

Social problems

These can include financial problems, debts and, perhaps more significantly, violence and intimidation for non-payment of debt, accommodation, employment, child care and legal problems.

Financial problems may be resolved by debt counselling, welfare rights advice or the provision of an alternative supply of prescribed

drugs. Many drug misusers also find themselves under considerable pressure from dealers to whom they owe money, leaving them -
vulnerable to intimidation. Financial advice may help. The same can be
said for support with accommodation.

The social environment in which a client lives may determine the level
of risky behaviour that is normal. Help with new accommodation may
avoid unhealthy, miserable conditions, where unhygienic drug misuse
and injecting and/or sharing may be the norm. Sharing needles and
syringes is more likely when clients are not in their own home. This type
of risk is increased among the homeless. Clients with legal problems often
require support in the form of court reports. Attending a drug agency for
help can be seen as a real alternative to custody. It is therefore worth professionals working closely with probation officers if necessary.

Psychological problems

These can include the effects of loss of control and chaotic drug misuse,
together with depression, anxiety, anger and aggression.

Drug misusers who present at drug agencies often do so for brief
periods in their lives when they seem to lose control of their drug misuse
and start to misuse chaotically. The data in Chapter 3 illustrate these problems. Generic professionals are also likely to become involved at this
stage. Life-saving interventions include health care and stabilization
prescriptions, but basic cognitive behavioural interventions can also be
useful. These are particularly appropriate for aftercare maintenance, helping drug misusers not to relapse back into old risky behaviour patterns
(see practical guidelines in Part Three). In addition, the most common
psychological problems in the general population, such as depression, anxiety and anger, can all be associated with drug misuse, either as direct
effects, withdrawal effects of drugs (Keene and Trinder, 1995) or as
precipitating factors leading to self-medication (see Chapter 8).

Methods derived from cognitive behavioural psychology can be used
for general problems of depression and anxiety or specific drug-related
problems, such as compulsive behaviour and loss of control. An
additional measure is the short-term prescription of drugs such as
benzodiazepines to stabilize drug misuse and/or treat anxiety and antidepressants for depression though there is, of course, a risk of a growing
dependence on benzodiazepines.

General behavioural techniques involve teaching clients to monitor
and understand specific behaviours in terms of what causes or motivates each action (antecedents and consequences). Clients can then learn
to modify their behaviour patterns by changing the controlling factors
and trying out new behaviours and coping strategies. The methods
include: assessment of antecedents and consequences: behavioural

monitoring; schedules of reinforcement and punishment; stimulus control and generalization; deconditioning and desensitization (cue exposure) and modelling and imitation.

Cognitive behavioural techniques are designed to modify both cognitions and behaviour. The essence of this work is again identification of problematic behaviours, antecedents and consequences. To this is added a comprehensive assessment of the maladaptive or controlling thoughts precipitating or following these behaviours. Clients are then taught to change and control their thoughts as a means of controlling behaviour. The methods include cognitive restructuring, self-instruction training, stress inoculation, thought stopping and basic self-control, self-talk methods and relaxation tapes. These methods are in widespread use for reducing anxiety and promoting relaxation, etc. Tapes and instruction booklets are usually available from psychology and health promotion departments in health authorities. These techniques are often effective when used in conjunction with practical development of new coping strategies and development of alternative activities such as physical exercise and sport. Changes in lifestyle can also help reduce stress and anxiety. Engaging in new occupations may provide new forms of interest and social life that militate against high-risk drug misuse.

Behavioural and cognitive interventions involve:

- assessment of individual problem behaviours and maladaptive cognitions together with the controlling conditions (antecedents and consequences);
- the development of structured interventions for cognitive and behavioural change aimed at achievable and measurable goals. These plans are carefully tailored to the individual;
- the active participation of the client;
- comprehensive baseline measures of the problem behaviour and consequent changes.

There are many texts outlining cognitive behavioural techniques for both clinical psychologists and generic professionals and the following may be useful: Goldfried and Bergin (1986), Kanfer and Goldstein (1986), Egan (1990) and Trower, Casey and Dryden (1991).

OUTREACH PROJECTS

The growing evidence that many risky drug misusers were not attending drug agencies led to the development of outreach projects. An outreach project involves individual staff members gaining access to networks of drug misusers who would not normally receive any service provision.

Outreach can have one of two purposes: it can be a means of reaching whole populations (public health measure) or it can function as a low threshold entry point into an agency. The public health approach fulfils a widespread need for harm minimization amongst those who do not want drug treatment; the agency-orientated approach expands the role of the old agencies by developing projects to access new 'clients' and possibly link them into the agency itself.

The agency-orientated approach to outreach work

This usually involves working with small numbers of drug misusers directly in their own homes or in their communities. It can mean seeing clients only once or twice but it is more usual for workers to get to know a core of clients and see them on a regular basis, so obtaining introductions to others. Workers often see part of their task as encouraging these people to attend drug agencies. For example, in San Francisco outreach workers gave treatment coupons to high-risk drug misusers (Sorenson *et al.*, 1993) and succeeded in attracting them into treatment. McCoy, Rivers and Khoury (1993) recommended that outreach programmes be incorporated into mainstream health and drug treatment programmes, to function independently and also as low threshold entry points.

Outreach work is difficult, dangerous and often unpopular. Outreach workers may take the more comfortable option of working closely with office-based services, which can be counterproductive. Bolton and Sellick (1991) found that using volunteers from the drug-using community itself proved a useful way of extending outreach work. Gilman (1992) points out that professionals may be more reliable but often have difficulties accessing target groups and may be vulnerable in ways that indigenous workers are not. He suggests that it is useful to involve indigenous workers to encourage a sense of social responsibility amongst drug misusers and to help change social norms related to risk behaviour.

The public health approach to outreach work

It has now been demonstrated that outreach projects reach different populations from those attending drug agencies, particularly attracting younger misusers who inject infrequently (Wechsberg *et al.*, 1993). This led to a change in purpose for outreach projects, as many workers focused on educating wide networks of misusers and linking these networks into a range of health and harm minimization services. An example is the peer education project implemented by the Wirral Drugs Service which developed to prevent HIV amongst drug misusers working as prostitutes (Hanslope, 1994).

This approach also utilizes social support by peers (Rhodes and Humfleet, 1993). Stimson (1995), discussing AIDS and injecting drug use in the UK between 1988 and 1993, states that:

The social networks through which HIV may be transmitted are the same social networks that may be co-opted for HIV prevention. Future outreach services must turn to these networks as a way of targeting and encouraging changes among broad populations of drug injectors. Such models might use indigenous advocates, working within social networks, supported by community outreach workers.

Rhodes and Stimson (1994) argue that there should be a change in focus from the individual drug misuser to large social groups and communities. Rhodes, Hartnoll and Johnson (1991), in a review of HIV outreach health education in Europe and the USA, propose that HIV prevention education should not be the province of specialist agencies but should focus instead on 'community change'.

The main problem with outreach is that it is by its nature difficult to evaluate without epidemiological studies of populations. Moreover, the difficulty in measuring outcome is increased by the tendency of outreach workers to attempt a range of different tasks without a clear set of priorities.

Practical guidelines for outreach work

The following guidelines will be useful when establishing and developing an outreach strategy. A useful practical account for those developing outreach projects is that of Gilman (1992), *Outreach*, published in the Drugs Work series by the London Institute for the Study of Drug Dependence.

The tasks inherent in the development of an outreach project are similar to those involved in the education project discussed in the previous chapter. The difficulties with outreach work are that the tasks are necessarily unstructured and much of the success of a project depends on the initiative of the individual workers in developing personal contacts and maintaining trust.

This need for flexibility should not over-ride the need for clarity in aims and knowledge of the target group. It is possible that the group actually reached may differ from the group originally targeted. It may then be necessary to start again or it may be constructive to change the original objectives to suit the new target group. It is not often constructive to become ambivalent about the aims and/or ambiguous about the messages given.

Flexibility involves having a very clear picture of the aims, objectives and target group and then closely monitoring the match of target group and project objectives. This allows appropriate changes based on a thorough knowledge of both.

Find out as much as possible about the population to be targeted

This is often referred to as developing a community profile, surveying the type and extent of drug misuse and the community norms (Fitzpatrick and Gerard, 1993). A good example of this type of approach is that developed in Edinburgh (McDermott and McBride, 1993). Here drug misusers and drugs workers formed a 'coalition' with local young people to help reduce drug-related harm and focused mainly on clubs (Gilman, 1992).

Determine target groups

As became clear in the case study in the previous chapter, the aims and the target group need to be carefully matched. This cannot be done unless both are carefully defined and understood.

Clarify aims

These aims can be public health aims, such as targeting particular groups within a total population with literature or syringes and condoms, or making contact with individuals in order to link them with local drug agencies. The aims will partly determine who is targeted and how, who is recruited and what is actually done. For example, outreach workers aiming at widespread harm minimization would attempt to provide as many people as possible with information and contact numbers for a wide range of generic services, whereas those aiming at low threshold entry into services might focus on one-to-one work and publicize the services that can provide more specialized care.

Clarify and prioritize objectives

Having identified aims it is necessary to clarify the practical steps necessary to achieve them. Many outreach workers fail simply because they do not have a clear idea about how to achieve their aims.

Clarify how day-to-day tasks will actually be carried out

It is important to decide early on how contacts will be made and networks accessed. The work can be difficult and risky. For example, particular aspects such as confidentiality and anonymity of clients need careful planning. It is therefore useful to contact experienced outreach workers on drugs and HIV projects for support and advice.

Monitor progress

This includes monitoring exactly what happens in order to be as flexible as possible in changing the tasks, objectives and even aims of the project as the worker gains more knowledge and experience of the target group.

The essence of good outreach is therefore both clarity and flexibility: clarity of aims and the practical means of achieving these and flexibility in changing them when necessary, to fit the target group accessed.

SYRINGE EXCHANGE: RESEARCH, CASE STUDY AND PRACTICAL GUIDELINES

RESEARCH

The method most clearly designed to reduce both HIV and drug-related health risk as a whole is needle and syringe exchange. This does not try to integrate harm minimization with drug dependency treatment and as a consequence is less confusing to professionals and drug misusers alike. Needles and syringes are offered free on an exchange basis or sold from pharmacists. The aim is to ensure that anyone who injects drugs will use a clean needle and syringe rather than share.

Syringe distribution and syringe exchange programmes operate on the assumption that drug injectors share syringes because sterile syringes are difficult to obtain.

Syringes have been supplied through pharmacies, vending machines, outreach workers and through special syringe exchange schemes. They work successfully in many countries including Holland, Sweden, New Zealand, Canada and Australia (Stimson *et al.*, 1990). Syringe exchange schemes (as opposed to syringe distribution schemes such as pharmacy sales) are basically concerned with dispensing sterile needles and syringes to people who inject drugs and providing facilities for their disposal. These schemes usually link dispensing and disposal, with the continued access to clean equipment being to some extent contingent on the return of used needles and syringes. This service is provided free and usually from a particular base at regular times during the week. Syringe exchange schemes are often based in drug services or have links with these services, which enables them to provide help and advice concerning drug problems, education about HIV risk behaviours, basic health care and access to other services (Keene *et al.*, 1993).

Generally, syringe exchange attenders report less injecting and less sharing of equipment. They are also likely to be more knowledgeable about HIV and make harm minimization changes in their behaviour (Stimson *et al.*, 1992; Keene *et al.*, 1993; Frischer *et al.*, 1993). Needle exchange programmes can also function as outreach programmes, firstly in the sense that they make contact with at-risk groups in order to attract them into treatment (Ginzburg, 1989) and secondly in terms of public health aims as they deal with drug misusers as a whole rather than just those individuals who present at treatment agencies. For example, individuals will often be given more needles and syringes than they need so that they can distribute them amongst other drug misusers. In some places drug dealers are given large numbers of needles and syringes to distribute when they sell drugs.

The report of the Advisory Council on the Misuse of Drugs (ACMD, 1988) gave support to syringe exchange as a means of preventing the spread of HIV among injecting drug misusers and recommended that resources be given to an expansion of the service. Syringe exchanges reach many people who are not reached by more conventional services (Donoghoe, Dolan and Stimson, 1991a). Syringe-sharing rates in the UK and the Netherlands are lower than rates in many US cities where legal sterile syringes are unavailable. In seven US cities, a National Institute on Drug Abuse tracking study found continuing high rates of syringe sharing (in four of the seven cities 70% or more had recently shared) and only modest risk reduction (Battjes, Pickens and Ansel, 1991). Negative effects of increased syringe distribution have not been reported. In Amsterdam, the total number of injectors has not increased (Buning, 1991) and participants have not increased their frequency of drug mis-use or of injecting (Van den Hoek, Van Haastrecht and Continho, 1989). The limits of syringe exchange are partly operational; they reach drug injectors but are not very successful in reaching women, younger injectors or those with a shorter history of injecting (Donoghoe, Dolan and Stimson, 1991a).

SYRINGE EXCHANGE BASED IN A DRUG AGENCY: A CASE STUDY

In the previous chapter practical guidelines were provided for drug education and/or prevention. A case study, 'Project Pitfall', was examined in order to highlight the problems inherent in that kind of work. In this chapter a similar format will be used. A case study of syringe exchange will be examined in order to pinpoint the opportuni-ties and constraints of service provision. This will provide guidance both for those involved in this type of service provision and for those whose clients may use it.

As in the previous chapter, it is necessary to emphasize the two most important factors when aiming an education or prevention service at a particular population – target population and basic task. Neither should be taken for granted. First, the nature and extent of the target population is not known: it should not be assumed that this population is equiva-lent to that presenting at drug agencies. Second, the task of providing an appropriate HIV prevention service for this population should not be assumed to be the same as drug treatment, nor that it requires specialist drug workers for its implementation.

These premises were taken for granted in the initial stages of the development of syringe exchange in England and Wales and as a consequence early service provision became the province of drug workers and was located in a drug treatment setting (Lart and Stimson, 1991). There are problems of limiting this work with drug misusers to drug agencies (Keene *et al.*, 1993).

The following section presents a case study of a successful drug agency-based syringe exchange and considers the limitations within the wider context of harm minimization and HIV prevention as a whole. First the advantages of drug agency-based services are outlined, then the limitations and constraints of such a service in the wider context of public health provision are analysed.

One of the most successful drug agency-based syringe exchanges in Wales was located in a non-statutory, drop-in drug agency. This agency developed a large caseload within a period of two years, many of whom had not previously been in contact with drug services and some of whom consequently went on to use other drug agency facilities.

The drug agency is near the town centre and is open from 10.00am to 5.00pm weekdays and manned by voluntary as well as paid workers. The setting and procedures are informal and the atmosphere relaxed and friendly. This syringe exchange is one of the most successful and least institutional in Wales. The main room in the building is used as a drop-in area where both client and staff sit. The furniture is old but comfortable and literature, information and client work (poems, letters, drawings) are displayed. The front door is open and if someone comes in they are greeted by staff. Clients are usually offered tea or coffee and people often stay to chat (even if they have not come to see a particular member of staff or to ask for help). The drug treatment service offered is:

> ... assessment and crisis work and structured ongoing counselling, this (counselling) is usually begun on an informal basis.

The syringe exchange service is separate to the treatment service, but staff attempt to provide syringe exchange clients with a counselling service.

> Though syringe exchange is very informal initially, clients gradually start to talk more to staff and the initial help and advice often leads to the development of a counselling relationship.

The syringe exchange attracted a large number of clients not previously in touch with health services and many of these have gone on to make use of other project facilities.

> There are advantages of it (syringe exchange) being in the drug team, it is easy to communicate with the team staff, referrals and contacts come from the drug project and there is staff support and cover.

The staff identified four significant factors in the successful development of the syringe exchange. First, the drug project had been established as a drop-in agency for seven years:

> (We) have gradually built up contact with networks of users over the years.

Second, the project was helped in the early days by several local individuals.

> A few older established users attached to the centre had stopped using but had respect and credibility within drug-using circles. They acted as outreach workers for the syringe exchange.

Third, they pointed out that they were based in a comparatively small, compact city, where:

> One centre could attract most of the population but the city is big enough for anonymity to be real.

Fourth, that the drug project itself provided an informal, harm reduction-orientated, drop-in service into which needle exchange could be easily integrated and that:

> There are no conflicts of philosophy and policy. I had been used to penalizing clients for injecting drugs. In my previous work in this field, methadone scripts would have been stopped if there was evidence of injecting. It had initially been difficult to adapt to the idea that clients would continue to inject drugs; however, I am reconciled to this.

There were three salaried staff on the team and although links appear to have been made with hard-to-reach groups, there was a lack of support to develop and work with these contacts. For example, the project had made contact with women who work in massage parlours, many of whom were misusing drugs, but did not obtain funding to develop these links. The disadvantage of a project with 'street credibility' for clients was the lack of respectability with other professional agencies. The project had difficulties obtaining local professional backing, though they have good relations with the local genitourinary medicine clinic.

> I think we are as efficient as we can get, our costs for exchange are practically nil. There are no costs for rent, staffing, admin, etc. The only costs are for equipment. But we need a full-time staff member in order to be fully effective.

THE ADVANTAGES OF DRUG-AGENCY BASED SYRINGE EXCHANGE

The early drug agency-based initiatives in Britain brought about foundational and almost revolutionary changes. It was largely as a consequence of the provision of these services in 200 agencies across England and Wales (Lart and Stimson, 1991) that attitudes to HIV prevention among drug misusers became more positive. The syringe exchanges first brought about a change in attitudes of the professional drug workers themselves and then stimulated a healthy change in attitude in their clients, who began to share less often. This caused a change in attitudes amongst the helping professions generally. Syringe exchange became a respectable, taken for granted part of everyday life and an integral, accepted part of public health service provision in all areas of social and health care (with the exception of the prison service).

> If you work in a drug agency, you see drug use in a different perspective to everybody else.

> We have perhaps helped towards a 'cultural change' in that it is becoming the 'norm' to use clean needles.

THE DISADVANTAGES OF DRUG AGENCY-BASED SYRINGE EXCHANGE

A serious problem for agencies in rural areas and those with widely dispersed populations was identified early on; many drug injectors did not attend the services. Those working in such areas were more aware of the need for a wide range of generic, community-based services (Keene and Stimson, 1992). It soon became apparent that the catchment areas of drug agency-based syringe exchanges were small. In Wales, where this case study agency was based, almost 42% travelled less than one mile and 40% travelled one to five miles to reach a scheme. Only 19% of attenders travelled more than five miles to reach a syringe exchange.

Research in Britain as a whole also indicated that large numbers of injecting drug users were not attending the agency-based syringe exchange services (Stimson *et al.*, 1988; Stimson, Dolan and Donoghoe, 1990; Glanz, Byrne and Jackson, 1990). Information from pharmacist surveys (Glanz, Byrne and Jackson, 1990) and other work (Keene *et al.*, 1993) indicates that many drug users were buying equipment from pharmacists rather than using the agency-based exchanges.

Drug workers themselves were among the first to recognize the limitations of agency-based provision.

> The constraints are that we can't reach everyone, there are not enough of us and there is not enough medical support.

> It is untenable to be doing needle exchange and acting as a counsellor, they need to be separate. By providing things people need you are in a position of power; this is incompatible with counselling. If someone is trying to give up drugs they won't want their counsellor to know they are getting needles.

> The disadvantages are that my clients cannot use the service without coming into contact with drug treatment staff: also some clients say that others want to come but won't come into a drug agency.

Drug agency workers, as a professional group, initially attempted to resolve this by expanding their own services. They attempted it in various ways, from sending outreach workers to introduce potential clients to the agency to setting up satellite clinics, yet were severely limited by both time and staffing constraints. Despite the effectiveness of some of these interventions, the specialists had in effect ensured (through their definition of syringe exchange) that the task remained within their own professional remit, effectively limiting the involvement of other professional or non-professional groups.

> Pharmacists are not trained for this kind of thing, it is not their role and it will clash with their everyday job.

> Ordinary general health workers and social workers don't have the knowledge or skills to deal with drug users.

> Private space is needed for confidentiality and pharmacists do not have the facilities for the necessary counselling.

Despite the potential problems inherent in limiting harm minimization to sympathetic drug workers, there were far greater problems where specialist drug professionals worked with incompatible treatment models. For example, where drug treatment professionals believed in the abstinence approach and also had total responsibility for service provision for drug misusers as a whole, they could be opposed to any form of syringe exchange or harm minimization at all. This was initially the case in north Wales, where an abstinence approach was the norm and syringe exchange was opposed (Keene, Parry-Langdon and Stimson, 1991).

AN EXTENDED MULTIDISCIPLINARY SERVICE: GENERIC PUBLIC HEALTH PROVIDERS, HEALTH CARE WORKERS AND PHARMACISTS

Within a few years it had become evident that, whilst drug workers had been extremely influential and instrumental in implementing syringe exchange in the first place (indeed, it is doubtful whether the service would have been introduced at all without their dedication), these influences combined to limit the development and range of this significant new initiative. This consequence was concealed perhaps by the rapidly growing expertise of professional drug workers.

As time passed it became clear that two of the most obvious health promotion providers had been virtually excluded. They were the community pharmacists and primary health care workers, professions with no conflicting ideologies or treatment priorities and an extremely widespread network of outlets. Initially, initiatives were developed independently of drug agencies by health workers and pharmacists primarily concerned with HIV prevention rather than helping people with drug problems. These services sometimes developed, with limited funding, in the face of opposition from the drug services and health authorities (Keene, Parry-Langdon and Stimson, 1991).

However, in 1991 there followed a change in government policy and pharmacists were remunerated for offering exchange schemes or disposal points. They were also encouraged to sell syringes to drug misusers and provided with free disposal tubes by health authorities (Keene and Stimson, 1991).

In the ensuing years health authorities and pharmacists themselves developed community-based pharmacy schemes across England and Wales, offering either sales and/or disposal points or more formalized exchange schemes. Some of these programmes ran alongside the specialist drug agency schemes and served as a support or fall-back for agency clients when the schemes were closed or inaccessible for other reasons. However, many schemes provided the only service in remote or rural areas and in the absence of a specialist drug agency syringe exchange. Services provided by pharmacists included syringe exchange; sale of needles and syringes; referral to drug agencies, genito-urinary clinics and HIV counselling services; literature on cleaning needles/syringes, safer drug use and safer sex; and condoms. Eventually specialist drug workers and pharmacists themselves came to see the extended provision offered by community pharmacists as indispensable.

> Pharmacists sell syringes, so clean equipment is available, easily accessible in a way that drug users will use, without there being disincentives.

> The logic of using pharmacies seems inescapable. Little is known of rural drug-using populations beyond the fact that they exist and will not travel far for help ... some users still prefer to stay outside the formal scheme by buying their equipment.

> Low key contact points, it is anonymous, professionals are available in rural areas, au fait with microbiology and familiar with aseptic techniques. We can't have a centralized depot as the county is too big, we need local access points.

WILL YOUR CLIENTS USE THE SUCCESSFUL AGENCY-BASED SERVICE?

The drug agency in the above case study was successful in providing a specific syringe exchange service (i.e. free needles and syringes in exchange for returned used needles and syringes) to those who were prepared to travel to the agency and return used syringes. It was also successful in offering a basic harm minimization and drug counselling service as part of the package of care for those clients who would accept it and functioned as a low threshold service for access to other agencies. As far as the funders, staff and clients of this exchange were concerned, the agency provided an efficient and well-used service. Evaluations of the service provided for clients who attended were extremely favourable (Keene and Stimson, 1991): however, certain questions remained unanswered. Would all injecting drug misusers use this service? That is, was this type of service the only effective means for reaching the target group as a whole with the most appropriate service?

PRACTICAL GUIDELINES FOR SYRINGE EXCHANGE

While links with drug treatment agencies are extremely useful for making contacts with some clients and for referring others, it may be more constructive to define the priorities and tasks of HIV prevention as a whole, independently of drug treatment (Keene and Stimson, 1996).

Separating syringe exchange from drug treatment

If the public health objectives of harm minimization and HIV prevention are separated from the specialist skills and philosophies of the drug treatment professions, it is possible to see a wide range of possible participants and methods for tackling the problem. Pharmacists provide only one example of a health care profession delivering a harm minimization service for drug misusers. The development of syringe exchange schemes provides a useful service for a certain proportion of injecting drug users, but the integrated package presents obstacles for other service providers and the customers themselves. It may be more practical to use a wider range of health and social care professionals to dispense needles and syringes and organize various community-based disposal points together with education concerning safe disposal at home, while maintaining the specialist drug agency services and developing links between them.

Separating dispersal of syringes from disposal of syringes

If the concept of syringe exchange is divided into two distinct tasks, dispersal and disposal of needles and syringes, the following practical developments offer an extension of agency-based schemes.

Dispersal points can be placed in a wide range of easily accessible places, from pharmacists who sell equipment to other health care agencies such as general practitioners and accident and emergency departments. These service providers could also offer bleach cleaning instructions and home disposal tubes. Disposal facilities can be provided by these and/or others, but need not be provided by all who dispense. This would increase the number of dispersal points, as many professionals are happy to provide syringes but not to become involved in complex syringe exchange schemes. Disposal collection facilities could be provided free by the statutory services. For example, a regular monthly, door-to-door collection and disposal service would prove invaluable to many pharmacists who could use the service to dispose of other returned medicines.

Many areas in England and Wales have adopted this approach, ensuring that pharmacists continue to sell needles and syringes. Clean equipment is then consistently available and easily accessible to drug misusers, without insisting on careful monitoring of returned equipment.

Different elements of syringe exchange

Whether professionals are involved in developing a comprehensive syringe exchange service, providing part of this overall service or simply encouraging clients to use these services, it is useful to understand the wider picture. The essence of effective syringe exchange is diversity. The following guidelines give a wide range of different types of task within the overall service provision. Neither agency-based nor community-based provision is appropriate alone: both types are essential parts of a comprehensive syringe exchange scheme, consisting of distribution points, disposal points and drug agency bases.

A comprehensive syringe exchange service will offer a service at different levels, with a focus on five particular areas: the distribution of syringes, the safe disposal of syringes, education regarding bleach cleaning, health care service provision and low threshold entry into drug treatment.

Professionals can refer their clients to one of these facilities and/or provide one or more themselves. This section provides information on what can be provided by different professionals and what is available.

Drug agency-based syringe exchange services

Most statutory and non-statutory drug agencies in Britain provide a syringe exchange programme (with the exception of abstinence-based services) (Stimson *et al*. 1988; Stimson, Dolan and Donoghoe, 1990). These programmes offer all the above services as part of an overall package. Because these agencies offer a far more comprehensive range of services than is provided elsewhere, they are often the first choice for both professionals and clients. They offer a wide range of facilities, most of which can be obtained on an anonymous basis. Agencies usually try to draw a distinction between clients attending the syringe exchange for harm minimization (anonymous) and those attending for drug treatment. The exchange staff work together with drug team staff and clients can be referred to the drug team for prescribing services (not anonymous). The exchange acts therefore as a low threshold entry point into a harm minimization prescribing service and clients may find themselves counselled by drug workers as part of the syringe exchange service.

Most syringe exchanges offer the full range of services outlined below, together with additional general harm minimization services (see also harm minimization guidelines). Many exchanges are based in health authority premises and employ nurses as part of the staff team, so are well placed to offer a comprehensive health care service. In addition, the attached drug agency teams often include psychiatric consultants and other doctors. Exchanges provide some or all of the following: a range of drug equipment including a (much appreciated) choice of needles and syringes; sterile water, swabs, wound cleansing packs; containers for disposal or return; condoms, spermicides and bleach/cleaning containers.

It is clear that if clients are prepared to attend agency-based syringe exchanges, they are the best possible option. Research indicates that many clients do not attend these agencies for practical reasons such as travelling distance and that those who do attend also need to buy needles and syringes from pharmacists (Keene and Stimson, 1991). They may well find themselves in situations when they need to share and are obliged to clean needles and syringes. It is therefore advisable to provide clients with alternatives to be used instead of, or in conjunction with, drug agency syringe exchanges.

Non-specialist syringe exchange: the wider picture

It can be seen that drug agencies provide a comprehensive syringe exchange and harm minimization service. Syringe exchange as a whole includes several different services, each of which can also be offered by an even wider variety of different professionals. There is no reason why various aspects of a service should be offered together; many different professionals can provide one or more of them, whether it is offering leaflets and lists of local pharmacists or ensuring disposal tubes and bins are available in residential facilities.

PRACTICAL SUGGESTIONS FOR PROVIDING SYRINGE EXCHANGE SERVICES

The simplest way to obtain needles and syringes is to buy them from a pharmacist. It is therefore useful to provide clients with a list of pharmacists prepared to sell needles and syringes to drug misusers. (Local drug agencies and health authorities can provide these lists.)

The most useful service almost all agencies can provide is disposal. This consists of a sharps box (safe plastic disposal bin). Some health authorities will also provide supplies of small plastic disposal tubes for safe keeping and home disposal.

These two elements provide the basic minimum syringe exchange service, ensuring that clients have ready access to clean equipment and the facilities to dispose of used equipment safely. The following suggestions offer ways of developing this basic service and list the various services available, provided by different professional groups. The significance of many of these services is that a client can remain anonymous. It is useful to find out how far anonyminity is guaranteed as this is likely to be the factor determining whether or not clients will use the service.

- Pharmacists can sell needles and syringes or provide disposal facilities or both. They may also take part in traditional organized syringe exchange programmes, where they offer free syringes conditional on their return when used. A coordinator is often appointed to facilitate and monitor progress.

- Whereas a basic disposal facility can be offered by a range of drop-in, day centre or residential facilities, it is also possible to set up full syringe exchange satellite clinics in non-statutory and health service community bases. These clinics can operate on an occasional but regular basis, such as one afternoon a week. If staffed by nurses these satellite services can also offer a basic health care facility.
- Outreach workers can carry needles and syringes to clients who do not attend services.
- Indigenous workers (current and ex-drug misusers and others with access to drug-injecting groups) can be involved in order to encourage a sense of social responsibility amongst drug injectors and to help change social norms related to HIV risk behaviour. (This may already be contributing to lower sharing rates among non-attenders.)
- Syringe exchange coordinators can be responsible for organizing distribution and will provide disposal facilities. These workers offer training, support and monitoring of generic professionals, outreach workers, volunteers and indigenous workers.
- Syringe exchange based in agencies can provide a range of services including health care aspects and social and welfare provision, accommodation advice and referral to other agencies.
- Drug treatment agencies can develop a distinct, separate harm minimization service and, as a consequence, design prescribing specifically for harm minimization and HIV prevention, rather than the traditional treatment withdrawal schedules.
- Many drug agencies and/or health authorities provide information leaflets listing local services (including dispersal and disposal points), addresses and phone numbers of all syringe distribution and disposal points in each area. It is useful for professionals to obtain these lists and make them available in agencies; they should include health care and local authority premises, pharmacies, police stations and magistrates' courts.
- It is essential to emphasize the importance of effective cleaning of equipment in emergencies. There is a residual risk to all misusers who may be without clean equipment occasionally. Literature and containers for bleach cleaning should be offered together with short simple information cards or stickers giving information on cleaning and safe disposal. Bleach-cleaning containers can be obtained from drug agencies and at syringe distribution points.
- Effective home disposal methods are needed when disposal points are not easily available; for example, safe destruction at home. This is particularly important in rural areas where lengthy storage of used syringes for later return to an agency may be impractical and may itself encourage reuse of syringes.

- As high levels of risk behaviour in prison are common (see Chapters 5 and 9), workers should develop links with prisons and pre-release prisoners.

PRACTICAL GUIDELINES FOR GENERAL HARM MINIMIZATION

There are many different tasks involved in a comprehensive harm mini-mization service and most generic professionals can be involved in carrying out one or more of them. It is important to stress that the drug treatment agencies provide the most efficient and comprehensive syringe exchange service and should be the first recommendation to drug misusers. The following services are available in most drug agencies and formal syringe exchanges. Generic professionals can be seen as offering a supplementary service to those clients who do not wish to attend drug agencies or cannot attend on a regular basis, for whatever reasons. The following services are also available from a wide range of non-specialist agencies and professionals. Some services provide elements free of charge, such as syringe exchange and syringe disposal points, others make a nominal charge for the purchase of equipment such as needles and syringes.

Targeting the relevant information and help

Griffiths *et al.* (1992), in their 'drug transitions study', found that different patterns of drug misuse were correlated with different routes of admin-istration and different types of health risks. Des Jarlais, Friedman and Casriel, (1990) identify three different target groups for preventing HIV:

1. preventing drug injecting amongst those at risk from starting to inject;
2. providing drug treatment for those who want to stop injecting drugs;
3. providing safer injections for those who are likely to continue.

Griffiths *et al.* (1994) carried out a study of 408 heroin users and found that there were two different types of use – 'chasing the dragon' (44%) and injection (54%). They suggest that interventions should take into account the different routes of current administration of drugs and consider the potential for future transmission within continued drug misuse.

The message of the previous chapter, that education interventions should be tailored and targeted at appropriate groups, remains true for harm minimization. Professionals should provide appropriate informa-tion depending on the type of activity of clients and the level of risk.

For example, the following aims can all be considered to be part of harm minimization. The aims and therefore the methods will depend on which group is targeted.

Aims

- To reduce the harm associated with all drug misuse.
- To stop people starting to inject.
- To stop people injecting.
- To stop people sharing (borrowing and/or lending).
- To stop people sharing with unclean equipment.

Methods

- Give information about dangers of drug misuse as a whole.
- Give information about dangers of injecting.
- Encourage other routes of administration and provide oral drugs.
- Provide clean equipment.
- Teach cleaning techniques.

WHERE TO GET DIFFERENT KINDS OF HARM MINIMIZATION SERVICE
AND WHAT CAN GENERIC PROFESSIONALS PROVIDE THEMSELVES?

The following harm minimization methods are available to clients, whether or not they inject. The main task of the generic professional is not necessarily to provide a harm minimization service, but rather to ensure that their clients have access to the full range of harm minimization and HIV prevention services provided by others.

- Information.
- Syringes and needles.
- Disposal points.
- Information about syringe cleaning.
- Regular basic health care.
- Prescribing facilities from a specialist agency or general practitioner.
- HIV information, testing and counselling facilities.
- Cognitive behavioural interventions for controlling chaotic behaviour.
- Social support for housing and financial problems.
- Low threshold access to drug treatment.
- Cognitive behavioural interventions for preventing relapse into old risky behaviour patterns (see practical guidelines in Part Three).

Information

The first guideline is simply to obtain as much information as possible about available services. Outreach, prescribing and syringe exchange are all examples of harm minimization services provided across Britain. They ensure that clients have easy access to information, clean equipment and disposal facilities, health care and prescribing services. Professionals can find out from health authorities, health promotion departments or drug agencies which of these facilities are available and where. This information should be made available to clients. There is seldom any system of referral, as the significant characteristic of such service provision is a greater level of anonymity than usual. Clients buying or disposing of equipment do not have to give a name and many syringe exchange programmes may be happy with only a first name or similar token identifier. Professionals can offer clients this information and other literature on safe sex and safer drug use. More detailed information is available from the Institute for the Study of Drug Dependence (1–4 Hatton Place, Hatton Garden, London EC1N 8NI).

Information on safer non-injecting drug use

Safety precautions for use of Ecstasy and stimulants are largely concerned with the risks of overheating and dehydration and therefore the need for cooling off periods and drinking water. Occasionally an individual will react badly to the drug or a mixture or substitute for the drug. The need for immediate access to medical support should be stressed and clients given information about basic first aid and the importance of the recovery position. Safety precautions for the use of depressants are concerned with the risks of overdose, particularly after periods of abstinence when tolerance will be greatly reduced.

Information on safer injecting drug use

This type of information is concerned with two areas – transmission of HIV and hepatitis and basic health risks. The risk of HIV and hepatitis infection can be reduced by ensuring that clients always have their own needles and syringes and that they do not share these or other drug use equipment such as spoons or water for flushing syringes. General health risks can be reduced by encouraging clients to use smaller needles and not reuse their own needles and syringes (see Part Three for outline of general health care issues).

Information is now available regarding safer and more hygienic methods of crushing tablets and preparing powders and drawing up substances into syringes. Information can also be obtained about safer injecting techniques and safer injecting sites (e.g. how to avoid arteries). The issue of how much information should be given to clients is controversial and depends largely on the training, experience and knowledge of the professional. Perhaps the most practical solution is again to ensure clients have access to this information, if they need it, through information leaflets. One of the best information leaflets for injecting drug users is *What Works: Safer Injecting Guide*, produced by the Exeter Drug Project (59 Magdalen Street, Exeter, EX2 4HY).

Distribution of new needles and syringes

The main agencies and professionals involved are drug agencies, syringe exchanges, hospitals and community pharmacists.

Safe disposal of syringes

The main provisions are containers for disposal, disposal bins and other disposal points, and safe home disposal containers.

Agencies and professionals involved include drug agencies, syringe exchanges, hospitals and community pharmacists. Disposal points or sharps boxes have been made available in hospitals, general practitioners' surgeries, probation offices, hostels, therapeutic communities, night shelters and police stations. These facilities are not yet available in prisons. Generic professionals can also provide disposal bins and safe disposal tubes themselves (available from health authorities).

Provision of information and equipment for cleaning needles and syringes

This includes the provision of information about syringe cleaning techniques, the importance of cleaning spoons, needles, etc., information on syringe decontamination, bleach/cleaning containers, sterile water and decontaminants (disinfecting/sterilizing tablets).

Agencies and professionals involved include drug agencies, syringe exchanges, hospitals and community pharmacies. Lists of addresses and contact names are available from local drug agencies and community pharmacists.

Disinfecting/sterilizing tablets have been made available in hospitals, general practitioners' surgeries, probation offices, hostels, therapeutic communities, night shelters, police stations and prisons.*

Health care provision

Many drug misusers find it difficult to obtain health care services as health professionals have often had negative experiences of drug misusers in the past. It is therefore useful for professionals to mediate on behalf of their clients, whether in accident and emergency departments or with GPs or family practitioner authorities. Some agencies will provide first aid equipment and basic health care equipment such as swabs and wound cleansing packs.

*Recently withdrawn from prisons for health and safety test.

Agencies and professionals involved include drug agencies, syringe exchanges, hospitals, community pharmacies and general practitioners.

Prescribed substitute drugs

The availability of prescribed substitute drugs is extremely variable. Some drug agencies and GPs are prepared to prescribe maintenance or long-term prescriptions of opiates or amphetamine substitutes, for harm minimization purposes. A small minority will even prescribe injectable drugs. Many GPs are willing to prescribe drugs if a social or health care professional agrees to monitor the client. However, some are unhappy to prescribe anything other than tranquillizers unless a specialist drug worker is involved.

Cognitive behavioural interventions for controlling chaotic behaviour

A small group of risky drug misusers will pass through periods when their drug misuse becomes chaotic and uncontrolled. These periods are usually brief, but clients are probably at greatest risk at these times. The cognitive behavioural interventions described in the section on social and psychological problems will be of help to clients in stabilizing their drug misuse and other behaviours.

Social support and help with general psychological problems

Social services, citizens' advice bureaux and housing associations can provide essential information about welfare rights, housing and financial problems. General practitioners and drug teams can provide access to psychological support and/or prescribe for anxiety and depression.

Low threshold entry into drug treatment and referral to other services

As many drug agencies are self-referral, it is not necessary for professionals to refer clients but they can give information and possibly contact names in the agencies. Clients may prefer to attend drug agencies informally at first and they can also use agency syringe exchanges without giving personal information; many agencies do not require a name but simply initials and date of birth.

Once clients are attending drug agencies the agencies can often help in obtaining a general practitioner and access to other services such as a psychologist or HIV counselling, to which they might otherwise have difficulty gaining access.

Cognitive behavioural interventions for preventing relapse (see practical guidelines in Part Three)

Part Three considers the practical difficulties of maintaining safer drug misuse behaviour over long periods of time with minimal professional input.

HIV prevention, testing and counselling facilities

These services are available from specialist agencies and genitourinary clinics. Drug agencies will have contacts with HIV agencies and will refer clients for testing and for pre- and post-test counselling.

Information on safe sex and condoms should be provided. There is now a wide range of information leaflets concerning safe sex and many agencies offer free condoms. There is always a danger that well-known facts about drug use and sex will not be available to younger drug misusers.

Agencies and professionals involved include drug agencies, syringe exchanges, hospitals and community pharmacies, genitourinary clinics, general practitioners' surgeries, probation offices, hostels, therapeutic communities, night shelters, police stations and prisons (not condoms). Health promotion departments will often provide leaflets and information to both professionals and clients.

Provision of sexual protection

Condoms, spermicides and female comdoms are available from drug agencies, syringe exchanges and community pharmacists.

Work with HIV positive misusers

There is an increasing need to consider HIV-seropositive misusers. It has been shown that these people are not receiving preventive care. It is therefore necessary to target this group, particularly for early treatment of asymptomatic HIV infection (Solomon *et al.*, 1991). Many drug workers now find themselves working with clients who are likely to die of AIDS and are having to develop new ways of working to reduce risky drug misuse and sexual behaviour (Des Jarlais, 1990). Des Jarlais suggests that drug treatment programmes have gone through four stages of response to this phenomenon: denial, panic, coping and potential burnout. It is, however, clearly necessary for professionals who wish to avoid this themselves to seek help and support from both specialist drug agencies and HIV services.

CONCLUSIONS

In conclusion, it seems clear that as many social and health care professionals as possible should be involved in harm minimization. The wide range of tasks make it possible for generic professionals to work with each other and drug agencies to provide a comprehensive service for all drug misusers.

The main points arising from this chapter are:

- There is no need for 'treatment' in order to prevent drug-related harm.
- There is no need for dispersal and disposal of drug use equipment to be a part of the same package.
- There is a great need for direct health care to maintain individual health and prevent HIV and hepatitis (individual health care and public health measures).
- There is no need for specialist knowledge of drugs, or experience with drug misusers, to provide each of the separate elements of harm minimization.

This chapter has given an account of harm minimization. The following chapter gives an account of drug dependence treatment. It should be emphasized that whilst a client who needs treatment for drug dependence is likely to need harm minimization before, during or after treatment, a client who requires harm minimization input does not necessarily need drug treatment at all.

It is important for professionals to decide which goals and methods are most appropriate for which clients. They may decide a client needs both harm minimization and drug dependence treatment, but this should not be taken for granted. This multiplicity of different needs is partly what makes working with drug misusers difficult, but this also arises when working with people with mental health or physical problems. The essence of efficient practice is practical assessment of different types of need that can be met using a range of resources.

WHAT IS HARM MINIMIZATION WITHOUT DRUG TREATMENT?

Harm minimization is often seen as a low threshold entry into drug treatment or as a short-term objective on the way to the ultimate aim of abstinence, rather than an end in itself.

These issues are confused in the field and as a consequence many professionals and clients are not clear about the difference between harm minimization and drug treatment. It is only when these categories are clearly defined that it can be decided which interventions are most appropriate. This definition is best determined by clarifying the needs of different target groups and the aims and methods of each service.

The aims of harm minimization are to reduce drug-related harm and the basic aim of drug treatment is to reduce drug misuse. The former aims to help clients stay alive, stay healthy and perhaps also out of custody, whilst continuing to misuse drugs; the latter aims to help clients cut down or give up drugs.

There is clearly an argument for providing harm minimization services for clients in drug treatment and there is an equally valid argument for providing them for clients who do not need drug treatment.

WHAT IS HARM MINIMIZATION WITHOUT SYRINGE EXCHANGE?

The basic harm minimization methods outlined above are necessary for risky and particularly injecting drug misusers. Harm minimization can be provided as an integral part of established syringe exchange programmes but this is a wide and diffuse target group, including many drug misusers who do not become part of a syringe exchange programme. In order to target this population effectively it is also necessary to provide services (such as needles and syringes and health care) separately from these programmes.

These services can be separated out to allow more professionals to be involved in provision and more clients to have access. The case study above has highlighted the problems of combining syringe dispersal, syringe disposal and drug treatment; each combination reduces the number of professionals qualified to provide the service and the number of clients prepared to accept the package. Whilst a range of harm minimization services may be an integral part of a syringe exchange or a drug treatment programme, neither package should limit the development of services by becoming identified as an essential aspect of harm minimization itself.

DOES HARM MINIMIZATION WORK?

It is clear from this chapter that reducing the harm associated with drug misuse is constructive. In the absence of a 'cure' or effective control of drug misuse, primary care workers should therefore focus on the promotion of less damaging behaviour among drug misusers (Robertson, 1989; Hopkins, 1991; Davies and Coggans, 1991). Harm minimization among drug misusers also reduces the risk of HIV. As a consequence the Advisory Council on the Misuse of Drugs (1988/1989) has clarified the priorities for professionals in the UK: 'AIDS presents a far greater threat to the individual and to society than drug use' and therefore the guidelines for working practice are equally clear: 'Prevention of HIV infection amongst injecting drug users must be a priority for all drugs work'.

The question, 'Are the aims of harm minimization achieved by the current methods and techniques?' can therefore be answered with a resounding 'Yes'. The difficulties lie not with its effectiveness but with moral controversies concerning the value of the aims and more practical questions about target groups and the most appropriate professionals to carry out the tasks. Many people would argue that the controversy concerning the morality of giving people information and equipment to make drug misuse safer is redundant in the face of the public health issues related to HIV. Others would assert that even without HIV it is still desirable to educate and equip people to protect them against unnecessary risks.

This chapter has not been concerned with moral controversies but rather with providing health and social care professionals with practical information on the most effective ways to reduce the health and social damage associated with drug misuse.

Treatment

8

INTRODUCTION: TREATING DRUG DEPENDENCE

The chapter is divided into six sections. The first, an introduction, examines problems of definition of dependence; the second analyses the three major treatment perspectives; the third considers the research issues; the fourth discusses the implications for practice; the fifth focuses on the change process, social factors and relapse prevention. This is followed by a case study of two treatment methods and finally practical guidelines.

WHAT IS IT AND WHAT IS IT FOR?

In Chapter 7 we examined the crucial distinction between harm minimization and treatment. Although in practice harm minimization and treatment of drug dependence are not necessarily clearcut categories, for the purposes of clarity, 'treatment' here will be used to refer to the professional response to physical and/or psychological dependence.

The chapter is therefore concerned with treatment of drug dependence and related drug problems, such as loss of control, rather than the wide range of problems that are associated in some way with drug misuse. Treatment generally involves therapeutic strategies for changing the drug misuser and/or their drug misuse behaviour, whether this takes the form of stopping, reducing or controlling drug dependence. Many drug workers (together with the author) feel strongly that harm minimization methods should always be used alongside treatment, but this does not mean that treatment methods should be used alongside harm minimization.

Having distinguished harm minimization from treatment, it is then necessary to clarify what exactly it is that is being treated. Much of the confusion in the field of addiction stems from both researchers and practitioners behaving as if there were general agreement about the concept of 'dependence', whereas this is not the case. First, scientific researchers erroneously presuppose there is consensus and that the phenomenon they call dependence can be easily defined and isolated, all peripheral variables controlled and experimental comparisons made. There is also a further presupposition that data inaccessible to this approach (concerning individual differences, subjective interpretations and the treatment process itself) are superfluous. This would, of course, be entirely valid if there was a general consensus about what dependence is and how to 'treat' it, but there is not. Second, practitioners and their clients may have completely different ideas of what they mean by addiction and this makes the concept to all intents and purposes useless for everyday communication between them.

A caveat

Although this chapter reviews work from the drug field, much of the literature and research on dependence and treatment is taken from the alcohol field. This is for several reasons: first, because theories of drug or 'chemical' dependence are largely derived from theories of alcohol dependence; second, because drug dependency treatment models are based on models of alcohol treatment, whether behavioural or disease models; third, most of the research and literature in the field of dependence is concerned with alcohol.

The following proviso is also necessary: to generalize about all drugs or to make generalizations on the basis of one (not necessarily representative) drug, such as alcohol, may be problematic. Theories of dependence assume that all drugs (including alcohol) share the same characteristics whereas, in fact, drug dependence encompasses an extremely wide range of different types of physiological dependence, some of which may not be similar to alcohol at all. Similarly, it could be equally risky to make statements about drug treatment as a whole and particularly to generalize, for example, from alcohol treatment to amphetamine treatment.

UNDERSTANDING THE PROBLEM

Dependence

Before outlining three very different perspectives of drug dependence, it is useful to give an overarching definition that attempts to amalgamate these varied approaches. This definition has been criticized as being too medical, too psychological and too general by various theoretical camps. In looking for a working definition we can do no better than rely on the WHO draft to the 10th revision of the *International Classification of Diseases* (WHO, 1989) where the dependence syndrome is defined as:

A cluster of physiological, behavioural and cognitive phenomena in which the use of a drug or class of drugs takes on a much higher priority for a given individual than other behaviours that once had higher value. A central descriptive characteristic of the dependence syndrome is the desire (often strong, sometimes overpowering) to take drugs ...

Physiological dependence, behavioural disorder or disease?

It is important to recognize that the experts do not agree about what dependence is, why it develops or what to do about it. Professionals cannot even agree about what to call it, some dismissing the word 'addiction' altogether. In deference to this latter camp the phenomenon will be referred to as 'dependence', but the term will be used to embrace a range of different conceptions of drug misuse based on three perspectives – the physiological, the cognitive/behavioural and the 'disease' models.

The physiological perspective focuses on the physical aspects of dependence and is concerned with increasing tolerance to a drug, withdrawals and craving. The psychological perspective is based on a cognitive/behavioural approach involving the control of drug-misusing behaviour, as in 'controlled drinking' regimes. The abstinence-orientated perspective of Narcotics Anonymous understands drug dependence as a 'disease of the whole person' and recovery in terms of abstinence. This use of the word 'disease' can obviously lead to confusion since it may be an essential part of any medical model. For the purposes of this chapter, the term will be used to characterize the Alcoholics Anonymous/Narcotics Anonymous perspective.

The British treatment of addiction has different historical roots from that in America and elsewhere. In Britain, instead of attempting to prevent drug misuse in the population and punish individual drug misuse, doctors attempted to control drugs generally and treat drug dependence in individuals. Before the Rolleston and Brain reports in the late 1960s and 1970s, GPs were entitled to prescribe both heroin and cocaine for drug misusers. It is only in the past 25 years that drug treatment has become the province of a small minority of consultant psychiatrists who prescribe the substitute drug methadone in short withdrawal schedules, with the intention of simply weaning heroin misusers off rather than maintaining safer drug misuse.

The 'British system' refers to the earlier acceptance of the 'grey' market and some inevitable 'malpractice' but also a recognition that what is 'malpractice' in medical terms could be seen as 'good practice' in social terms and vice versa. Doctors supplied liberally to misusers and the black market was in effect controlled by GPs and the early clinics. But by the early 1970s prescribing was far less liberal. By 1977 a small group of clinic consultants (psychiatrists) decided that in terms of individual treatment it would be better to limit prescribing and contract withdrawal schemes became the norm. This did not take into account social or public health considerations, as these were not their remit.

It was only with the emergence of HIV and hepatitis C that the public health perspective became dominant. However, alongside the withdrawal clinics in the late 1970s other services started to develop, largely non-statutory with few qualified practitioners (Stimson and Oppenheimer, 1982). These agencies developed from early social and welfare initiatives and provided housing, welfare rights, etc. for clients.

As with attitudes towards harm minimization, current controversies about drug dependence bear a strong resemblance to those at the turn of the century. Opinion is divided between those who believe that drug misuse is a matter of personal choice and habitual behaviour (the cognitive/behavioural approach), those who see the problem in terms of loss of self-control and personal pathology (NA) and those who regard it as a form of physiological dependence. It will become apparent that there is hardly any common ground and little communication between the three points of view. Differences between doctors, psychiatrists and psychologists are focused on the relative importance of physical and behavioural factors. The physiological perspective stands in contrast to the disease model of NA and the two approaches have very different underlying theoretical premises and knowledge bases. The psychological approach emphasizes the importance of the cognitive and behavioural aspects of dependence, even to the extent of suggesting as Stockwell (1994) does, that physiological dependence and withdrawal are '... an adaption to heavy drinking of no practical significance'.

Drug treatment within the statutory professions in Britain and Australia is based largely on physiological and psychological perspectives and consequently takes little account of social factors or long-term aftercare. This is in direct contrast to America, Canada and many European countries where there is general agreement among laymen and professionals alike that dependence is a disease which influences all aspects of a person's social and spiritual lifestyle. As a consequence, the approach of Narcotics Anonymous and the Minnesota Method is usually limited to voluntary and private residential services in Britain, whereas it is far more widespread in other countries, particularly in North America.

There are lessons to be learned from both the American and the British approaches to treatment. For example, the British psychological approach offers simple practical cognitive behavioural interventions that have been shown to be cheap and effective for a majority of people with substance misuse problems. The American approach offers important lessons about treatment process and the relevance of social factors. Perhaps more importantly in the light of relapse rates, the Americans have much to teach the British about the importance of after-care. In Britain, aftercare has not been seen as a significant and separate aspect of treatment and the importance of social and life skills is often ignored in the research literature, if not in the everyday practice of professionals.

It is also important to examine literature that does not fit conveniently into any one of the three camps, concerning treatment process, aftercare, social factors and relapse prevention. It should be emphasized that the author does not advocate any one perspective or approach: instead, the benefits of each will be outlined in order to allow a greater understanding of all methods available. This chapter will outline the basic positions within different professions and examine theoretical controversies in the literature.

It is doubtful whether the three perspectives can be reconciled into an encompassing theory. As we shall see, they contain essentially contradictory and irreconcilable components. However, several attempts have been made to amalgamate them into a 'biopsychosocial' model. Orford (1985) and Galizio and Maisto (1985) give comprehensive accounts of such models. There are problems with these overgeneralized understandings in that they are limited in their clinical usefulness and difficult to test within the framework of a behavioural science, as has been pointed out by Schwartz (1982).

The working definition given at the beginning of this chapter is taken from the most comprehensive integrated approach which has been developed internationally; this is the concept of a 'continuum of dependence' which is now used for both clinical and screening purposes. This approach has been developed for alcohol problems and it is therefore necessary to be cautious about the practicalities of extending it to drug problems, but a brief overview of the concepts will give greater understanding of the notion of dependence itself. In this approach, dependence is understood as a biopsychosocial continuum along which anybody can travel under certain circumstances. This is the model created by Griffith Edwards to guide clinicians and practitioners (Edwards and Gross, 1976).

Although until recently the WHO supported a disease model of drug dependence, it does not now support either the disease or cognitive behavioural approach, stating that the syndrome defined a 'cluster' of physiological, behavioural and social phenomena. For alcohol this is specified as: narrowing of drinking repertoire, salience of drink-seeking behaviour, tolerance, withdrawals and craving, relief drinking and the rapid reinstatement of the syndrome after abstinence (Edwards and Grant, 1976, 1977).

The syndrome includes characteristics from physiological, behavioural and disease models – physiological characteristics such as tolerance and withdrawal, behavioural characteristics such as narrowing of the drinking repertoire, together with characteristics identified by the Anonymous Fellowship (including both Alcoholics Anonymous and Narcotics Anonymous), particularly the salience of drinking and reinstatement of the syndrome (which the WHO refers to as 'remission'). The syndrome therefore can be seen as incorporating aspects of all three models. Its value for drugs other than alcohol will be apparent. However, the terminology is largely medical and as such is seen by those who are not doctors as too closely related to the medical or disease models, being criticized by both NA advocates and cognitive behavioural psychologists. The former state that dependence is a disease and not a medical problem (Lefevre, n.d.) and the latter assert that it is based on a physiological perspective and is therefore an ineffectual attempt to support the disease perspective in the face of the evidence of behavioural research (Heather and Robertson, 1981).

THE THREE PERSPECTIVES

THE PHYSIOLOGICAL PERSPECTIVE

Introduction

For heuristic purposes, this perspective (and the two that follow) are treated as 'ideal types'; that is, their characteristics are isolated, accentuated and expressed as abstract depictions which are not intended to represent reality. They are ideal in the logical sense. This method of analysis allows us to see the essential character of each approach. In this first case, it is most unlikely that any clinician faced with a drug misuser would concentrate entirely on physical symptoms to the exclusion of all other considerations. Indeed, the standard texts for general practitioners in Britain cover psychological and social issues as well as narrowly clinical ones (see, for example, Banks and Waller, 1988). Nevertheless, clinicians can be expected to give priority to clinical matters and to deal with physical evidence in the way that their training has taught them to do.

The research on the physiological effects of drugs on human beings is rigorously scientific and can obviously be incorporated into a model of scientific medicine without difficulty. The results of such research can be deployed to produce a physiological perspective of drug dependence which permits the construction of a treatment regime. In order to understand the nature of such an approach, it is necessary to summarize recent findings in the field of psychopharmacology.

It was not so long ago that drug dependence was understood as a function of the unpleasantness of withdrawal symptoms; that people continued to misuse drugs because they could not face withdrawal. It was only when the researchers emphasized high relapse rates and a growing dependence on drugs with no obvious withdrawal syndrome (as with cocaine) that people began to consider other reasons for dependence.

Research reports now offer some understanding of the pleasant or hedonist functions of drugs, rather than simply detailing the withdrawal effects. The 'dopamine hypothesis' is generally held to offer the most useful explanation of all drug-induced pleasure. Much current research in the psychopharmacology of drug misuse is now focused on the effects of drugs on dopamine in human biology.

Littleton and Little (1994) propose that the 'neurochemical basis for the rewarding effects of alcohol may be the potentiation of GABA at GABAA receptors (causing relaxation) and release of dopamine from mesolimbic neurones (causing euphoria)'. They add, however, that in terms of physiological dependence, 'The adaptive changes in these and other receptors are unclear'. These authors suggest that a simple model of reinforcement can be useful, where the rewarding effects of drug misuse can be seen as positive reinforcers of behaviour, whereas withdrawal is seen as providing a negative reinforcer.

These authors point out that research into depressants including alcohol has focused largely on withdrawal syndromes, which are easily measured. In contrast, research on stimulant use has concentrated on the hedonistic side of drug use, where effects on the central nervous system can again be measured. This has led to an imbalance in our knowledge of the positives and negatives of both types of drugs.

The idea that dependence may be due to physiological changes in receptors is controversial (Balfour, 1994). Although the reinforcing effects of drugs are seen as due to stimulation of the mesolimbic dopamine system of the brain, this may be only part of a complex reaction where a range of different receptors become sensitized and desensitized to particular drugs.

This explanation probably fits better with the experience of those who work with drug misusers themselves. Drug misusers vary the amounts of drugs they misuse, they may stagger drug-misusing sessions to ensure that they do not build up a tolerance or lose their sensitivity to a drug. They may regulate their misuse of a drug, leaving short drug-free periods between each dose in order to increase sensitivity or reduce need, or detoxify completely for similar reasons.

One thing is certain: it is rare for drug misusers to steadily misuse a drug every day on a continual basis until they are physically addicted. Instead, they are likely to control their drug misuse in various ways over time to increase the positive effects and decrease the potential costs of misuse.

Withdrawals

Much of the research on the physiology of dependence is limited to alcohol. The emphasis is on the negative effects of withdrawal. Physiological dependence is seen as causing adaptive changes in the brain which cause unpleasant effects when withdrawing. It is therefore suggested that it is the need to stop these physiological withdrawals that motivates people to keep on misusing the drug.

Traditionally the depressants, especially the opiates, were seen as the only seriously 'addictive' drugs (the exception being nicotine). A withdrawal or abstinence syndrome for cocaine was first recognized in 1976 (Caldwell, 1976). Depression is a common feature of withdrawal; it is suggested that this is related to the physiological effects of both amphetamines and cocaine blocking the uptake in the brain of the neurotransmitters noradrenaline and dopamine (together with changes in the beta-adrenergic and dopaminergic receptors) (Banks and Waller, 1988). As a consequence, tricyclic antidepressants have been shown to be useful in the treatment of cocaine dependence (Gawin and Kleber, 1984).

This will come as no surprise to practitioners who may have heard stimulant users complaining of depression or at least an inability to feel good after giving up their stimulants. Those who work with opiate users will know that they complain of the same thing for several months after giving up. Dackis and Marks (1983) have demonstrated that depression in opiate users more than doubles in the weeks following withdrawal. It is also possible that depression occurs after benzodiazepine withdrawal (Petursson and Lader, 1984).

Substitute drugs

The main substitute drug for opiates is methadone which is a direct substitute opiate and has a similar effect to heroin. As such, it is valued by heroin misusers although it has less hedonistic effect. It is also as addictive, if not more so. Methadone prescribing in the UK and the US has increased in the 1980s and 1990s. In the UK, government policy has supported flexible prescribing, as HIV and harm minimization influence the need to attract and retain heroin users in treatment. In the US it is perhaps the increase of drug-related criminal activity that had kept prescribing policies active.

Research shows that methadone prescribing can reduce the amount of injecting and sharing of needles and syringes. A prescription for drugs can also reduce criminal activity as it removes the need to buy drugs (Bean and Wilkinson, 1988; Jarvis and Parker, 1989; Hammersley, Forsyth and Lavelle, 1990). The role of methadone in maintenance and withdrawal is less clear. Recent work from Strang and Simpson (1990), Caplehorn *et al.* (1993), Strain *et al.* (1993), Novick *et al.* (1993) and others gives some indication of the positive effects of methadone prescribing. Caplehorn and his colleagues have shown that the greater the methadone script, the less chance that the patient would use heroin. Maremmani *et al.* (1994) have also demonstrated a positive correlation between methadone dose and treatment compliance. They conclude that methadone can have a rehabilitative effect and prevent relapse, if high enough doses are prescribed.

Strain *et al.* (1993) have demonstrated that patients prescribed methadone show improvements over time on scales of psychological and psychosocial functioning. The Amsterdam Methadone Maintenance programme results in clients who take better care of themselves, misuse less heroin or methadone post-treatment and, perhaps more significantly, are more likely to develop new social networks with non-drug users (Reijneveld and Plomp, 1993).

Part of the difficulty is that a prescription may involve the professional in a controlling role. This is distasteful for some professionals, particularly those concerned with psychodynamic aspects of treatment. Summerhill (1990) highlights some of these difficulties.

The use of methadone is, however, still controversial. Another heroin substitute, buprenorphine, relieves withdrawal symptoms and helps prevent the hedonistic effects of any heroin misused afterwards. This is also the case for naltrexone, which effectively stops heroin misusers deriving pleasure from the drug.

There are at present no similar 'blockers' or antagonists for cocaine and amphetamine. The problem with all antagonists is that in effect they block the body's natural pain control and pleasure receptors and consequently may cause unnecessary pain and detract from ordinary everyday life.

See Chapter 11 for prescribing guidelines.

Useful sources

Perhaps the best overview of the physiology of dependence can be gained from a special issue of *Addiction*, 'Comparing Drugs of Dependence' (November 1994), particularly the summary of biological processes (Grunberg) and of the withdrawal effects of different drugs (Gossop). Useful texts are also provided by Bowman and Rand (1980) and Balfour (1990). A readable issue of *New Scientist* (October 1994) covers similar ground.

The most useful British clinical text is that compiled and updated by the Department of Health, Scottish Home and Health Department and

Welsh Office (1991); *Drug Misuse and Dependence: Guidelines on Clinical Management*. This outlines the withdrawal effects of different drugs and guidelines for prescribing practice. For a more informal booklet for clients and professionals ISSD have published *The Methadone Handbook* (Preston, 1993).

THE PSYCHOLOGICAL PERSPECTIVE

Cognitive and behavioural change methods derive from social learning theory and cognitive psychology. As explained in the previous chapter, these can be very effective for certain conditions ranging from depression to phobias and, more significantly, for obsessive/compulsive behaviours (Bandura, 1977; Ellis, 1987; Beck, 1989). These methods have also been demonstrated to be effective in helping people control their drinking and drug misuse when reducing substance-related harm.

Similar methods based on social learning theory can also be used to control drug dependence. Social learning theory is a general theory offering an explanation of behaviour formation and maintenance. It assumes that behaviour can be either positively or negatively reinforced and that the sooner the reinforcement occurs, the more effective it will be. (Therefore short-term positive reinforcement could be seen to outweigh long-term negative reinforcement). The development of behavioural habits can also be influenced by 'modelling' and other social factors such as the need to conform (Bergin and Garfield, 1978; Bandura, 1977; Bolles, 1979; Catania and Harnad, 1988).

The advantage of placing the phenomenon of dependency disorders within the context of social learning theory and behavioural methods is that they are empirically testable and can be disproved. Thus if the maladaptive behaviour is not changed after 'treatment' in a controlled setting, then the chosen method can be shown to be of no use in dealing with the phenomenon in question. A great deal of research has now been completed in this area which sheds some light not only on the effectiveness of behavioural techniques in dealing with behavioural disorders, but also on the usefulness of defining dependence in behavioural terms. Cognitive and behavioural techniques of dependency treatment have been gradually refined in outcome studies with clients with diverse characteristics (Sanchez-Craig, 1990). These techniques are based on the notion that dependent drinkers can control their drinking, clients are capable of self-control and of taking responsibility for much of their treatment. Raistrick and Davidson (1985) give a useful account of the application of psychological techniques to alcoholism and drug dependence.

The influence of cognitive psychology is demonstrated in the work of Orford, Heather, Stockwell, Hodgson, Robertson and Marlatt. This work on the psychological interpretation of alcohol problems and dependence has to a greater or lesser extent adapted psychology to the particular problems of dependence generally. These theories of drug-dependent behaviour (while developed from rather generalized theories of behaviour) have integrated this with a cognitive approach to encompass the phenomenon of dependence as a subject in its own right. Perhaps the most clinically useful of these approaches is Marlatt's theory of relapse prevention (see Part Three). One of the most comprehensive is that of Orford, *Excessive Appetites – A Psychological View of Addictions* (1985). He develops a theory of the addictive process, taking into account biological influences and ideas of restraint (the counterbalancing of incentive and disincentive) and considering the social issue of moral conflict between deviant excess and conforming moderation. He develops this theory within the context of the social learning paradigm, but provides a comprehensive and coherent alternative model to that of the disease perspective, rather than simply interpreting dependence in behavioural terms and advocating general behavioural techniques.

The psychological approach sees substance misuse problems and dependence as lying on a continuum, rather than being divided into 'either/or' categories as in the disease approach. The issues raised by psychology are dealt with in a special edition of the *British Journal of Addiction* (**82** (4), 1987). The focus of attention is again alcohol, but the development of methods of early or brief intervention offer some guidance to those working with drug misuse.

The behavioural approach can be criticized because of the limitations of its theoretical base and cognitive psychology can be criticized for its lack of scientific testing but there seems little doubt from the accumulated evidence that psychology has a good deal to offer in the field of dependence. The most severe criticism is that while psychological theories and methods of treatment are useful in dealing with psychological dependencies, they have not yet been shown to be effective in serious cases of combined physiological and psychological dependence.

Behavioural interventions

Behavioural interventions are concerned with the accurate description of behavioural problems and change in the specific behaviours identified. The approach usually involves close monitoring of concrete changes in behaviour.

Initial assessment

Beliefs about personal drug misuse and the drug misuse behaviour itself are likely to have been strongly reinforced over long periods of time. It is necessary to discover which variables are important in continuing this reinforcement.

Positive and negative reinforcement

The aim is then to change the reinforcing properties of drug misuse behaviour, either by offering positive reinforcement for non-drug-using behaviours or negative reinforcement for drug misuse. For example, it is possible to use prescribed drugs (e.g. naltrexone) to block the physiologically reinforcing effects of opiates (although there is little evidence for effective post-treatment [long-term] results).

Cue exposure or extinction

The typical cues for drug misuse are produced without the reinforcing antecedents. The client is gradually and then repeatedly exposed to drugs or drug use situations which elicit craving, without actually using the drugs or without feeling positively reinforcing effects. Over time the craving induced by such cues is reduced.

Self-monitoring and self-control training

In effect, the terms 'self-monitoring' and 'self-control' refer to the basic behavioural techniques of identifying the antecedents and consequences of any problem behaviour and learning to alter behaviour by changing either or both. So the client is taught to identify those situations and cues that are likely to stimulate drug-misusing behaviour and to respond by avoiding the situation, refusing drugs and/or developing alternative coping strategies.

Cognitive interventions

Most contemporary behavioural psychologists also include a cognitive element in both their theory and practice. The way people define or interpret their experience is considered to have a strong influence on their behaviour (Ellis, 1962; Orford, 1977; Beck, 1989). Cognitive processes are now considered influential in both loss of control and relapse (Brown, Goldman and Christiansen, 1985). Cognitive findings are also reviewed by Marlatt (1985) who demonstrates that the influence of expectation and social cues can be significant in loss of control and relapse.

As with behavioural interventions, cognitive interventions are concerned with the accurate description of cognitions and beliefs, followed by identification of maladaptive beliefs and consequent changes to more constructive alternative beliefs.

Accurate initial descriptions of relevant beliefs about drugs and drug misuse are therefore essential, followed by continuing assessment, self-monitoring and feedback. The first step in the helping process is to examine the beliefs, thinking processes and patterns of behaviour which lead to and maintain drug misuse.

This is followed, first, by demonstration of inaccuracies in beliefs and, second, by identification of inconsistencies in personal belief and value systems. First, information about the effects of withdrawal and tolerance are very useful here. For example, a client attempting to withdraw from benzodiazepines may believe that they are inherently depressive or anxious, without realizing that these are the withdrawal effects of these drugs. Similarly, clients unaware of the decreased tolerance following a period of abstinence may misinterpret their increased vulnerability to the drug as a sign of complete lack of control (see attribution theory below). Second, it is also useful to highlight any inconsistencies and conflicts between the client's short- and long-term aims and priorities and help the client to weigh up the evidence for the pros and cons of drug misuse. It can then be demonstrated how beliefs are linked to emotional and behavioural consequences and alternative beliefs.

Clients can also be taught new skills and coping strategies, from a decision matrix to reminder cards and self-talk or cognitive restructuring.

Attribution theory and application

Psychological approaches to addiction include the application of attribution theory. The first attempts to examine the influence of the client's beliefs about dependence on their consequent behaviour was that of Eiser and colleagues in the late 1970s (Eiser and Sutton, 1977; Eiser and Gossop, 1979; Eiser, 1982). The most comprehensive exposition of the application of this theory to drug misuse as a whole is that of Davies (1993). Davies argues convincingly that drug misusers misuse drugs because they like them and that if they are taught to attribute difficulties in giving up to a 'disease' it will simply make it a lot more difficult to give up. Davies used data similar to those in Part One to clarify the influence of the subject's own beliefs on their behaviour, within a psychological model. He too identified different categories ('chronic' and 'sporadic') of drug misuser from subjects' subjective understandings (Davies, 1993, p. 135).

Part One has used these data within a different methodological framework (a sociological phenomenological perspective) to gain a greater understanding of the phenomenon of drug misuse itself. Both sets of research data can be seen to emphasize the relevance of the client's subjective beliefs. It is unclear how far the subject's subjective understanding contributes to the social reality of their own drug misuse (phenomenology) and how far it can be seen as a function of the inter-action between attributional style and other aspects of social cognition (cognitive psychology), but there can be little doubt that the individual's beliefs influence future drug misuse behaviour.

Brief interventions

Minimal or brief interventions have been used for people with alcohol problems for many years. Robertson and Heather (1986) and Miller, Sovereign and Krege (1988) provide the best manuals for use. Miller has shown that this intervention can produce a 20% reduction in the amount of alcohol used and, perhaps more significantly, a 37% reduction in 'peak intoxication levels'. Miller *et al.* suggest that the essence of effec-tive minimal intervention is an objective assessment of problems, an empathic quality of intervention and feedback to highlight discrepan-cies between desired goal and present state. There is no reason to suppose that similar brief interventions would not be effective at an early stage for drug misusers.

Brief interventions vary from self-help pamphlets to structured cognitive behavioural interventions by a psychologist. Heather (1989) has argued that the personal contact is an important aspect of brief interventions, but there is little evidence for the necessity of psycho-logist input. Psychiatrists use mainly simple advice whereas psychologists have developed cognitive and behavioural techniques. Within the field of brief interventions there is some confusion about what is cognitive behavioural psychology and what is simply giving information. Information and advice giving can involve cognitive behavioural methods such as behavioural monitoring (Heather, 1989) whereas basic education techniques can be described in cognitive behavioural terms.

There is also debate about the aims of brief interventions. If 'early intervention' methods, such as information and cognitive behavioural techniques, are seen in a preventive light this itself raises methodologi-cal issues. It should be asked whether epidemiological population measures of incidence and prevalence might not be more appropriate to assess early intervention methods rather than studies of treatment out-come. Although research indicates that non-serious drinkers can be helped to control their drinking at an 'early stage' (Edwards, 1986), the research cannot distinguish between those who will progress to a 'later' stage and those who drinking would not become harmful at all. Edwards himself is increasingly unsure of the comprehensiveness and validity of outcome data (1988) and the wisdom of excluding many aspects of dependence in favour of those emphasized by the now popu-lar cognitive behavioural perspective (1989a).

However, despite the growing influence of the cognitive behavioural perspective, the notion of sequential stages of alcoholism, which seems integral to behavioural notions of early intervention, is not derived from social learning theory itself. It is here, as might be expected, that psychologists would question the integrated approach. Hodgson, in an article entitled 'The alcohol dependence syndrome: a step in the wrong direction' (1980), criticizes the notion of a syndrome as too medical (although he does concede that the concept of a dependence state is valid) and yet if there is no sequential syndrome then the rationale for early intervention is less apparent. Not all academics feel that the concept of a continuum of dependence is necessary to prove the worth of brief interventions. Heather (1989) argues that if these techniques serve to reduce drinking at any time then they serve a useful purpose, whether or not dependence was a likely outcome.

Relapse prevention

Cognitive behavioural techniques have also been developed to help prevent relapse. These include self-monitoring, relapse fantasies and behavioural assessment methods to assess the relative risks of particular moods and situations. Clients are informed about the immediate and delayed effects of substances. Information is gained from clients through descriptions of past relapses and rehearsal of possible future relapses. In addition to teaching clients to be self-aware and able to assess risks, relapse prevention techniques also offer skills and coping strategies including relaxation training, stress management and efficacy-enhancing imagery.

THE DISEASE PERSPECTIVE

This perspective has developed from the belief that alcoholism is a physical and spiritual disease, which was the cornerstone of the creation of Alcoholics Anonymous in the United States in the 1930s. The concept was later enlarged and expanded to include all chemical dependence and the model is now encapsulated in the philosophy and methods of the Narcotics Anonymous groups in Britain and America and has been developed into what is called the Minnesota Method of therapy. The essence of this approach is that the affected individual has to experience a spiritual conversion and a moral transformation to cure the 'disease'. The cure can only be achieved through total abstinence.

Some 30 years later, Jellinek, in his book *The Disease Concept of Alcoholism* (1960), argued that alcoholism could be regarded as a physical disease and treated accordingly: although his analysis was not entirely consistent with the philosophy of AA, it was incorporated into their perspective. Jellinek is therefore often understood as providing the evidence for the spiritual disease model. He categorized some forms of alcohol problems as diseases but also stressed that there were many different kinds which did not fit into these categories. Jellinek's notion of the 'Gamma' form of alcoholism became most influential in the field, together with the idea that alcoholism was a unitary disease. These concepts reinforced and maintained AA, but the success of AA itself also helped; the mutually reinforcing elements of Jellinek's work and AA are of interest as it is these that kept the disease model alive. Some of Jellinek's work was taken further by Glatt (1972) and incorporated into a wider treatment programme.

The basic beliefs of the Anonymous Fellowship can be found in two texts, *Alcoholics Anonymous* (Anonymous, 1976) and *The Little Red Book* (Anonymous, 1970). Narcotics Anonymous is one of five self-help groups within the Fellowship. The other four are Alcholics Anonymous (for alcohol addicts), Alanon (for families of alcoholics), Alateen (for the teenage children of alcoholics) and Families Anonymous (for the families of drug addicts). Alcoholics Anonymous is by far the largest and has a longer history (founded in 1935); others have developed more recently as offshoots from the original Fellowship.

Dependence on any type of drug is referred to as 'chemical dependency' and Narcotics Anonymous is recommended by American authors for a variety of different types of drug misuse. For example, Millman (1988) recommends this approach for cocaine abusers.

The premise underlying this approach is that the 'addict' is not normal, that they suffer from an illness of the whole person. The addict is seen as having three characteristics: a physical allergy to alcohol and/or other drugs, an 'addictive personality' (described as immature and self-centred) and a spiritual sickness. The client is seen as an addict rather than a person who has an alcoholic disease. Chemical dependency is defined as a primary disease which causes other problems; it can involve dependence on one or a range of drugs.

Members of NA are recommended to attend regular meetings of the Anonymous Fellowship, to read regularly its approved literature and to work the Twelve Steps for recovery. As noted above, the disease model of addiction can be applied to families. It is important to emphasize that addictive disease is conceived of as different to physical addiction and therefore also identified in different ways. Only a certain proportion of the population are considered vulnerable. Various authors have attempted to describe the programme; perhaps the most accessible account is that of Kurtz (n.d.) *'Not God': A History of Alcoholics Anonymous.*

The Minnesota Method is founded on the basic premises of Alcoholics Anonymous and Narcotics Anonymous. This method is usually applied in residential settings and only five of the Twelve Steps are used before the clients are discharged into the community, where they then attend Fellowship groups in order to complete the remaining seven. There are slight differences between the emphasis of the two approaches but these are largely concerned with the intensive residential elements of the Minnesota Method. Information concerning it can be found in Lefevre (n.d.) *How to Combat Alcoholism and Addictions: The Promis Handbook on Alcoholism, Addictions and Recovery,* and Anderson (1981) *The Minnesota Experience.*

The basic assumptions are that addiction or chemical dependency is a disease and that all addicts 'should receive treatment within the same basic programme framework' (LeFevre). The disease concept of addiction is considered relevant to gambling, risk taking, relationships and some forms of behaviour. The addict is not held personally responsible for having the illness but is held responsible for their own recovery. Initial denial is seen as symptomatic of the illness. This illness or 'addictive disease' is assumed to 'originate as a disease of the human spirit' and 'this predisposition may be genetically determined'. This disease of the spirit 'leads to a disorder of mood' and 'affects intellectual processes' (pp. 45–7).

The philosophy of the Anonymous Fellowship emphasizes the contribution that non-professional ex-addicts make to the recovery of others in both residential and day care facilities. There is much controversy about the working relationship between professionals and non-professionals in non-statutory agencies and in residential treatment. Kostyk *et al.* (1993) found that non-professional ex-addicts can be effective co-leaders in therapeutic groups and provide useful role models by offering hope and optimism.

The Twelve Step treatment process

The following gives a brief description of the Twelve Steps of the therapeutic programme based on this model. The quotes are taken from various anonymous booklets and pamphlets published in connection with Narcotics Anonymous and the Anonymous Fellowship, generally published by Hazelden, City Centre, Minnesota.

Step One: Powerlessness

Step one is probably the most significant part of the process. The essence of the step is that clients should believe or accept that they are an addict and that they are powerless over their lives as a result (i.e. admit that their lives are unmanageable). If clients do not accept this interpretation of their experience the treatment process cannot proceed. This stage is often referred to as 'breaking denial'. The beliefs of clients in the initiation stages of treatment appear to be of particular relevance to later attitudes and behaviour (Keene, 1994b).

Step One refers to 'the foundation of recovery' and involves members coming to an understanding of their own powerlessness. 'We admitted we were powerless over drugs and that our lives had become unmanageable.' 'Reluctance to examine our powerlessness is as much a symptom of our illness as liver damage, withdrawal, or digestive disorder. Social pressures centred around the myth that 'willpower is all that is needed to control a drinking or drug problem can result in unwillingness to study our powerlessness.'

This emphasis on acceptance of personal powerlessness is an integral part of the notion of addiction as a disease and an essential starting point for the programme.

> As we develop a thorough understanding of addiction ... we will not be ashamed to admit that we are powerless over it, just as we would be powerless over any other disease. We will also learn that we will not be able to adapt our lives to the disease of addiction unless we have a thorough, ongoing program of recovery in the same way that a diabetic or heart patient has an ongoing program to keep the disease in check.

The notion of disease and powerlessness is also linked to the notion of responsibility for one's own recovery. This may appear to be paradoxical, but is explained within the NA philosophy in the following way: 'personal responsibility for addiction occurs when we have recognized it in ourself, or others have pointed out the symptoms to us, and we realize we are afflicted with a disease. It then becomes our responsibility to start a recovery programme'.

Step One then integrates the concept of unmanageability with the notion of powerlessness. 'Personal unmanageability relates to our attitudes and beliefs ... In many cases, personal unmanageability was present many years before chemical addiction.'

> The N.A. philosophy is that putting the cork in the bottle is not enough. We need to rejuvenate our personalities. We have to learn about ourselves on an intimate level. We have to discover what the N.A. programme calls our character defects and shortcomings in order to accept ourselves as human beings with strong and weak points just like everyone else.

The concepts of powerlessness and unmanageability are linked to the idea of an 'addictive personality' which can only be controlled through adherence to the programme and abstinence.

There are some character weaknesses that chemically dependent people do seem to have in common. One is self-centredness. This defect has to be present in each of us for our disease to prosper. Selfishness seems to need a direct assault to break our denial system and rebuild trust in and concern for other people. Another area of common personal unmanageability is the basic immaturity that seems to be prevalent among chemically dependent people. Immaturity may not be obvious. A person may be able to function very well when sober, but the least amount of agitation or disruption of the normal pattern will cause extreme reaction. Over-reacting is definitely immature. Some other examples are temper tantrums, not sharing feelings and emotions honestly with others, insisting on having one's own way and the like. Such behaviour patterns enlarge and gradually take over a large part of one's personality.

Step Two – Hope

This step involves recognizing and 'coming to hope' that one has the need to and ability to change, with the help of the programme and, more importantly, with the help of a 'higher power' or 'power outside of oneself'. This step is a necessary follow-up to acceptance of powerlessness, as without hope and faith in outside help, little progress can be made. In effect the first step in isolation could be worse than useless.

The notion of a higher power does not necessarily involve Christian or religious beliefs. Some solve this problem by defining their higher power as being the Anonymous Fellowship itself or their therapeutic groups, that is, without recourse to ideas of a spiritual god.

Step Three – Commitment to change

This step involves not only making a commitment to change, but also 'handing yourself over to others in the programmes, that is, learning to *trust* others'. This stage involves an effort on the part of the client to stop 'controlling or manipulating others', that is, relinquishing control over both themselves and over others. The idea of 'getting out of the driving seat' is also an important part of this process.

Step Four – The moral inventory

This stage involves the client writing a list of their previous character or personality defects prior to treatment. This list is referred to as a 'searching moral inventory'.

Step Five – Confession

This stage involves sharing this 'moral inventory' with someone else. This is often referred to as the 'confessional' step. The benefits felt by clients are often described as very similar to religious practices.

The final steps (6–12) do not form part of the inpatient or daycentre treatment programmes, but they are part of the continued experience of all those who complete the programme successfully. They take place in self-help groups after the structured treatment period. In effect, the Twelve Steps can be split into two stages: the first five steps concerned mainly with changing or converting the client, the last seven steps concerned with maintaining this change or conversion. The programme of treatment is not therefore complete without the second stage (i.e. the Narcotics Anonymous equivalent of relapse prevention).

Step Six involves emphasizing the willingness to change and to have the higher power remove defects.

Step Seven involves putting the initial commitments into action.

Step Eight involves becoming willing to make amends to people one has harmed.

Step Nine involves actually making amends, as far as this is practicable.

Step Ten involves learning to take a personal moral inventory on a regular basis, that is, continually monitoring and evaluating what happens in one's life, admitting when one is wrong and changing if necessary.

Step Eleven involves building and improving on faith and improving one's relationship with one's higher power.

Step Twelve. The final step is considered of extreme importance, as it demands helping others, consistently reminding oneself of how bad one was and reminding oneself of the principles of the programme by teaching them to other addicts. 'Twelfth Stepping' is thought to give members 'a purpose, an identity and self-respect' and is seen as an integral part of recovery and sobriety.

Although this description attempts to give an overall picture of the Twelve Step process, at present we have little understanding of what actually happens in these programmes or why an approach at odds with both the methods and beliefs of contemporary British academic and professional consensus should survive and flourish alongside its scientifically credible opponents. Information is usually acquired experientially by 'working the programme', with the consequent internalization of the belief system on which it is based. Reports by 'insiders' are often more concerned with promulgating this philosophy than objectively analysing the perspective. Those who have not gone through this process themselves are often critical, citing lack of research evidence, but have little understanding of how the treatment functions. The exception is the work of Denzin (1986, 1987a, b) who carried out intensive qualitative studies of Alcoholics Anonymous. The outcome research carried out by 'outsiders' has done little more than show that this approach is no more or less effective than any other (Cook, 1988).

The disease perspective incorporates an essential commitment to abstinence and there is some evidence to indicate that commitment to abstinence influences likelihood of relapse; Hall, Harassey and Wasserman (1990) followed up subjects for 12 weeks after treatment for alcohol, opiate and nicotine dependence and found that commitment to complete abstinence was related to lower risk and a longer time until relapse.

The wider interest of this perspective is, however, that it deals specifically with the factors known to be correlated with relapse: emotions, relationships and social norms. It also takes into account the social factors correlated with successful treatment and long-term maintenance of change, such as relationships with family and social support systems. The programme also emphasizes the importance of the sequential nature of the individual recovery process. (These issues are discussed in the following section.)

It is important to point out that the Twelve Step model combines those components which have been found to be useful with those which have not. For example, the ideology of the Anonymous Fellowship has been unacceptable to many of those who have sought its help. The author's research indicates that the 50% drop-out rate can largely be attributed to this source (Keene and Rayner, 1993) and the moral/spiritual content can also prevent initial self-referral and non-compliance during treatment (Keene, 1994b). In addition, the evidence for the effectiveness of the following aspects of the programme is limited: group therapy (Frank, 1974); individual counselling (Miller and Hester, 1986a); lectures/videos and attendance at AA/NA self-help groups (Miller and Hester, 1986b). The success of AA and the Minnesota Method is therefore, as might be expected, no greater than any other.

The consensus of professional opinion in Britain is similar to those professionals and public health campaigners of the late 19th and early 20th centuries, who disagreed with the developing moral/pathological approach to addiction. The situation is different in America where the 'spiritual disease' model and the Minnesota Method provide the dominant theory and method with regard to alcohol problems, but are not so readily used to explain drug misuse (Sanchez-Craig, 1990).

There is a substantial body of literature which criticizes this particular disease perspective, e.g. Heather and Robertson (1981, 1986), Fingarette (1988) and Keene (1994b). Robinson (1979) has discussed AA self-help groups in detail and Cook (1988) has reviewed the literature. Frank (1974, 1981) and Antze (1979) have taken a more independent view of the therapeutic processes. Kurtz (n.d.) and Denzin (1987a, b) have attempted overviews, but these texts can seem to be biased in favour of the underlying assumptions of the perspective itself.

As the notion of alcohol dependence as a unitary disease is increasingly questioned by both the psychiatric profession and behavioural psychologists, Jellinek's theory and AA therapy have become less influential in some countries and are seldom referenced in contemporary British journals. As Lindstrom (1992) points out, 'Jellinek's working hypothesis had a salutary effect on alcohol research. Today many researchers regard it as a principal obstacle to the advancement of knowledge and the emergence of interdisciplinary approaches to alcoholism' (p. 54). This is unfortunate as Jellinek had identified important aspects of alcohol dependence and treatment which have been largely ignored in Britain. He focused on the sequential aetiology of the disease and the importance of individual differences and he stressed the 'need to establish criteria for the suitability of any given method to a given patient', stating that the 'criteria so far established are superficial' (Jellinek and Bowman, 1946, quoted in Lindstrom, 1992, p. 3).

It is interesting that much of the literature criticizing this approach is largely preoccupied with examining the theoretical reasons why it should not be effective, rather than considering the reasons why it is effective in many cases. The approach is poorly represented in British journals or edited collections and only mentioned in passing in literature reviews. Yet the Anonymous Fellowship and the Minnesota Method underpin much American treatment and AA has established self-help networks in most countries (Alcoholics Anonymous, 1981). It therefore seems strange that the concept has been subject to so little study. As Bratt (1953) stated, 'writing about alcoholism and leaving out what the Americans have to say on the matter would be like writing about Communism and ignoring the lessons from Russia '(quoted in Lindstrom, 1992, p. 10). Twelve Step advocates are themselves equally dogmatic in their criticism and dismissal of alternative approaches, particularly behavioural psychology, as Marlatt and Fingarette found out to their cost in the USA.

THE RESEARCH ISSUES

It has been noted earlier that two of the most important characteristics of the field of dependence are how much is taken for granted and the fact that an enduring lack of consensus has led to confusion. It is not surprising, therefore, that an examination of the issues which have provided the agenda for research during the last 20 years or so cannot be directly related to the three main treatment approaches described above. In the following review of the more significant areas, an attempt will be made to relate the findings to the three different approaches. As explained earlier, much of the substance misuse research has been focused on alcohol, but the findings concerning the development of dependence, self-medication for disabilities, treatment, relapse and aftercare are similar across different substances.

Research in the drug and alcohol field indicates that a range of treatment models are relatively successful in reducing severity of dependence and/or other drug misuse or drinking behaviour, but that these changes are often not maintained at follow-up. Researchers have not yet succeeded in distinguishing between the efficacy of different treatment perspectives, in identifying the effective components of successful short-term treatment or in explaining the reasons for relapse (loss of treatment gains) in the one-year period following treatment.

However, recent reviews of outcome studies suggest that 'social treatment' programmes seem more successful than others in the short term. In the alcohol field, Holder *et al.* (1991), Lindstrom (1992) and Hodgson (1994) identify aspects of treatment that appear to be more useful than others, particularly social factors and improved social functioning, which prove significant in terms of pretreatment characteristics and treatment outcome. In the longer term, relapse within the two years following treatment has been shown to be correlated with social and emotional factors. When the reasons for loss of treatment gains are examined (Marlatt and Gordon, 1985; Marlatt, 1985; Wilson 1992), the main precursors of relapse are identified as negative emotional states, interpersonal conflict and social pressure. This raises the possibility that maintenance of treatment gains may be correlated with psychological and social factors rather than treatment variables themselves (Moos, Finney and Cronkite, 1990; Lindstrom, 1992).

Pretreatment and during-treatment variables do not predict post-treatment functioning: this suggests that other variables may have a significant effect on post-treatment functioning (Catelano *et al.*, 1990/91). They report that pretreatment factors have accounted for 10–20% of variance in post-treatment relapse, that during-treatment factors have accounted for 15–18% and that post-treatment factors have accounted for 50% of the variance in outcome. (These findings highlight the important work of Cronkite and Moos, 1980; Finney, Moos and Mewborn, 1980; Simpson and Sells, 1982; and Simson and Marsh, 1986).

As Catalano points out, if post-treatment factors are important, 'it will be necessary to focus more research on the post-treatment predictors if we are to address these potentially important factors ... which may be critical to post-treatment success'. Catalano and his colleagues identify the following factors as important in relapse: drug cravings, lack of involvement in active leisure and in education and an inability of treatment clients to establish non-drug-using contacts in work and educational settings. They suggest that cognitive behavioural skills training will help clients develop the interpersonal skills necessary in developing new social networks.

Unfortunately there is a dearth of longitudinal research examining the relationship between pretreatment, during-treatment and post-treatment factors. This has resulted in a lack of knowledge about the treatment process as a whole.

The importance of process in treatment has been emphasized by DiClemente and Prochaska (1985), Davidson and colleagues (1991) and Prochaska and DiClemente (1994), who argue that sequences or stages in the process of individual change over time may be significant in determining effectiveness of interventions. Attempts have been made to identify and measure stages of this process (Rollnick *et al.*, 1992). The negative outcomes of treatment studies and the absence of conclusive outcomes in comparative studies have led to much controversy about the relevance of the methodology which limits the subject matter to treatment outcome. The major proponent of British treatment research in the past two decades, Griffith Edwards, has himself questioned the 'usefulness of predictors of treatment outcome and indeed of the useful-ness of treatment outcome as a measure at all' (Edwards, 1988).

In the light of these findings (or lack of them), it may be important to adopt a different methodology within the treatment field itself that is appropriate to research into the subject areas that look most promising: identifying clinically useful therapeutic components, social factors within treatment and post-treatment recovery processes.

At present research is limited in three main ways: first, there are difficulties in distinguishing between the effects of different models as all appear equally effective. There is a need to identify the influential treatment components in each. Second, little work has been done to assess the influence of social variables. Although social and emotional factors have been shown to be correlated with relapse rather than physiological factors (Marlatt and Gordon, 1980, 1985), there is little research on these factors. Third, treatment gains are lost at follow-up one to two years after treatment. This raises the possibility that maintenance of treatment gains may be correlated with psychological and social factors rather than treatment variables (Lindstrom, 1992) yet there is again little research in these areas. While the distinction between treatment change processes and long-term maintenance of change is useful, much research indicates that social variables may not only be important in long-term maintenance of change but also for within-treatment change. Recent reviews suggest that social programmes focusing on individual functioning, such as social skills, family relationships, etc., seem more successful than other programmes (Holder *et al.*, 1991; Hodgson, 1994). Similarly, community reinforcement programmes seem successful (Fishbein and Azrin, 1985).

THE INFLUENCE OF THEORY IN TREATMENT

At present research demonstrates little or no difference between different treatments and what work there is, is largely in the alcohol field. Edwards and Orford (1977) had found little difference between different treatment approaches with alcohol problems and this work is often cited as the best example of an outcome study illustrating no difference between the two options of 'treatment' and 'advice'. This illustrated, as they pointed out:

> a reservation which should properly attach to any too absolute interpretation of a clinical trial is that conclusions based on the averaging process and simple comparisons of group means may fail to bring out important patient type–treatment interactions.

Orford, Oppenheimer and Edwards (1979) carried out a two-year follow-up of this study which confirmed the authors' initial hypothesis, finding that the long-term results were different for different categories of client. Those clients considered more seriously dependent achieved better results in the treatment group and others (suffering only disability) did better in the group receiving only advice. There was an interaction between the types of client and the goals they achieved, with the seriously addicted clients achieving better results through abstention and the others tending to achieve good results through controlled drinking.

However, as Lindstrom (1992) points out, variations of outcomes tend to be large within treatments and small between treatments, Although there is some indication that certain methods may be more or less effective (Hodgson, 1994), there are few data to indicate that one overall approach is better than another.

Researchers have succeeded in identifying and measuring many important variables in the outcome of alcohol treatment, such as severity of dependence (Edwards, 1988) and client characteristics (Glaser, 1980). However, a major difficulty remains. The problem lies in determining the interactions between individual variables and treatment model variables in order to clarify their relative significance in treatment outcome and follow-up (Glaser, 1980; Lindstrom, 1992). There has been less success in defining and comparing different treatment perspectives than in measuring other variables, largely due to an inability to identify the core components which differentiate between them. Although much is known of the theory and practice of cognitive behavioural models, very little is known about the core components of both 'treatment' and continuing 'recovery' in the Twelve Step model. Because of this gap in basic knowledge, very few comparative studies have yet been carried out. The exception is the MATCH study in America (Project MATCH, 1993), which itself suffers limitations in design partly due to difficulties in identifying and isolating the relevant differences between treatment models. A solution at present is seen to lie in this direction, the aim being to examine interactive effects of different treatment variables and to develop testable 'matching hypotheses'. In order to facilitate this future research it is necessary to identify and define core treatment components in each model.

As Marlatt (1988) explains, there are 'various conflicting positions put forth to explain the aetiology of addiction, along with corresponding recommendations for treatment'. He goes on to state that:

> A one-to-one correspondence between models of aetiology and the effectiveness of various treatment modalities may not exist in the area of addiction treatment ... there is no consensus of opinion or convergence of evidence concerning the underlying cause of addiction.

He states, in true behavioural tradition, that 'treatment' itself can be effective even when the causative factors are unknown or considered irrelevant to treatment outcome. There is much disparity in the field generally between models of aetiology and treatment. This is also true for psychotherapy generally, where a non-directive, reflective type of therapy based on what is loosely called a 'humanistic' or 'holistic' theory uses behavioural techniques of reinforcement and modelling, and where behavioural and cognitive theories can lead to therapies which involve the 'unconditional positive regard', warmth and empathy considered central to the more holistic theories (e.g. gestalt). Therefore it is not perhaps surprising to find that 'behavioural' and 'cognitive' techniques are used by therapists who believe firmly in the disease model.

A distinction can also be drawn not only between aetiology and treatment but also between initial change and maintenance of change. Marlatt himself is particularly concerned with prevention of relapse (1985). Working in America where the disease concept of alcoholism has widespread professional acceptance, he has utilized these arguments to adapt a behavioural theory of relapse prevention to fit with the disease concept of treatment.

Whatever the praticalities determining the focus of Marlatt's work, it does seem possible to distinguish the process of relapse itself from that of initial change or 'stopping' use. Relapse prevention techniques can be used to reinforce abstinence or controlled drinking strategies.

THE INFLUENCE OF SOCIAL AND EMOTIONAL VARIABLES

The study of individual characteristics before and after treatment would be incomplete without a consideration of individual functioning in the social environment. Perhaps the clearest exposition of the relevance of social factors in the overall picture has been that of Rudolph Moos and colleagues (1990). Moos illustrated the influence of social functioning and social environment in both the outcome of alcohol treatment and the maintenance of change. Several research reviews also give an indication of the importance of individual differences in social circumstances to treatment outcome and post-treatment recovery. Baekland, Lundwall and Kissen (1975), in a review of 400 studies (1953–1973), noted higher socioeconomic status and social stability among clients who succeeded. Fink and colleagues in 1987 and Babor and colleagues in 1988 carried out studies indicating that a correlation between continued heavy drinking and disabilities in other areas tends to increase with a longer follow-up period, especially with regard to physical and psychological health (Lindstrom, 1992).

Perhaps more significant are the reasons given for relapse among misusers themselves. The study of post-treatment relapse highlights the factors connected with treatment failure over time. Although the medical definition of dependence involves tolerance, withdrawals and craving, research has provided little evidence that any of these factors are correlated with relapse or more generally with attempts to stay off alcohol and drugs (Somers and Marlatt,1992; Wilson, 1992; West and Gossop, 1994). Craving itself has not been found to be a primary pre-cipitant of lapses (Ludwig, 1972; Marlatt and Gordon, 1985; Drummond, Cooper and Glautier, 1990). Instead, the main precursors of relapse have been identified as negative emotional states, interpersonal conflict and social pressure (Marlatt and Gordon, 1980; Marlatt, 1985).

EFFECTIVE TREATMENT METHODS: SHORT-TERM GAINS

Of the wide range of interventions, the socially based seem most successful. Hodgson (1994) gives an overview of different approaches used in the treatment of alcohol problems.

There is good evidence to suggest the approaches directed at improving social and marital relationships, self-control and stress management are effective. There is at present little evidence to suggest that aversion therapies, confrontational interventions, educational lectures or films and group psychodynamic therapies are effective, nor is there any evidence that use of psychotropic medications is effective. (p. 1529)

One of the major recent reviews is that of Holder *et al.* (1991). Hodgson draws the following conclusions about it.

> Approaches which are directed towards improving social and marital relationships are very effective. If we combine the work on social skills training, behavioural marital therapy and community reinforcement there are 21 studies, all of which are positive. Furthermore, investigations of those factors which predict relapse and recovery suggest that family stability, cohesion and social support are among the most important (Orford and Edwards, 1979; Billings and Moos, 1983). There is good evidence that interventions directed towards self-control and stress management training are effective (18 out of 27 studies were effective). (pp. 1530–1)

SOCIAL TREATMENT: EFFECTIVE METHODS FOR SHORT-TERM CHANGE AND MAINTENANCE OF CHANGE

Social skills

Lindstrom also reports favourably on the outcome of social skills training over time, remarking that positive effects were demonstrated at one-year follow-up by Erikson, Bjorns and and Gotestam (1986). He suggests that the social skills needed by moderate drinkers and excessive drinkers will differ. He refers to studies by Valliant (1983) and Edwards *et al.* (1987) to illustrate the view that, over time, 'Excessive drinking is likely to become less mouldable and more independent of others' reactions, excessive drinking resulting in changing reactivity to alcohol and disabling consequences to personality'. He proposes that rehabilitation of the highly dependent drinker 'seems to require a more fundamental psychological shift including a reappraisal of self and others' and suggests that 'specific skills, such as those prescribed by Sisson and Azrin (1986) and Al Anon (self-help support for families of alcoholics), may be useful in modifying the behaviour of the highly dependent drinker' (p. 102). Work in this area is extremely sparse but Longabaugh and Beattie (1985) have carried out a preliminary study which indicates that the greater a person's social investment, the more influential the social environment will be.

Social context and community

Work in the area of community reinforcement is also extremely limited, but early evidence suggests that it may be very effective (Hunt and Azrin, 1973; Azrin *et al.*, 1982; Mallams *et al.*, 1982; Sisson and Azrin, 1986). The community reinforcement approach involves the modification of the client's social environment; in effect, making an assessment of social resources, or 'reinforcement contingencies', available and using them, rather than taking the individual out of their environment to treat them (Azrin *et al.*, 1982). Azrin evaluated his project and followed up treatment and comparison groups at six months and two years. Drinkers in the community reinforcement group continued to abstain from drinking at least 90% of the time, whereas drinkers in the matched comparison group drank at least 50% of the time (in a three-month period following treatment).

Azrin's initial success in developing and evaluating this type of programme indicates that long-term gains can be maintained if significant people in the alcoholic's natural environment remain in a position to influence their behaviour after therapy. Azrin's work suggests that the problem is the client's inability to deal with the problem on their own and that they need social support to learn to control themselves. Other authors have also focused on the need for 'governance of self'; Lindstrom cites the work of Mack (1981) and Khantzian and Mack (1989).

It can be seen that social treatment approaches offer a possible way forward in the drug and alcohol field. However, because this type of approach is rare, there are few programmes to evaluate (Azrin had developed his own for this purpose). What has not been recognized is that the Anonymous Fellowship and the Twelve Steps offer a means of examining this type of programme, as they are in effect utilizing the social environment to change behaviour and maintain change over time.

Relapse and follow-up

The importance of social factors in follow-up has been stressed earlier. There is, however, little research in this area and much of this is unreliable, partly because it is difficult to control for social and relationship variables or to generalize. Ravndal and Vaglum (1994) found a correlation in women patients between 'destructive relationships' and failure in treatment. Birke, Edelmann and Davis (1990) found that negative affect and interpersonal conflict were predictors of relapse, but not social pressure or cognitive attributional styles. (See also Part Three.)

IS TREATMENT INEFFECTIVE IN THE LONG TERM?

It is well documented that addictive problems are highly susceptible to recurrence following successful intervention (West and Gossop, 1994; Somers and Marlatt, 1992). Lindstrom, in his recent review of alcohol treatment, states that, 'Controlled studies of outcome criteria have generally demonstrated only weak and short-term effects of alcoholism treatment ... with virtually no effects remaining after one or two years' (p. 30). He concludes, 'One cannot expect to find, one year after discharge from an alcoholism programme, any significant difference in drinking outcome between a treatment group and a control group' (p. 294). He explains that this does not indicate that remission of alcohol problems is impossible, simply that research tells us little about other factors that may be important in this process: 'The interpretation of this result should not be that treatment of alcoholics is ineffective, but that any long-term remission is primarily due to circumstances over which the therapist has little or no control'.

Lindstrom points out that the studies of treatment with positive results are those with little treatment follow-up. He lists a range of studies illustrating loss of treatment effect over time (e.g. Valliant, 1983). It also becomes clear when comparing review data that little has changed in the past 20 years, despite apparent advances in research. A comparison of studies from two distinct time periods (1952–1972) and 1978–1983) by Riley *et al.* (1987) could find no evidence in either period, amongst old or new treatment models, of long-term treatment gains.

Generally effective non-methadone drug treatment programmes have been shown to work best if they have adequate staff levels and quality assurance procedures, but most significantly, if they have adequate follow-up (McCaughrin and Price, 1992).

Treatment gains therefore appear to be only short term. However, while within-treatment processes are qualitatively different from maintenance of change processes, it is important to emphasize that the short-term benefits of treatment could be kept up if that treatment were continued indefinitely in some form. Maisto and Carey (1987) found that staying in treatment seems to be associated with benefits that may be maintained for some time.

It may also be useful to ask which clinical strategies interact best with what follow-up post-treatment strategies. For example, Saunders and Kershaw (1979), Valliant (1983), Nordstrom and Burglund (1986) and Edwards *et al.* (1987) have presented findings that indicate that social relations are a major precondition for the persistence of whatever the severely dependent drinker may have gained from treatment.

Feigelman and Jaquith (1992) have argued that day care should be structured to treat drug abuse as a family problem, in that it requires behavioural change in all family members. Galanter (1993) emphasizes the effectiveness of involving family and peers in a cognitive behavioural programme. Gibson *et al.* (1992) have demonstrated that family problems as a whole decrease in the first few months of drug treatment of one family member, when the patient relates better to their family and there are fewer difficulties and problems within the family. This is substantiated by Spear and Mason (1991), who compared medical insurance claims for families two years before and after treatment of one member. They found that there was a significant decrease in health claims in the whole family.

DIFFICULT AREAS AND BOUNDARY DISPUTES

The review of research above has demonstrated the controversial and in many ways unsatisfactory state of knowledge of drug misuse and its effective treatment. There is simply a very poor fit between research and practice. This is perhaps most evident in those issues which challenge the autonomy of the different treatment perspectives and the notion of dependence treatment itself. While it is true that outcome research on drug misuse where behavioural hypotheses can be tested demonstrates that these techniques can be effective in changing behaviour, it does not follow that other therapeutic regimes are less effective or ineffective. It is possible that both professionals and non-professional helpers tend to rely on what works rather on the evidence of one particular theory. Clinicians may develop their own understanding from experience rather than academic theories and scientific research.

HIV

HIV has greatly influenced attitudes to drug treatment and raised questions about the weight given to treatment priorities as a whole in contrast to harm minimization. This is because HIV is more likely to be prevented through harm minimization methods than many treatment methods. The exception is the use of methadone maintenance treatment for those at risk of HIV. The risk of infection has led many GPs to respond to drug dependence with more flexible long-term prescriptions for a range of different drugs rather than simply heroin withdrawal schedules. This change in attitudes is a result of the need to establish and maintain therapeutic contact with patients. These aims obviously have increased significance if the patient is HIV positive. These issues have been covered more fully in the previous chapter on harm minimization but it is important to deal specifically with the implications for drug dependence treatment here.

These implications can be summarized in one guideline: that generic professionals should attempt to link more drug misusers into harm minimization services of any kind, whether they are dependent on drugs or not. If a drug misuser is referred to an agency for drug treatment, this does not necessarily include a harm minimization service if the treatment is abstinence based, nor does it guarantee a continued harm minimization service if drug treatment is not accepted by the client. The professional should therefore ensure that the treatment agency also offers a harm minimization service independently of drug treatment and/or put their client in contact with pharmacists, syringe exchanges and a general practitioner who will provide harm minimization and health care respectively.

Harm minimization information and clean syringes should be made available whether the client is receiving treatment for drug dependence or not. This can create problems if clients appear to be receiving mixed messages from different staff in agencies. For example, it may be difficult for a client to collect a withdrawal prescription prior to exchanging needles and syringes. Many agencies attempt to keep syringe exchange and harm minimization facilities separate from prescribing services to facilitate the use of both. For a prescribing GP it is important to do the same or at least to give patients precise information about where clean needles and syringes can be easily obtained.

The guidelines for all professionals are broadly similar: to ensure that clients receiving treatment for drug dependence also know exactly where harm minimization services are available or, failing this, at least know the address of the nearest pharmacy or chemist where needles and syringes are available. (This information is available from drug teams or health authorities.) The same is necessary for general health care facilities. Many drug misusers are not registered with a GP and it can be part of the generic professional's task to liaise with GPs to develop a working relationship in respect of particular clients. Doctors and other professionals should be able to offer referral for HIV antibody testing and counselling and should be aware of the need for health care and immunological monitoring for people who are HIV positive. Professionals should also be aware that the early symptoms of HIV may mimic drug dependence or withdrawal. These and further details are available in the Department of Health booklet, *Drug Misuse and Dependence: Guidelines on Clinical Management* (Department of Health, Scottish Office and Welsh Office, 1991).

The issues regarding the treatment of drug dependence in people with HIV and AIDS are described by Brettle et al. (1994). They describe how a multidisciplinary outpatient medical team functions to deal with the medical, social and public health implications of working with this group. The particular implications of HIV for the aims of methadone treatment are discussed by Summerhill (1990), who highlights the public health function of prescribing for clients with HIV to prevent further spread of the infection.

Controlled drug use or abstinence?

There is very little research concerned with the controlled misuse of drugs, so we are therefore reliant on what we can learn from alcohol misuse. Much of the research into psychological interventions has been conducted in the alcohol field, where controlled use is more acceptable (Hodgson, 1976; Heather and Robertson, 1981, 1986; Chick, Lloyd and Crombie, 1985; Babor *et al.*, 1988; Hester and Miller, 1989). As Lindstrom (1992) states, 'There is ample evidence demonstrating that brief interventions to help to cut down and control alcohol use may be modestly but reliably effective in helping the less serious type of problem drinker'. There is little controversy over whether cognitive behavioural methods are useful for heavy drinkers and those who are psychologically dependent; but people disagree as to how far the methods are useful when dealing with the biopsychosocial phenomenon of 'dependence'.

Despite criticisms of the narrow and self-limiting nature of the behavioural approach, there is no doubt that much of the most useful recent research in the alcohol field has been carried out by cognitive behavioural psychologists. This work is listed in a thorough review of early intervention strategies provided by Babor, Ritson and Hodgson (1986). This review suggests that 'Modest, but reliable effects on drinking behaviour and related problems can follow from brief interventions, especially with the less serious type of problem drinker'. The types of intervention evaluated include information, advice, self-help manuals and groups and periodic monitoring of progress by a health worker. The authors conclude that 'Low intensity, brief interventions have much to recommend them as the first approach to the problem drinker in the primary care setting'. This work is corroborated by Chick, Lloyd and Crombie (1985), who conclude that nurses can effect significant change in patients' drinking behaviour, if intervention is made before the problem has become too severe. There are criticisms of the type of controlled experimental/quasi experimental method used in many of these studies, both in terms of the need for epidemiological data and the bias introduced by a behavioural analysis of problems. But as Babor *et al.* (1988) point out, the data collected have proved extremely useful, as the research indicates that both light and heavy drinking can be brought under individual control and drinking levels maintained or cut down.

The research and discussion concerning the possibility of severely dependent clients returning to controlled drinking, using the behavioural techniques of early intervention, are less optimistic (Lindstrom, 1991). Though there is much controversy over this issue, there is little research to indicate that controlled drinking is a viable option for most of those who are severely addicted. The discussion in this area centres around the measures used to assess severe dependence and how to identify those individuals who come into this category.

In the 1970s Griffith Edwards and his colleagues (Edwards and Grant, 1976, 1977) concluded that this type of difficulty is compounded by confusion between 'dependence' and 'disability'. Edwards states that because these terms are confused, the problem of definition becomes much worse. His work led to clarification of the notion of disabilities in the broader sense which referred to physical disabilities (such as liver disease), psychological disabilities (for example, depression) and social disabilities (such as unemployment). It will be obvious that while most disabilities may be drug dependent in origin, others may not. The difficulty is that, for example, if a drug misuser is depressed, the depression may be a consequence of the drug misuse or vice versa. Moreover, the relationship between them may itself affect the balance between dependence and disability. The greater the misuse of drugs to counteract depression, the greater the depression: the deeper the depression, the more need to take drugs. As we shall see, this has significant consequences for treatment.

The problem of deciding whether dependence has caused certain disabilities or whether certain disabilities have led to and aggravated dependence is common in the drugs field. It seems very difficult, if not impossible, to discover simple objective criteria for assessing this, for professionals and clients alike. This difficulty seems inherent in the phenomenon itself. There is no doubt that there is a correlation between severe dependence and disability but which causal direction is accepted often depends not on research (of which there is surprisingly little), but on the initial theoretical structure and beliefs of the observer. An observer believing in the disease model would be more likely to define disabilities as a sign of dependence, whereas a psychiatrist or a psychologist may be more inclined to define dependence as a sign of underlying disabilities.

This situation explains to some extent the confusion of both professionals and clients themselves. In an attempt to remedy the lack of research evidence in the drugs field, it is again useful to look to research in the alcohol field.

Griffith Edwards defined different patient types in terms of disability and dependence. He believed that if the dependence syndrome were severe, then the chances of returning to controlled drinking were low. If the syndrome were mild, then controlled drinking would be possible. He felt it essential, but problematic, to group subjects in terms of degrees of dependence rather than in terms of disability. Edwards concluded:

> For people who are disabilitated simple strategies of social adjustment or psychotherapy (coping mechanisms) might be useful in reversing the 'dependence syndrome'. There is an important contrast here with the benefits of such general strategies for severe dependence, where similar approaches may be expected to lead to remission. (1986, p. 149)

That is, what is good for someone mildly addicted may be a disaster for someone who is heavily addicted and vice versa (see also Edwards, 1988, 1989a).

The significant point raised in discussion of this study was the difficulties in assigning subjects to groups in terms of whether the drinking caused the disability or the disability caused the drinking. Edwards is attempting to overcome differences between different theorists and develop 'a shared model of understanding' by defining two types of alcohol problem. We can see how this useful distinction can be applied to drug misuse generally.

The question is now, if brief interventions and controlled alcohol use can have a benevolent influence, why should drug misuse be any different? Earlier sections in this book have considered the sociological research examining recreational and non-problematic drug misuse. Much of this work suggests that drug misuse can be controlled. The issue is in large part concerned with the type of drug misuse, its potential for creating dependence and the patterns of misuse.

The physiology of dependence

A further area of difficulty arises when interactions between physiological and cognitive variables appear to influence 'loss of control' and relapse. From the medical perspective, one of the problems inherent in the notion of simply using behavioural techniques is that drugs have both short- and long-term physiological effects. It can, for example, be demonstrated that physiological factors influence the significance of psychological expectancy factors. Research with subjects who are physically dependent on alcohol demonstrates that they respond to the alcoholic content in drinks, whereas the response of less dependent subjects is determined mainly by their beliefs and expectancies about drinking. The effect of an initial drink on a severely dependent subject's craving to drink further is not merely the result of cues, beliefs and rationalization; it occurs without the drinker being aware that they have consumed alcohol. It has been demonstrated that there is a statistically significant increase in craving three hours after a 'primary dose' (Hodgson and Stockwell, 1977). The authors state that this can be explained using the behavioural perspective, but only if it is possible to find out what pattern of reinforcement history differentiates the severely dependent from the less dependent drinker. Hodgson and Stockwell continue:

> Such a learning experience involves a set of discriminative stimuli and also a powerful reinforcer so that, after hundreds of repetitions, the severely dependent alcoholic will tend to experience a compulsion

to drink when exposed to the cognitive and physiological cues associated with stopping drinking.

They suggest that this altered physiological state is a component of craving but only for seriously dependent alcoholics for whom it has become a discriminative stimulus. For less serious alcoholics this physiological state is seen only as a component of craving when the alcoholic believes they have consumed alcohol.

Marlatt, Demming and Read (1973) carried out an experiment in order to test whether people addicted to alcohol lose control of their drinking (a premise integral to the disease concept and opposed to the cognitive behavioural approach). They found no difference between 'alcoholics' and social drinkers in alcohol consumption based on actual primer drinks (physiological stimuli), but found increases based on expectation.

It is also necessary to include an understanding of physiological variables in illegal drugs. Heroin has been shown to produce more severe dependence than either cocaine or amphetamine (Gossop, 1992). Gossop demonstrated that severity of dependence was also influenced by route of administration, amounts misused and length of time misusing. (He also showed that dependence problems were common among heroin misusers who had never attended treatment.)

There is therefore conflicting evidence regarding the interaction of physiological or behavioural factors and subjective expectations and cognitions. It is impossible to research these interactions 'scientifically' because subjective interpretation is not observable. Much research is based on apparently positivist methodology: examining observed behavioural and physiological changes invokes both cognitive processes and the notion of craving as variables in an experiment. Craving is not as yet defined in physiological or behavioural terms; instead, it is often based on vague cognitive and emotional concepts far removed from any scientific approach, but the concept remains a dominant one in the field due to its clinical usefulness.

Dual diagnosis

In the case of dual diagnosis it is clear that drugs prescribed for mental health reasons can lead to dependency problems. For example, Landry *et al.* (1991) suggest that patients attending treatment for anxiety disorders are commonly prescribed psychoactive drugs which can lead to dependence if not carefully monitored.

In the same way, illicit drug misusers may be self-medicating (knowingly or unknowingly) for problems of anxiety or depression. Khantzian (1985) suggested that individuals misuse drugs adaptively to cope with intensive adolescent-like anxiety. He formulates a self-medication hypothesis, proposing that drug misusers will select drugs which will 'medicate' their dominant painful feelings. It is also possible that illicit drug misusers may be self-medicating for behavioural problems such as aggressive and/or violent behaviour. Again there are problems in determining whether behavioural problems are an effect of drug misuse and withdrawals or whether they predate them. Powell and Taylor (1992) showed that hostility and anger were common among opiate misusers after withdrawal: they suggest that anger management would be a useful intervention at this stage.

It is perhaps the relationship between depression and substance misuse that is most significant to practitioners (Paton, Kessler and Kandel, 1977; Deakin, Levy and Wells, 1987). Many clients will say that they used drugs (or alcohol) because they were depressed and that therefore it is the underlying depression that needs to be treated, whereas others will see depression as a consequence of their drug misuse. It is of course likely that each contributes to the other in a worsening spiral of problems but in terms of practical assessment, it is very important to distinguish between the two as the type of problem will determine the type of intervention.

Although there are difficulties in determining the causal relationship, many professionals are concerned at the incidence of 'co-morbidity' (Kandel, 1982) and the increased incidence of depression. Deykin, Buka and Zeena (1992) have shown that among adolescents in drug treatment the rate of depression is three times higher than in a control population and that this is linked to a very high incidence of physical and sexual abuse and neglect in the histories of depressive drug misusers. They have demonstrated that the high rate of depression amongst substance misusers can be explained in terms of two distinct groups: those with primary depression and those with depression subsequent to dependence. The authors suggest that these different types of development of depression require two different types of treatment approach.

> Patients with primary depression who are chemically dependent might well profit from treatment directed at their depression rather than their chemical dependence. Conversely, those who become depressed after they have developed chemical dependence might show greater improvement when treatment is targeted at their alcohol and drug misuse rather than their depression.

Deykin *et al.* also found that males were twice as likely to have depression following drug misuse than vice versa, while females were more likely to be self-medicating.

Magruder-Habib, Hubbard and Ginzburg (1992) have shown that those at highest risk of suicide were women misusing a range of non-narcotic drugs. They found that if clients felt suicidal tendencies soon after treatment these would be more likely a year later and that it was related to relapse.

Dual diagnosis itself has practical implications in terms of therapeutic intervention, as what is believed to be good practice for dealing with dependence may be at odds with treatment of depression and other mental health problems where prescribed drugs are considered beneficial. There are also particular treatment issues concerned with the psychiatric models of care and the disease/abstinence approach to dependence. Dermott and Pyett (1994) recommend that conflicts over conceptualization should be resolved; while this is most improbable, it is important to recognize the likely conflicts between psychiatrists and disease model adherents and negotiate working relationships that take the differences in belief and practice into account. In addition to these problems, Kofoed (1993) outlines the difficulty of working with these complex patients over time, suggesting that different types of treatment intervention should be planned in a structured way through the treatment and recovery process.

Fresh complications are also introduced by those clients who misuse a variety of different drugs (polydrug misuse). Although it is clearly the case that some clients will prefer certain types of drugs, there are also an increasing number of polydrug misusers. The problems of dependence are then compounded. For example, it is fairly common for certain drugs such as depressants to be misused interchangeably, e.g. heroin, methadone and alcohol. Work by Stastny and Potter (1991) and Carrol, Rounsaville and Bryant (1993) indicates that heavy alcohol use is not uncommon among opiate addicts and cocaine misusers respectively. A longitudinal study of heroin addicts 12 years after treatment shows that a quarter of a group of 298 were classified as heavy drinkers, indicating a form of substitute drug taking (Lehman, Barrett and Simpson, 1990). Iguchi et al. (1993) have also identified illicit benzodiazepines and sedative misuse amongst methadone maintenance clients. San et al. (1993a) found that benzodiazepine misuse (particularly flunitrazepam) was common among heroin addicts. However, these authors report elsewhere (San et al., 1993b) that clients on a drug-free programme consumed more of a variety of illicit substances than those in methadone maintenance.

It seems likely that certain types or combinations of drugs are preferred by different people. This seems clear from the qualitative data earlier in the book. Other researchers have also identified this phenomenon. For example, Kidorf and Stitzer (1993) found that patients had preferences for particular groups of drugs and not others, e.g. methadone patients did not particularly like cocaine. Craig and Olson (1990) compared cocaine and heroin misusers and found that the latter were more likely to have anxious personalities and evidence somatic distress, whereas cocaine misusers demonstrated antisocial personality traits. The authors conclude that there were clear personality differences between the two groups. Unfortunately it is difficult to determine causal direction, that is, to distinguish between the effects of long-term misuse and the reasons for misusing.

EFFECTIVE TREATMENT – THE IMPLICATIONS FOR PRACTICE: TREATMENT PROCESS AND AFTERCARE

Research has shown that any of the three main treatment interventions can be effective in the short term. These methods can be used in conjunction with each other but may be more effective if client and treatment method can be matched appropriately.

There are two substantive implications for practice:

Process

The efficacy of treatment is improved if the stage of a process of change is identified and monitored for each client, as each stage requires a different intervention. It is important in assessment and treatment itself to respond appropriately at different stages in the process of developing and dealing with drug problems (see also Part Three).

Aftercare

It will be evident that some kind of aftercare service is essential for preventing relapse, by developing psychological and social skills together with a range of social support systems in the community for follow-up maintenance of treatment gains. Social factors and the social context are particularly relevant in aftercare. It is necessary to consider social theories of drug misuse and the implications for treatment in terms of social factors, including the influence of old and new networks of family, friends and colleagues.

INDIVIDUAL DIFFERENCES AND MATCHING CLIENTS TO TREATMENTS

Work on individual differences and matching has been limited to the alcohol field (Glaser, 1980; Schwartz, 1982; Miller and Hester, 1986b; Marlatt, 1988; Moos *et al.*, 1990; Keene, 1994). It has been largely inconclusive.

It is often argued that the problems inherent in the field of service provision are caused by the lack of one generalized theory of dependence and exacerbated by conflict between professionals with differing beliefs as to both theory and practice. There seems little likelihood of consensus developing. It may be that the breadth and variety of service provision would suffer in some respects if this ideal were to be achieved. There are practical advantages to different approaches.

Although the disease perspective requires more intensive input than a behavioural or social learning approach (see Mash and Terdal, 1976; Goldfried and Bergin, 1986), ongoing Narcotics Anonymous support is free, in the form of a countrywide network of self-help groups. On the other hand, the cognitive behavioural approach is likely to be more accessible to a wider range of clients with drug problems and to generic professionals such as health care workers and social workers, as a social learning approach to drug and alcohol abuse and treatment often forms part of their basic training.

Glaser (1980) considered the implications for treatment and the possibilities of matching clients to different services, stating, 'The question is not whether similarities and differences between individuals exist – they do – but which is prepotent in determining the outcome of treatment?' (p. 178). He makes the point that:

> In most circumstances at the present time, the similarities between individuals are assumed to be prepotent, and hence all individuals presenting are dealt with in the same way ... which is perhaps why the matching hypothesis in most fields remains a hypothesis.

Research has added extra weight to these arguments, several authors stressing the influence of individual differences in treatment process and outcome (McCrady and Sher, 1983; Caddy and Block, 1985). However, discussion of client matching has remained largely abstract and the practical implications for treatment allocation remain vague. Research in the 1990s is as yet inconclusive, though the results from the large American study of alcohol treatment, 'Project MATCH', should give much needed information about the possibilities of matching clients to the different treatment regimes.

PROCESSES OF CHANGE

It is only very recently that researchers have focused on the **process** of therapeutic change. We therefore know very little about it. This may be simple because process research cannot be easily undertaken within the confines of behavioural science or medicine.

However, the 'scientific' data obtained by controlled studies of human behaviour are not necessarily acceptable to the medical community of clinicians. It is possible that while scientific data are useful to researchers, in the area of clinical practice itself the quantitative outcome data of the sciences are less useful than the qualitative material derived from individual case studies. As much clinical knowledge is built on the accumulated information of casework experience, many clinicians may come to view both diagnosis and treatment as professional crafts, dependent on qualitative data concerning process and interpretation, rather than scientific procedures.

This difference between the preferred knowledge base of researchers and practitioners is probably best illustrated by a recent phenomenon in the addictions field, that of 'motivational interviewing' (Miller, 1983) and the related idea of 'stages of change' (Prochaska and DiClemente, 1983, 1986) (see below). Methods of interviewing and counselling in the field were developed in the late 1980s using notions of individual change processes, apparently in the absence of any coherent theory or testable hypotheses.

The popularity of these methods among practitioners spread rapidly in the early 1990s. These ideas were popularized and now form a base for specialist and non-specialist training in Britain (see Aquarius, 'Managing Drink' training pack) and counselling techniques themselves (Robinson, Rollnick and MacEwan, 1991).

As a consequence of the gap between psychological researchers and Twelve Step advocates, the 'stages of change' model of Miller and Prochaska and DiClemente has been hailed as an entirely new concept in substance misuse treatment. In effect the Twelve Step approach incorporates a theory of stages of change in the therapeutic process. Many academics were unable to explain the popularity of these two unproven methods but their relevance to practitioners can be easily understood in terms of the importance of individual differences and individual change processes to clinicians. Data concerning individual change process cannot be gained using positivist methods, therefore scientifically orientated research can do little to provide the practitioner with guidance for work with individual clients. The motivational model offers a means of integrating notions of long-term individual change with more general theories of aetiology and treatment. The practical methods developed from these ideas are discussed in Part Three.

The notion of the importance of individual change processes was highlighted in 1983 in Prochaska and DiClemente's much quoted paper 'Stages and processes of self-change of smoking: toward an integrated model of change'. Prochaska had also published a more complex account, 'Systems of psychotherapy: a transtheoretical perspective' (1979), and, with DiClemente, built on the idea of transtheoretical therapy (Prochaska and DiClemente, 1982).

This work was paralleled by W.R. Miller's development of the concept of motivational interviewing. Miller's work was published in the journal *Behavioural Psychotherapy* in 1983 and again (with Sovereign and Krege) in 1987.

The ideas of change process and motivational interviewing have been criticized on the grounds that they have no underpinning theory and that they are not testable (see *British Journal of Addiction*, letters and articles 1992/93). Academics from the psychological and medical fields have expressed surprise at the widespread popularity of the approach amongst practitioners. It is therefore interesting to examine the enthusiasm for clinical methods not accessible to positivist-based research.

It seems clear that within the field at the present time there are theories of the aetiology of dependence, treatment and relapse prevention. It cannot be taken for granted that practitioners use corresponding theories for each stage. It may be that they can select several different theories for each client, describing the development of the problem in terms of a particular theoretical perspective, and then, perhaps independently of this initial understanding of the aetiology, choose a particular understanding of a treatment process to structure their practical response. While it is no doubt true that academics would require some degree of 'fit' between a theory of aetiology and one of treatment, it is not necessarily true for either practitioners or clients in the field. It seems possible for clients to work both with therapists who take a cognitive behavioural approach and those who work with the disease perspective: it is also possible for 'eclectic' practitioners to use both. The cognitive dissonance which would be generated in many academics is also problematic for many practitioners and clients but for a certain proportion of individuals these discrepancies are not important, if they are recognized at all.

In summary, it may be useful to conceive of three layers of theory: first, that of the aetiology or development of the phenomenon; second, that of initial change or treatment; and third, that of maintenance of change or relapse prevention. These considerations have some relevance in considering the possibilities of matching clients to different treatments. In practice, perhaps the two significant stages in treatment are those of initial change and maintenance of change. It is only very recently that the second stage of maintenance of change has been given any attention at all.

MAINTENANCE OF CHANGE: AFTERCARE AND THE SOCIAL
CONTEXT OF DEPENDENCE

The implementation of coherent aftercare models is often not an important consideration for many treatment agencies. However, research indicates that most people relapse within two years of treatment (see below) and that the major influences on successful maintenance of gains are of a social nature. Cognitive behavioural treatment regimes are effective in the short term but the effects are not long-lasting. This lack of focus on the post-treatment period may be because that approach neglects both the social psychological theory of small groups and the sociological theories of social interaction and wider social networks.

As Barry Brown points out in his introduction to a special edition of the *International Journal of the Addictions* (1990/91):

> The ideology and tradition of drug user treatments, borrowed from the fields of psychiatry and psychology, emphasizes a development of the individual's coping skills ... to allow the individual to change his/her way of functioning in the community. The emphasis is on changing the individual, so that he/she can change their environment ... it has not included the service provider working to change the individual's environment beyond the limited effort of contracting community agencies for services on the client's behalf. (p. 1082)

He suggests that the following would be useful:

- work to change role demands and expectations placed on the client by the family, school or employer;
- efforts to develop new and prosocial networks;
- initiatives to help existing clients structure their free time activities to avoid relapse.

He points out that activity in the first two areas has been practically non-existent while the third, though starting to generate interest, is still largely undeveloped. He suggests that this is partly a consequence of a lack of funding for these initiatives and partly a reflection of the professional theories and remits of those presently involved (p. 1083). Once research has identified the importance of such factors as stigma, subculture, social role, social and life skills and social networks, it makes sense to consider what information we have about these areas in the relevant literature.

The importance of establishing social networks can be seen as particularly important for illicit drug misusers, as sociological research indicates that there are subcultures of drug misuse and that misusers may be socialized into this way of life (Becker, 1964; Young, 1971; Parker, Baker and Newcombe, 1988; Bloor *et al.*, 1994). If drug misusers are to stop misusing they will need to move out of these social networks and corresponding occupations and develop a complete change in lifestyle. It follows that old social networks will need to be discarded and new ones developed. Individuals will have to change their own reactions to things and develop new skills. It is at this most crucial time that treatment ends and the help stops.

As emphasized earlier, we have something to learn from each of the different treatment perspectives. The contribution of the disease perspective of NA is knowledge about the importance of social context in aftercare. In America the Twelve Step programme of the Anonymous Fellowship integrates aftercare into the actual treatment process and extends the period of 'treatment recovery' over several years (the average length of attendance at AA is 5–6 years). This period of recovery includes the provision of structured social support networks in the aftercare or post-treatment period. In Britain we lack any structured pattern of aftercare provision and need alternatives to the self-help aftercare of the Anonymous Fellowship, which will fit better with the cognitive behavioural and harm minimization approaches.

As we are now aware that relapse after drug treatment is the norm rather than the exception, research is now developing into the identification of the conditions associated with relapse, focusing on areas such as employment, relationships, social settings, psychological stress and crossaddiction (Leukefeld and Tims, 1989). Aftercare therefore consists of remedial education, vocational counselling, life skills training, family support and cognitive behavioural relapse prevention techniques. In its most extreme form, stages of treatment can be seen as preparation for aftercare rather than the other way round.

The two main elements of aftercare are individual cognitive behavioural skills (Marlatt and Gordon, 1985) and the development of social support networks with social skills training (Catalano and Hawkins, 1985). These methods rely on an accurate assessment of each individual's needs and resources post-treatment and the consequent ability to help with a range of social, psychological and health needs. These can range from teaching clients to recognize drug-taking cues to improving housing and employment prospects, from building up social networks to dealing with stress or avoiding stressful situations.

Risk of relapse is highest in the first 6–12 months after treatment, particularly in the first three months. There are different kinds of relapse, from one slip to a full return to old patterns of behaviour and social groups. There is more chance of success with swift reintervention after the relapse (Brown, 1979; Brown and Ashery, 1979; Hawkins and Catalano, 1985). There is also evidence that self-help groups, networking and skills training help minimize relapse (Nurco *et al.*, 1983; Hawkins and Catalano, 1985).

It is not possible to develop effective aftercare without reference to the wider society in which the individual lives. Researchers who take no account of social context will remain oblivious to many drug misusers' lack of contact with everyday social networks, as a consequence of which they may lack understanding of certain forms of social etiquette. As an additional handicap, they may be unaccustomed to social intercourse when sober and lack the necessary skills. The Anonymous Fellowship apparently best grasps the enormity of this task for some people and provides a self-help subculture which functions as an interim support group or even as a bridge back into non-drug-using society. It is not suggested that other treatment perspectives should replicate the Anonymous Fellowship, rather that they should be pre-pared to learn from other approaches and consider the advantages of providing alternative forms of aftercare support.

Catelano and Hawkins (1985) describe a project that attempted to develop a social programme derived from theory and research. It was called 'Project Skills' and involved cognitive behavioural skills training and social network development using volunteer partners for each client to act as a bridge to the non-drug-using world. Catelano and Hawkins proposed that this programme would reduce the factors associated with relapse and strengthen those associated with the main-tenance of (in this case) abstinence. They identified the following factors as connected with relapse:

- the absence of strong prosocial interpersonal networks (including family and friends);
- pressure to use drugs from drug-misusing peers and family;
- isolation;
- lack of productive work or school roles;
- lack of involvement in active leisure or recreational activities;
- negative emotional states;
- physical discomfort.

They developed four main tasks to lessen the effect of the factors identified: first, to develop and enhance informal social supports in the community; second, to increase involvement in productive roles in the community (whether in work, school or in the home); third, to facilitate involvement in drug-free, active recreational and leisure activities; fourth, to develop and practise (cognitive behavioural) skills to maintain treatment gains (including skills to become involved in social and leisure activities, skills to cope with stress and 'negative emotional states', skills to recognize and avoid situations with high risk of relapse and skills to cope with slips without having a full-blown relapse).

Rounsaville (1986) also considers the clinical implications of recent research on relapse. He points out that relapse is the main problem for clinicians and researchers alike and suggests that treatment should focus on the likelihood of lapses or relapse and consequently encourage clients to return to treatment soon after lapse or relapse for help. In order to facilitate this he proposes that specialist agencies should develop poli-cies and programmes that explicitly encourage re-entry into treatment after relapse. He suggests that clients should be taught to anticipate and cope with feelings and conditions that are a high risk for relapse and that, while in treatment itself, clients should be encouraged to form self-help groups in order to enhance the supportiveness of post-treatment social environments. He suggests that programmes offer these groups facilities within the treatment project buildings and emphasizes the importance of support for family and spouses along the lines of the Anonymous Fellowship groups, Al Anon and Al Ateen. He proposes that there should be continuing contact after treatment, though less frequently, by meetings, phone calls or drop-in and that there should be a crisis intervention service.

There remains the problem of how to obtain aftercare without the disease perspective. In effect, many cognitive behavioural and methadone maintenance programmes lack a cohesive aftercare component. In Britain, professionals often disagree with the disease perspective, having received training in a psychological perspective. They nevertheless refer clients to Narcotics Anonymous because it provides a socially orientated, self-help network lacking in any of the more mainline statutory services.

In America the problem is slightly different. The majority of service provision, particularly the residential rehabilitation of Minnesota, links directly and cost-effectively into the self-help aftercare provision of the Anonymous Fellowship. However, methadone maintenance programmes are in conflict with Narcotics Anonymous which insists on abstinence.

Therefore, in both Britain and America there are needs for the development of comprehensive aftercare programmes which incorporate social and individual relapse prevention strategies and link with treatment programmes themselves. (For detailed practical information concerning the implementation of psychological relapse prevention programmes, see Marlatt and George (1984) and for social relapse prevention programmes see Catalano and Hawkins (1985).

This chapter has concentrated on the theoretical and research implications for three areas of practice: process, social factors and aftercare. These were, until very recently, the three most neglected areas in the literature.

The work of professionals themselves seldom follows academic theories, scientific research or textbook methods. Instead, it develops from personal experience, accumulated practice wisdom and common sense. So, they make assessments of health needs and health risks and they offer health care and medication for underlying mental health problems. They will appraise psychological deficits and provide relaxation techniques and cognitive behavioural methods for psychological problems, and perhaps also social skills or assertiveness training. They will make assessments of social needs and then provide welfare rights advice, housing and other forms of social work help. They will offer individual counselling and support groups.

It is possible to use cognitive and behavioural methods irrespective of the theoretical interpretation of the problem. This is particularly true for the aftercare period of relapse prevention, whether for controlled use or abstinence. In the same way it is possible to give social and life skills training and help in developing social support networks independently of any theoretical beliefs about dependence.

Within the field of addiction, as in the field of psychotherapy generally, the actual methods used by different types of therapist are often similar, though the interpretations may be diverse. For example, the practice of the Minnesota Method could be seen as involving a large cognitive behavioural element. To some extent, any therapeutic technique can be seen to involve cognitive and behavioural elements in terms of restructuring cognition and reinforcing behaviours. Likewise, any 'behavioural' technique can also be seen to incorporate other therapeutic elements, resulting from a therapeutic relationship with a 'powerful, helping' figure.

FASHIONS IN THEORY CHANGE, COMMONSENSE METHODS REMAIN MUCH THE SAME

Edwards (Edwards and Grant, 1977) is perhaps one of the most respected critics of the disease model, arguing against the notion of total loss of control and the unitary concept itself, yet he points out, 'In the very recent past the received wisdom was exactly the opposite of what we accept today' (Edwards, 1989a). He reminds us that in the 19th and early 20th centuries withdrawal symptoms were recognized as a clinical reality and continues: 'At some point in the first half of the 20th century, clinical awareness of withdrawal phenomena faded ... an inspection of a few contemporary texts certainly suggests that over the decades the dependence potential of alcohol was generally downplayed' (*ibid*, p. 449).

Edwards attempts to examine the reasons why the questions of the physiology of alcoholism (neuroadaption, tolerance and withdrawal) are considered 'redundant' today. He laments the absence of 'The full history of the way in which medicine and science have regarded alcohol withdrawal symptoms'. He describes the rise of medical or 'disease' theories in America in the 1950s, quoting the work of Victor and Adams describing the time course and clinical spectrum of alcohol withdrawal and the WHO report of 1955, comparing alcohol withdrawal symptoms to those of opiate withdrawal and suggesting a common syndrome.

Edwards considers that things changed after the war when the newer theories 'of the founding fathers of post-war alcoholism' stressed that 'tolerance and withdrawal were matters only very peripheral to the view of alcoholism which they were at that time developing' (*ibid*, p. 450). The work of Jellinek was in direct contrast to this, defining particular types of 'alcoholism' as diseases and consequently withdrawal symptoms and tolerance as among the most significant factors in the study of alcohol.

Edwards sums up: 'How is the ebb and flow to be explained?' He adds that, 'One might speculate on the possible impact of changes in the relationship between the world of clinical authority and the high ground of scientific authority' (*ibid.*, p. 451) but draws no general conclusions, except strongly to emphasize the importance of 'clinical experience', even though it 'is a notoriously unreliable arbiter on complicated issues'.

THEORY TO PRACTICE ... OR PRACTICE FROM EXPERIENCE?

Before describing practical interventions for generic practitioners, it will be useful to examine in detail what the specialist drug agency professionals actually do in practice. The following case studies add weight to Griffith Edwards' remarks that 'clinical' experience may lead to very different conclusions from scientific research.

The case study below illustrates that scientific research and theoretical directives may be less influential among practitioners than academics. The professionals here have very different theoretical orientations but, in effect, very similar practice experience. The end result is not necessarily what one would expect ...

A CASE STUDY OF TWO DIFFERENT PRACTICE MODELS: A COMMUNITY DRUG TEAM AND THE MINNESOTA METHOD

The previous two chapters have used case studies to give guidelines for developing drug education and HIV prevention services and information for professionals wishing to use this type of service. The case studies have outlined the aims and advantages of certain services, while highlighting the possible problems and difficulties of professional involvement and the constraints of ideology. This chapter gives two examples, the first of a community drug team and the second of a day centre based on the Minnesota Method. The data give information about the service provided for professionals wishing to refer clients or co-work with these agencies in the UK (for details see Keene, 1994b, and Keene and Trinder, 1995; Trinder and Keene, 1996).

THE COMMUNITY DRUG TEAM: PROS AND CONS

This case brings together the views of a community drug team in order to describe the essential qualities, and to highlight opportunities and to constraints on specialist involvement. The issues are, again, how to target those populations most in need and what service is most appropriate. The main problems are also similar to those of syringe exchange: regional specialist teams have limited caseload capacity relative to numbers of problematic drug misusers and many drug misusers wish to avoid any formal contact with statutory drug agencies due to the illicit nature of drug misuse itself.

The answer for syringe exchange providers was to use pharmacists and generic professionals to provide a basic service to the wider community of injecting drug misusers. The answer for community drug teams is to encourage social and health care professionals to provide a service for their clients with drug problems. Team staff therefore spend much of their time acting as consultants to generic professionals, offering advice, information and training. They will carry out initial assessments for social and health workers, they will co-work with other professionals and provide specialist counselling alongside other professional input (e.g. in conjunction with the prescribing schedule of general practitioners). The team staff also took direct referrals, but some preferred to deal only with the most serious problems and/or play a partial or temporary role in overall care.

The team comprised a psychiatric consultant, a manager, two nurses and a social worker. Until recently the service was referral only with no facility for clients to self-refer, in keeping with health authority traditions. This has recently been modified to include a drop-in facility (i.e. self-referrals).

Aims

The aim is the treatment of dependence, to reduce drug misuse and to help the client give up drugs completely when they are ready. Three intermediate aims are to help resolve the problems underlying and contributing to drug misuse, to help the client gain more control and misuse drugs less chaotically and to reduce the harm caused by drug misuse (i.e. harm minimization).

Staff definitions of the problem

Staff seldom use the word 'addiction', but have various different concepts of physical and psychological dependence and emphasize the importance of the reasons underlying drug misuse, often describing it as a form of 'self-medication'.

> I would rephrase the question 'What are the problems of drug use?' and ask myself 'What problems do people have that cause them to use drugs and what problems do they have whilst they

are using drugs?' ... these would include (a) problems directly related to drug use; access to drug of use, behaviour while intoxicated, finance, and (b) general life needs and problems (often coloured by their drug use). Drugs form such an integral part of some clients' lives that it is difficult to separate them out.

Drugs used in one way can be beneficial, but can also lead to difficulties.

I don't think addiction as a concept is accepted by anyone here, they believe in people being dependent on the use of a substance, but this is a very flexible concept.

I don't want to condone drug use and I don't want to condemn it either.

Staff explanations of their job

Very few statutory drug teams in the UK are entirely abstinence based, although many will have policies which present controlled drug misuse and harm minimization as interim goals on the road to the ultimate long-term aim of abstinence. There is a growing agency consensus that drug dependence, together with drug misuse as a whole, can be best dealt with through controlling or stabilizing it, dealing with underlying problems and teaching new coping skills. This is reflected in the attitudes of staff in the CDT.

Prescribing is an integral part of a drug treatment service, whether for harm minimization, stabilization or withdrawal or maintenance.

There is agreement here that punitive measures are not a solution.

What clients can expect on referral

The major aim of helping clients to stop misusing drugs and the immediate objectives of dealing with underlying problems and preventing drug-related harm are evident in the work carried out by community drug teams. This work includes:

- assessment and care plans;
- practical help;
- individual counselling;
- self-help support groups and developing new social networks;
- teaching new coping skills and diversionary activities;

- aftercare and relapse prevention;
- prescribed drugs.

Assessment and care plans

After referral, the client will usually be seen for an assessment interview. This can be carried out by any of the staff – a community psychiatric nurse, a social worker or a consultant psychiatrist. Assessments focus on health and social harm as well as degree of dependence. The purpose of the assessment is to obtain information about patterns of drug misuse, other problems and the relationships between the two. The first session will also be used to assess the client's levels of HIV and general health risk and give information aimed at reducing these risks.

> The aim of the assessment is to get as much information as possible about what the client feels the problems are – see what they want. We may never see the client again so it's also a way of doing some education on harm reduction and reducing the spread of HIV.

The staff will take a detailed case history of the client, including history of drug misuse with changes in patterns of misuse and previous treatment, mental and physical health; this will then be placed in its social context, including details of social situation, family background, legal situation and educational and employment history. The assessment of the client's problems leads to identification of particular needs and the development of an individual care plan tailored to each client.

The assessments carried out in the community drug team are not dissimilar to those outlined in Part Three.

Practical help

Practical help takes the form of welfare rights, child care, housing and legal help and may well include referral to other agencies. Team workers may also be involved with negotiating with the client's GP or probation officer. These practical issues are often seen as a precursor to dealing with possible psychological problems underlying drug misuse.

> You need to sort out the chaos before you can get at the person. Until they're sorted out, the person just goes round in a circle.

> There's no point in sorting out the drugs problem if you don't deal with other problems, as they will just go back to the drugs.

> Sometimes getting reasonable accommodation away from the old drug-using friends is enough.

Individual counselling

The term 'counselling' is often used to describe both practical problem solving and indepth examination of drug misuse and drug-related problems. Both forms of individual counselling constitute the major component of treatment in the community drug team. It runs parallel with any prescription and is normally carried out on a regular weekly or fortnightly basis. Group work is less important and those groups that have been set up serve an educative rather than therapeutic role.

Therapeutic counselling techniques follow a client-centred approach, leading to goal setting and task-centred work. These educational and psychological interventions are aimed at helping clients to understand reasons underlying drug misuse, teaching the misuser new skills for coping with psychological problems and removing the need for drugs to cope.

> (It's like) taking the lid off the pot, seeing what's going on and dealing with what comes out.

The basic counselling structure common to much social and health care is used by all staff (see Chapter 10): this involves assessment, listening to the client's view, questions and challenging, negotiating goals and agreements and teaching new coping skills.

Cognitive and behavioural interventions are used to help the client learn to control their behaviour. An important aspect of cognitive behavioural work with clients is goal setting. Clients will be asked about their short-term drug-related goals, their short-term life goals and their long-term life goals. Staff will then help them to draw up manageable targets. Reaching such targets gives clients a sense of achievement. Staff emphasize that it is essential that goals are not too ambitious.

> It's important you get it right. If you set people up for failure, they'll fail. You have to break it into manageable goals then when they've achieved one, they think, 'I can do something else'.

The team use a non-directive, reflective, client-centred counselling approach, though staff often strongly encourage clients to make changes in their drug misuse and lifestyle or alternatively restrain them from making changes until they are sure that the clients themselves have the personal and social resources to do so.

Staff acknowledge the importance of the social context of clients in maintaining drug misuse and clients are taught to recognize the pressures placed upon them in certain situations. The team also encourage clients to develop new interests and social networks, including a range of different sports and leisure activities.

Self-help support groups and developing new social networks

There were no self-help support groups attached to the team during the research period, but the staff themselves continue to see clients after 'treatment completion' if necessary. Treatment completion itself is a vague and ill-fitting concept for the staff, as few clients actually stopped attending for positive reasons. The community drug team offers support on a fairly regular basis, clients can return at any time and many continue to keep in touch for years.

Learning new (drug-free) coping skills and diversionary activities

Cognitive and behavioural methods are used to teach coping skills; these can vary from relaxation training to assertiveness training to basic social skills. Diversionary activity covers the whole range of leisure activities and voluntary work. These not only provide outside interests to occupy the clients' time, but also help in the formation of new social networks.

Many clients experience problems with anxiety and aggression. Staff follow a programme of anxiety or anger management and relaxation with them. This involves identifying physical and emotional aspects of anxiety or anger; identifying situations which can trigger these emotions and learning methods of coping with them; methods of physical and mental relaxation using tapes and breathing exercises; practical problem solving to reduce the occurrence of stressful situations and, sometimes, controlled exposure to feared situations.

Aftercare and relapse prevention

There is no formal system of aftercare. The period of treatment with the community drug team is not time limited and clients remain in treatment for as long as they want to, with some attending for over three years. As clients become more stable, the frequency with which they attend the centre is reduced to once a month or once every six weeks. When a client is discharged, it is either their own decision or it is because they have been breaking rules or missing appointments.

Clients are taught specific relapse prevention skills whilst in treatment. Relapse prevention is based on cognitive behavioural techniques and involves teaching the client techniques to increase tolerance to relapse-inducing situations, learn coping skills or avoid the situations where relapse would be likely.

Events or feeling which trigger use can also be identified. The worker can then help the client to develop ways of coping with these triggers.

Prescribed drugs

The aim of prescribing as analysed in the previous chapter was to reduce drug-related harm. The CDT prescribes drugs for two purposes: the first, for harm minimization (discussed in the previous chapter) and the second to facilitate withdrawal as part of treatment. The drug team will prescribe non-injectable opiates (usually methadone), benzodiazepines and/or dexamphetamine. They also arrange for GPs to prescribe these drugs. A contract is drawn up with clients who receive a prescription on the understanding that they attend for counselling on a regular basis. If a client breaks the contract or abuses the system (e.g. by selling the prescription), they are likely to be given a warning; if they continually break their contract, the script will usually be reduced and then terminated.

Staff describe a range of aims for prescribing: initial stabilization, basic harm minimization, short- and long-term withdrawal regimes and long-term maintenance. Most staff see prescribing as an integral part of drug treatment and many are disparaging about the notion that non-medical staff alone could deal with drug misusers.

> That's why we need to give maintenance scripts – to stabilize the chaos, so that we can get them to the stage where they can begin to look at the underlying problem.

> It's the drugs that have done it. He has a diazepam script and he's taking it spaced through the day. He was buying a bit more, but he's stopped that now. It's the drugs that are stabilizing him.

Summary

It can be seen from this case study that counselling and therapeutic change are integral to the work of the community drug team, who will help reduce drug-related harm, prescribe drugs and assist with underlying problems as part of the process of encouraging clients to come off drugs.

Harm minimization is offered to clients as an integral part of the drug treatment package offered by this agency, which also had a separate syringe exchange service attached. This aspect of the work is seen as a means to the end of treating drug misuse (see previous syringe exchange case study) rather than an end in itself.

As discussed in the previous chapter, there is much controversy in the UK about the aims and roles of non-statutory drugs agencies and community drug teams. This controversy arises largely from confusion about what constitutes harm minimization and what constitutes drug treatment. It is only when these categories are clearly defined that it can be decided which interventions are most appropriate for each.

There is clearly a place for harm minimization alongside drug treatment, if only to keep clients alive, healthy, safe and out of custody whilst they are in treatment, but there is also clearly a place for harm minimization for clients who do not need drug treatment. To blur the boundaries may help obtain harm minimization services for those drug misusers in treatment, but it does not help to provide harm minimization services for the majority of drug misusers not in drug treatment programmes.

This is particularly apparent in the contractual relationships between agencies and funding bodies, where the funder may emphasize the treatment of drug dependence whereas the agency staff may prefer to prioritize harm minimization. As a consequence, the aims and objectives of agencies are often blurred and access to harm minimization services such as clean equipment, prescribed substitute drugs and health care services is available but only as part of a drug treatment package.

This situation has created a great deal of confusion in the UK. An attempt to resolve the problem was made in the previous chapter, by defining the basic minimum necessary for harm minimization without drug treatment. The agency in this case study provides both drug treatment and harm minimization and there is little attempt made to distinguish between the two. As a consequence, clients attending the agency, perhaps in the hope of a basic health harm minimization service, are likely to receive drug treatment counselling and a range of therapeutic interventions to deal with underlying psychological problems and develop new individual and social skills.

A CASE STUDY OF A TWELVE STEP PROGRAMME CENTRE

Twelve Step programme: pros and cons

Confusion does not arise for abstinence-based agencies who do not attempt to provide harm minimization facilities, but it should be remembered that harm minimization services are not available at all unless drug agencies provide them.

Twelve Step non-residential services are rare in the UK. This case study brings together the views of the staff of a Minnesota Method day centre in order to highlight the essential qualities of an 'ideal type' service. As with previous case studies, there are opportunities and constraints of specialist involvement and professional ideology. The problem of small specialist teams for diverse populations is resolved partly by co-working with generic professionals as for the community drug teams, but the service also makes much use of the network of self-help Narcotics Anonymous groups to offer support whilst in treatment and, perhaps more importantly, to carry on the support for several years in the crucial period after treatment.

The centre staff, in common with the CDT, work closely with social and health care professionals, spending much time discussing clients, offering advice, information and training and carrying out initial assessments. They receive direct referrals from health, social services and probation, but in contrast to the CDT, the Twelve Step programme takes many more self-referrals (almost a third of its client caseload).

The staff comprise a manager and four staff who are all ex-addicts who had worked through the Twelve Step programme themselves. They have no formal professional qualifications.

Aims

The primary aims of treatment are twofold – abstinence and ongoing recovery. This involves dealing with underlying problems and the development of coping skills to facilitate the social and moral changes necessary to maintain an abstinent lifestyle.

As discussed earlier in the chapter, the ideology differs radically from that of the community drug teams, yet in effect both are concerned with similar issues, underlying problems and drug-free methods of coping. The difference in philosophy is reflected rather in the sequence of treatment steps; the Twelve Step method insists on abstinence first, whereas the community drug team attempts to deal with problems and develop new skills prior to changes in drug misuse.

The ideology and treatment process of the Minnesota Method and Anonymous Fellowship are integrated in the day centre which fully embraces their philosophy and ideology. The centre does not attempt to reproduce the self-help format of the former nor the residential community concept of the latter. Instead, it combines the advantages of both NA and Minnesota in a day centre with a structured programme of regular weekly day-long sessions.

Staff definitions of the problem

Staff definitions of the problem again reflect the primary importance of 'addiction', but staff focus largely on the reasons underlying it, emphasizing that recovery was impossible without dealing with the emotional, social, spiritual and moral reasons for drug misuse. The basic premises underlying the programme are explained in detail earlier in this chapter: the following quotations illustrate these beliefs.

> Clients are not normal, but addicts, it is a disease of the whole person, not something they can get rid of ... they have an addictive personality.

They have a physical allergy to alcohol, an 'alcoholic personality' (immature and self-centred) and a spiritual sickness.

The disease of addiction is the primary cause of all other problems or 'symptoms'.

The addict has a disease, but is also responsible for his own cure.

Staff explanations of their job

Staff did not understand their job as distinct from the Twelve Step programme itself. When asked to be explicit about their role in the treatment process they defined themselves as guides and saw their job as guiding or helping other addicts along the road to recovery. The staff's personal experience of the recovery process is therefore seen as an important factor in their ability to help others.

Staff should have themselves worked the programme before they are able to help other addicts. The job is to guide people through an experience. It would be difficult for a counsellor to do this guiding if he had not been through the process himself.

The programme at the Centre is the same as the first four Steps of the Twelve Step programme. The Centre gives clients work to do on the first four steps.

My role is supporting clients, when necessary, to deal with denial, but mainly to deal with the problems of everyday living, for example, the need to work at relationships, to talk honestly about problems, overcome pride, accept the reality of a situation and then do something about it ... enabling patients to face reality; to cope with stress without chemicals. To help patients become happy, useful members of society.

What clients can expect on referral

- Assessment and care plans
- Practical help
- Individual counselling and group therapy
- Self-help support groups and developing new social networks
- Teaching new coping skills and diversionary activities
- Aftercare and relapse prevention

Clients will experience a different reception at a Twelve Step service from that they can expect at a CDT. First, they will be acquainted with the 'disease' model and assessed within these criteria. If they are diagnosed

as an addict, they will be informed that the only means to recovery is
through abstinence from all drugs and that this is an essential first step.
Attempts to query the diagnosis are interpreted as 'denial'.

At first glance, then, reception appears very different in each agency.
However, if the client accepts their initial diagnosis, the service actually
provided is very similar to that in the community drug teams. The two
important exceptions are the absence of prescribed drugs and syringe
exchange facilities. These services would of course conflict with the belief that
addiction is a progressive disease and abstinence the only route to recovery.

Assessment and care plans

Assessments focus on health and social harm as well as degree of
dependence: they also include moral and spiritual deterioration and a
clearcut definition of addictive disease, which differs radically from the
vaguer notions of dependence used in the community drug teams. The
assessment includes the identification of a range of problems and pin-
points the needs of each individual client.

> The counsellor goes through the diagnostic interview on the
> assessment form. Loss of control is a key factor. Then the client is
> shown the Jellinek Chart and asked to judge where they think they
> come. The diagnosis carries more weight if clients do
> it themselves.

Practical help

As in the community drug teams, this takes the form of welfare rights,
child care, housing and legal help and may well include referral to
other agencies.

> We help people to find somewhere to live, find a GP and gener-
> ally sort their lives out. We help them in any way we can.

Individual counselling and group therapy

Individual counselling is an integral part of treatment, but the main
treatment tool is group therapy and the emphasis is not on the counsellors but the therapeutic influence of group peers. Group therapy is
intensive and emotional. This is in marked contrast to the counselling
offered in the community drug team, which focuses on cognitive and
behavioural issues.

The basic counselling structure and process differ slightly to that
offered in the community drug team: whilst the focus in the drug team is

on reflective, non-directive client-centred counselling, that provided in the Twelve Step programme is more directive, confrontational and structured. The emphasis is on emotional issues rather than cognitive and behavioural ones, though the cognitive and behavioural change techniques used are not dissimilar to those employed by the community drug teams.

The role of the counsellors at the centre is to ensure that clients have assimilated the ideas of NA at each stage, each counsellor 'guiding' the client through the first five steps in a way that is less likely to occur at Anonymous meetings. The client is asked to read the relevant literature and write down their understanding of the ideas and discuss them with the counsellor. In this way the counsellor can assess whether the client has understood and then accepted (believes) the new ways of looking at themselves. A range of therapeutic processes are involved. It appears to be important that a client fully understands and accepts each stage in turn before progressing to the following stage in the sequence. Problems are likely to occur if a client does not fully participate in any one part.

> The therapeutic process involves encouraging people to take emotional risks within a group environment which is based on gradually developed feelings of trust, honesty and integrity. Within the security of the group, through the method of experiential learning, clients learn how to handle emotions appropriately.

> Re-occurring themes in the sessions include emotions, needs and wants and constructive means of dealing with them. The commonest problems that arise are those concerned with interpersonal relationships and conflicts; depression; denial and manipulative strategies or games.

> You must sacrifice your individuality to the group and receiving back their interest in you.

> Clients are asked to consider personal examples of their own powerlessness and unmanageability, to write these down and discuss them both with their centre counsellor and in their therapeutic group.

Self-help support groups and developing new social networks

Self-help groups reflect the emphasis on the therapeutic influence of the peer group. This is the clear advantage of the Twelve Step model, the active support of voluntary non-professionals over prolonged periods of time.

A new client becomes an NA member and will often be advised to stay away from old friends or acquaintances if they misuse drugs and to develop instead a new social life around NA meetings. Many successful clients continue to socialize mainly with other NA members, their social environment as a whole acting as a reinforcer of the new beliefs. The programme at the centre is seen by staff as likely to be effective only when the client attends regular NA meetings and keeps in close contact with their NA sponsor. Clients will therefore usually be asked to leave if they do not attend NA meetings during treatment at the centre.

The network of support groups usually meet on a weekly basis and there is often more than one group available. In common with all other Anonymous Fellowships, such as Alcoholics Anonymous, these self-help structures usually allocate a sponsor to each client. This person will act as a voluntary individual counsellor. In effect, these self-help networks offer social contacts and support on an ongoing basis, 24 hours a day for as long as the client feels it necessary. Successful clients socialize more and more with other NA members, rely a good deal on their NA sponsor and try to get their families involved in similar groups. It is important to emphasize that although the important treatment element seems to be the process of individual change, it takes place within a social context which reinforces and maintains this change as well as playing an active part in its initiation (Keene, 1994b).

> By phoning someone and sharing with them, you are helping that person too.

> It is usual for the client to have developed firm links with NA groups by this stage (second/third step) and to have been allocated an NA sponsor who acts in a supportive, advisory capacity to the client/member.

Teaching new coping skills and diversionary activities

Whilst the Twelve Step centre will teach a range of coping skills, diversionary activity is specifically limited to the activities of the self-help network which serves to occupy the client's time and develop and maintain new social networks.

> The Twelfth Step is a fundamental part of the recovery programme, it consists in helping other addicts. This is very important, the helper will be working in a specific way (the way that had worked for him). These were particular ways of helping that are as rewarding for the helper as for the receiver.

Aftercare and relapse prevention

The Twelve Step model is a long-term recovery programme which continues after the end of formal treatment, often for many years. The

self-help groups function as social support networks to maintain abstinence and prevent relapse.

This model provides two specific forms of relapse prevention. These
differ from those taught within the cognitive behavioural model of the
community drug team. The first involves recognizing one's inability to
control drug misuse and secondly recognizing the importance of
emotions in provoking relapse. Clients are taught to recognize the
danger signs and act to avoid potentially dangerous situations. The
Twelfth Stepping described above is seen as an important means of
remaining abstinent.

> People learn about their own weaknesses, they are responsible
> for making sure these don't cause them to slip. Clients are taught
> to understand themselves so they can't deceive themselves, so
> they can't rationalize using again.

> Twelfth Stepping is what keeps you on the straight and narrow,
> maybe for the rest of your life.

Summary

The Minnesota Method is controversial in the UK. The agency in this
case study is one of the few Twelve Step agencies funded through
statutory sources and also one of a small minority of non-residential
services. However, many private residential services, together with a
nationwide self-help network (Narcotics Anonymous), are based on this
approach. The community drug teams and nonstatutory agencies are
often therefore obliged to work in conjunction with residential and aftercare self-help services using this approach. Finally, it is clear that whilst
the community drug team can integrated syringe and harm minimization into its work, this agency cannot do so without compromising the
basic principle of abstinence.

DISCUSSION

The basic difference in treatment of drug dependence is identified in the
aims of the two agencies above, either abstinence or controlled use. This
emphasis will determine the type of service provided. The two case
studies illustrate an agency that believes in treating drug dependence by
prescribing drugs, either for maintenance or withdrawal, and an agency
that believes that abstinence is the only way to recovery. There are other
models of drug treatment, such as the medical model, discussed earlier,
which also emphasize the necessity for abstinence but may be more
flexible in terms of prescribing practice. However, the basic similarity of

most forms of drug treatment lies in the ultimate aim of reducing or stopping drug misuse altogether. Drug treatment agencies that use harm minimization goals will therefore usually integrate them within a therapeutic drug treatment programme aimed at reduced drug misuse and/or abstinence in the long term.

Drug treatment or harm minimization or both?

The decision to try to cut down or give up drugs and engage in drug treatment is often not straightforward for clients, many of whom will refuse to accept the diagnosis of addict and/or refuse to become abstinent. As the majority of drug-misusing clients are neither addicted to drugs nor ready to give them up, this should not come as a surprise. These clients do not need treatment for drug dependence at all. Instead they need harm minimization services and prescribed drugs, to protect their health and prevent the transmission of HIV and hepatitis. The basic harm minimization service appropriate to these drug misusers is described in the previous chapter. The Twelve Step service is clearly not for those clients who need harm minimization facilities.

However, this basic health service will in all probability not yet be available outside a drug agency. In the absence of straightforward harm minimization support and health care, the drug treatment service offered by community drug teams and non-statutory drug agencies may be the best place to obtain these services. Not only do they offer harm minimization services, but they provide access to essential health care, as a key role of drug agencies is to negotiate with GPs and other health care providers on behalf of their clients.

Most statutory (not disease model) drug workers in the UK now integrate harm minimization goals with counselling and/or therapy with a long-term aim of reduced drug misuse and/or abstinence. The consensus is that drug misuse can be best dealt with, or treated, by initially controlling or stabilizing misuse, dealing with underlying problems and teaching new coping skills. These elements are seen as precursors or parallel interventions alongside counselling for therapeutic behaviour change and abstinence.

The only answer for most drug misusers seems to be – if you can't get harm minimization services without drug counselling treatment then it is advisable to take both.

This chapter and the previous one give an indication of the different goals and methods of each approach. It is important for professionals to decide which goals and methods are most appropriate for which clients. They may decide a client needs both harm minimization and drug dependency treatment, but this should not be taken for granted.

Differences between harm minimization and drug treatment

Table 8.1 A comparison of harm minimization and drug treatment

Harm minimization	Drug treatment
Helping drug misusers reduce drug-related harm	Helping drug misusers reduce or stop drug misuse
Maintenance prescribing and flexible prescribing	Flexible and withdrawal prescribing regimes
No change in drug misuse necessary	Change in drug misuse (reduction or abstinence)
No change in lifestyle necessary	Change in lifestyle
No counselling or therapeutic input	Counselling or therapeutic input
No therapeutic change	Therapeutic change
No examination of underlying reasons for drug misuse	Dealing with underlying problems
Syringe exchange	No syringe exchange (though this may be available nearby)
Drug use equipment, sterile water, bleach, swabs, etc.	No drug misuse equipment
Health care	Health care
Social support	Social support
HIV information, condoms and support services	HIV information, condoms and support services

It can be seen that whilst both harm minimization and treatment services offer health care, social support and HIV prevention, they have a great many differences. For example, the rationale for prescribing drugs for harm minimization is entirely different to that of prescribing drugs for treatment. The former will prescribe on a continuing basis to help the drug misuser reduce the health and social problems associated with drug misuse, the latter will not prescribe independently of a treatment regime.

Because people can be both risky drug misusers and dependent, most drug agencies do both harm minimization and drug treatment as a complete package, thereby offering harm minimization services to help their clients at any stage before, during or after treatment. This is a significant advance for most agencies and has undoubtedly reduced a great deal of drug-related harm and prevented HIV infection among drug misusers in treatment.

However, it should not be assumed that because dependent drug misusers need harm minimization, risky drug misusers will need drug treatment.

Harm minimization would also be very useful to the great majority of drug misusers who do not want drug treatment. In essence, a harm minimization service should provide basic information, equipment and a health care service for the majority of risky drug misusers who do not want drug treatment at all.

Drug treatment: different models, common methods?

For those clients who do want to reduce or stop misusing drugs and are looking for drug treatment, any of the three main approaches will provide a great deal of help and support in the form of cognitive behavioural, medical or Twelve Step interventions. Those clients who want to learn to control their drug misuse and to reduce their dosage over time will prefer a psychologically based or medical service; those who can accept that they are addicts and become abstinent may prefer to comply with the demands of the Twelve Step programme and receive the long-term social support offered by the Narcotics Anonymous self-help networks.

The aims of agencies are clear from conversations with staff, Those in statutory services in the UK seldom use the word 'addiction', but are concerned instead with the reasons underlying drug misuse and often describe the drug misuse itself as a form of 'self-medication'. This is in direct contrast to many self-help and other services in the residential and private sectors which are usually abstinence orientated and may utilize a Twelve Step or similar abstinence model, dealing with the 'primary' problem or disease of 'addiction' itself.

This strange coexistence of contradictory approaches is better understood when the short-term objectives and methods of both are examined. Most drug counsellors of whichever persuasion have an eclectic approach to treatment, using a range of different counselling methods. The above case studies have illustrated how the objectives and methods are in essence very similar, though the sequence of steps will vary, the abstinence-orientated models insisting on abstinence first rather than as the ultimate goal.

The aims and beliefs of each service are obviously of some interest to clients, but whether they are of practical significance is less clear. Drug workers will be well aware that many clients have used a wide range of different types of service and will often use two types one after the other (for example, moving from day centre to residential agency) or two types together (if they need drug treatment from a community drug team once a week, but also want the 24-hour support offered by self-help NA networks). This is a curious situation which has for a long time passed unquestioned by workers in the field.

The answer is clearly illustrated in the two case studies presented here; the service clients receive is actually very similar whichever agency they attend.

- Both services encourage the client to identify drug misuse as a serious problem and to recognize that it causes other problems.
- Both services examine the problems underlying and perpetuating drug misuse.
- Both services use a process model of change (either the Twelve Step process or the motivational change model of Miller and Prochaska and DiClemente).
- Both services encourage the client to weigh up the pros and cons of drug misuse.
- Both services encourage the client to change drug misuse behaviour and other aspects of their lifestyle.
- Both services use recognized counselling and therapeutic techniques.
- Both services have a structured programme involving the setting of a series of concrete goals.
- Both services use cognitive behavioural techniques to change attitudes and beliefs and modify behaviours.
- Both services offer practical help and support concerning health, housing, welfare rights, child care problems, etc.
- Both services teach skills and strategies for coping without drugs.
- Both services encourage clients to develop new leisure interests, occupations, social contacts and support networks.
- Both services provide some kind of relapse prevention support.

Referral or 'do it yourself' or both?

In summary, if generic professionals choose to refer, they must ensure that they match the client's aims, and perhaps also beliefs, to those of each agency. The first step is clearly to determine if the client simply needs harm minimization or if they want drug treatment. It is then necessary to decide if the client is likely to benefit from the more intensive therapeutic intervention and aftercare support of Twelve Step or the counselling and prescription drugs of a community drug team. Having made these decisions, they can be fairly certain that the methods used will be fairly similar.

Alternatively, they may choose to work with clients themselves or co-work with other agencies. In this case they must again first determine the need for harm minimization services; if there is a need, ensure that the client has access to syringes, needles and health care and prescribing facilities from a specialist agency or general practitioner. Only then will the professional be in a position to treat drug problems themselves. However, once these basic health and safety precautions have been taken, they can utilize the range of **methods** that both models have in common. Practical guidelines are outlined in the following chapters.

CONCLUSIONS: USING THE METHODS DIFFERENT AGENCIES HAVE IN COMMON

There are two kinds of knowledge and therefore two ways for professionals to learn: first, by reading academic research and second, from professional and clinical experience. Both routes led to similar common-sense interventions. Despite their different theoretical persuasions, the actual methods used by professionals in the field are in effect very similar. The sequence in which problems are dealt with may vary and the theoretical interpretations are clearly different but the methods derived from clinical experience are not dissimilar.

Social and health care professionals can use these methods together with generic skills to deal with drug dependence, if they have specialist knowledge and skill in motivational interviewing and/or motivating clients, together with assessment and relapse prevention skills.

The following methods and approaches can be used by health and social care professions to treat drug problems; many of these are part of their own professional training and experience and they will find themselves already equipped with the basic counselling and helping skills necessary. As with all problems, particular professions will need to consult each other for specialist expertise. For example, probation and health care professionals would need to consult social workers about implications of drug misuse for child care and social workers will need to consult doctors about health risks.

SUMMARY OF COMMON SHORT-TERM OBJECTIVES AND METHODS FOR TREATING DRUG DEPENDENCE

The methods outlined below are preceded by those for harm minimization in order to emphasize its importance as a corollary to drug treatment. It should be remembered that whilst harm minimization may be an integral part of drug treatment, there is no reason to suppose that drug treatment should be an integral part of harm minimization.

Harm minimization: before and during treatment

The treatment of drug dependence involves harm minimization before and during treatment, and again, if and when the client relapses. See previous chapter for details.

- Provide information about dangers of drug misuse and means of safer drug misuse.
- Ensure that the client has access to clean syringes and needles.

- Ensure that the client is receiving health care.
- Ensure that the client has access to prescribing facilities from a specialist agency or general practitioner.
- HIV testing and counselling facilities.
- Ensure that client has access to hepatitis inoculation.
- Social support for housing, financial problems, child care, etc.

Treatment for drug dependence

- Ask about drug problems (see Part Three).
- Motivate (see Part Three).
- Assess the client's drug misuse and other problems (see Part Three).
- Help the client to understand the links between their drug misuse and other problems.
- Help the client to weigh up the pros and cons of drug misuse and come to a decision about change.
- Encourage the client to change drug misuse behaviour and other aspects of their lifestyle.
- Help the client solve the underlying problems contributing to or maintaining drug misuse.
- Teach the client new coping skills to enable them to manage without using drugs.
- Teach the client relapse prevention skills (see Part Three).

Generic methods

- Client-centred counselling.
- Task-centred counselling.
- Relaxation, stress management and anxiety and anger control.
- Cognitive behavioural techniques for monitoring behaviour, behaviour change and relapse prevention.
- Diversionary activities and the development of new social skills and social networks.

It can be seen that most of the methods and skills involved in the treatment of drug dependence are part of the everyday work of most social and health care professionals, including basic helping and counselling skills together with health care and social support. The more specialist skills are discussed in Part Three.

The chapters in Part Three deal with the three major differences between working with drug misusers and working with other clients and outline the methods for specialist interventions in these areas. First, assessment of drug problems. Second, motivational interviewing and, third, relapse prevention.

SUMMARY

It can be seen from research findings that dependency is a recurring condition rather than a one-off illness. The combination of a physiological vulnerability (through previous periods of dependence), a psychological need and a stressful social environment, can cause individuals to become dependent on drugs at intervals throughout their lives. Social and Health Care professionals can offer help and support before, during and after these intervals.

Institutional control 9

INTRODUCTION

In Chapter 5 the views of drug misusers about life in prison were illustrated. In this chapter the research evidence is set against the views of staff in the prison as a means of identifying the major issues that arise in attempting to reconcile the priorities of the criminal justice system with those of public health. Prisons, as 'total institutions', constrain the lives of inmates in ways that drastically interrupt the behaviour, including their drug misuse, that characterizes their lives in the community before and after custody.

POLICY ISSUES

The prison service in England and Wales is the responsibility of the Home Secretary, while in Scotland and Northern Ireland it comes within the remit of the respective Secretaries of State. There are significant differences among the three in prison policy, but all of them subscribe to the recent White Paper *Tackling Drugs Together* (Department of Health, 1995). This document sets the framework for the period 1995–98 and includes a number of specific references to drug misuse in prisons. The ministerial *Foreword* sets the tone: among the main proposals to which great importance is attached is 'tough controls on drugs in prison'. This is followed by sets of objectives which include, under crime, 'to reduce the level of drug misuse in prisons' and, under public health, 'to ensure that drug misusers have access to a range of advice, counselling, treatment, rehabilitation and aftercare services'.

Later in the White Paper, it states that these broad policy aims are to be implemented by 'the reduction of drug misuse in prisons as a key performance indicator. Compulsory drug testing, improved security ... and effective treatment services will be among the measures introduced' (p. 15). And 'The launching of a new Prison Service drug strategy further commits the Service to vigorous measures against drug dealing and the misuse of drugs, as well as the provision of effective support for drug misusers.' Compulsory drug testing is now in place. The rationale for these measures is that, 'The misuse of drugs, particularly hard drugs, in prisons is an increasingly serious problem' (p. 34). There are several references to HIV and AIDS in the document but they do not include mention of prisons.

The difficulty of reconciling the crime and public health objectives has been mentioned briefly in Chapter 5. As we shall see, it runs as a recurrent theme through the statements of prison staff when they reflect on the nature of their responsibilities for the supervision of drug misusers in custody. However, before analysing these views, it is necessary to review briefly the extent of drug misuse in prisons as revealed by recent research.

DRUG MISUSE IN THE PRISON POPULATION

It is only relatively recently that research has been undertaken in an attempt to determine the extent of drug misuse among the prison population. As a consequence, there is little substantial statistical evidence available. Almost all the studies are of male prisoners (Fletcher, 1990). Many recent investigations are of particular subgroups and are largely concerned with dependent or problematic drug misuse and/or injecting drug misuse and HIV (Turnbull, Dolan and Stimson, 1991, 1992; Power et al., 1992). One of the exceptions is a recent Home Office survey of 25 custodial institutions (Madden, Swinton and Gunn, 1992), showing that 34% of a sample of 1751 prisoners had misused cannabis before prison, with 9% having misused heroin or amphetamine and 3% sedatives. The study was based on formal interviews in a custodial setting and excluded drugs such as LSD, Ecstasy and benzodiazepines. Other research indicates that many drug misusers spend time in custody and that many continue to inject in prison (Carvell and Harb, 1990; Dolan et al., 1991; Turnbull, Dolan and Stimson, 1991; Keene, 1994a; Power et al., 1992). (See also ACPO Annual Drugs Conference 1992 and Reports of the Chief Inspector of Prisons (Tumim, 1992–95). The evidence from Scotland seems unequivocal: 'The Scottish prison population contains a high proportion of prisoners who are drug users' and '... drug use is a significant part of prison life...' (Shewan, Gemmel and Davies, 1994). The study reported in Chapter 5 in a Welsh prison appears to confirm this view.

It is necessary to remember that drug misuse is notoriously difficult to quantify; drug offences *per se* are not an accurate reflection of misuse (Parole Release Scheme, 1989; Fletcher, 1990; Tumim, 1992). Only a small proportion of drug misusers will inform their probation officer of drug misuse (ACMD 1991), and less will inform prison authorities. The range and type of drug misuse examined in the research reported below is therefore wider than much previous work in which the focus has often been on particular types of drugs and drug misuse.

DRUG MISUSE IN A CUSTODIAL SETTING

The views of the drug misusers in prison reported in Chapter 5 are also reflected in many ways in the attitudes of uniformed staff. In brief, benzodiazepines and cannabis were seen as harmless and useful whereas those inmates regularly using hard drugs such as heroin and cocaine were perceived as needing treatment. Amphetamine occupied a middle ground but there was concern that inmates would share infected needles and syringes. In contrast, the non-uniformed staff resembled the non-drug-using prisoners in their views, seeing drug misuse as a homogeneous problem that required an all-encompassing solution – prevention or treatment.

CONTROL

As might be expected, the over-riding requirement of maintaining control of prisoners featured prominently in staff perspectives. The issue was not a straightforward one. Drugs were seen by many officers as a solution, not only to institutional problems but also for individuals who could not cope with various aspects of prison life. Although prison staff were preoccupied with discipline and the atmosphere in the prison, opinions varied as to the best solutions. Some, in common with the inmates, emphasized the calming effects of drugs such as cannabis and prescribed drugs such as Valium, whereas others feared the development of a black market creating a culture where intimidation and violence were more likely.

> I think that cannabis is generally accepted and tolerated. It makes inmates quiet, they have a smoke, go to bed and are quite happy and no problem to anyone... The difficulty is that if we do tighten up on cannabis, what sort of problems would this create for landing staff and for security?

Similarly, many felt that a humane regime which tolerated the misuse of prescribed and other drugs (particularly cannabis) was conducive to effective control.

> Prescribing drugs is humane for many prisoners... there is always a compassionate misuse if the circumstances dictate... all prison

officers have agreed at sometime that prescribed drugs are the humane answer when a prisoner is under stress.

Cannabis stops people needing other drugs in prison. It does seem to keep them peaceful and happy.

There could, however, be problems.

One of the objections to drugs is the wide swings of mood, it also introduces the element of unpredictability, ranging from exaggerated attempts at self-harm to assaults on peers or staff.

If a first-time prisoner, perhaps slightly subnormal, comes onto the wing he may be told to come to the hospital staff, 'tell them you want Valium and we will give you tobacco in return; if you don't do it we will beat you up'.

However, in this small local prison, large-scale dealing was not regarded as a serious problem.

Not a lot of drugs are ever found at one time, though regular amounts come in for daily misuse.

The problems of identifying misusers remain despite the introduction of urine testing. Although cannabis can remain in the urine for four to six weeks, other drugs misused occasionally are not so easily identified.

There is a 'hidden curriculum' of drug misuse in prisons, unseen because it is difficult to identify occasional misusers who were not addicts.

Most officers felt that security and control were clear priorities in prison but were ambivalent about the effects of tightening up on drugs as they felt this might have an adverse effect on control.

I think what we see is the tip of the iceberg. I'm sure that a lot of cannabis comes through visits every day. The only way to find it would be to strip search all the inmates after visits. I disagree with laxity at visits but if an emphasis is placed on maintaining the family unit, as is prison policy, controlled visits destroy the ambience. There are practical constraints, there are hundreds of visits a week.

The problem is where to draw the line between people's rights and how we tackle drug problems. It is also practically very difficult. It is easy to say we could strip search all prisoners at each contact with the outside world, but this would present an awful lot of problems and be an infringement of their rights. Is it logical to go to these lengths? These measures would completely undermine one of the institution's main priorities, we aim to maintain links with the family to improve quality of life for inmates on their release.

HEALTH

Uniformed staff tended to agree with inmates about the need for treatment for those with serious drug problems.

If someone is ill they should be treated, if there was a health unit people would be likely to identify themselves as having drug problems.

To some extent there is more need for sedative drugs for this group of people than for the population outside. We do have many people in here who are either disturbed or distressed.

Officers were nevertheless concerned about the misuse of hard drugs and particularly injecting practices and needle sharing. They were keenly aware of the risks of HIV and the difficulties of preventing the misuse of needles and syringes on the wings, emphasizing that diabetics needed needles and that home-made needles were easily constructed. Particular officers suggested that cleaning decontaminants should be available, if not syringe exchanges.

I am aware of injecting drug misuse, generally of amphetamine and benzodiazepines. I am told that intravenous drug misuse is greater than we think, I don't see it... maybe because my eyes are shut.

The indications lead to the inescapable conclusion that if somebody has reached the stage when they inject, that they would be quite happy to share works... by this I mean not just injecting together in a group but actually using the same needle and syringe. I would not sanction needle exchange in prison, but I think it almost essential that we instruct intravenous drug misusers in the hygienic misuse of works... we could also make cleaning agents available and should explore the possibilities of doing this further.

I don't think HIV in prison is made as public as it should be.

REHABILITATION

Officers were aware that drug misuse in prison had to be seen in many ways as a reflection of drug misuse in the community.

Drug misuse has mushroomed inside as it has outside, amongst young people in the past ten years.

Many inmates misuse drugs rather than alcohol outside prison where they are cheaper and this is reflected in prison where benzodiazepines are misused as if they were a bottle of whisky, they all have a swig.

The drug problem in prison is simply a reflection of the extent of the same problem outside prison. If people think that drug taking is reaching epidemic proportions in prison, this should be seen in the context of drug misuse amongst people outside, for example in pubs. Part of our problem in prison is that they are being supplied from outside by parents, siblings and relatives. The question of problematic drug misuse therefore needs to be addressed to the community as a whole and not just a specific part of it.

There is an extraordinary tolerance to drugs in some prisoners.

Some of the more innocent young offenders come in never having touched drugs and quickly learn about it.

DISCUSSION

Perhaps the most interesting aspect of these data is that a significant distinction is drawn between more and less harmful types of drug rather than their legal status. So cannabis and tranquillizers are grouped together as useful, fairly harmless drugs, whereas addictive or injectable drugs are seen as causing serious problems for individuals and the institution as a whole. The prison staff comment that 'Drug use relieves boredom and feeds addiction' sums up the difficulties: it does serve a purpose but can cause problems.

However, non-uniformed staff, in common with the general public, were less likely to distinguish between sedative drugs and drugs that cause more volatile behaviour. Violence is often seen either as a result of drug misuse or a consequence of not prescribing drugs, rather than a distinction made between the different effects of different drugs.

However, as we have demonstrated, the majority of staff, like the inmates, did distinguish between the more and less harmful drugs, recommending the use of the latter and the treatment of the former. This may be because many inmates had experience of drugs and all inmates and staff had direct knowledge of prison wings where drug misuse was common practice. Amphetamine, and perhaps also cocaine, occupied a middle ground where there was less concern about dependence but a concern that inmates would share infected needles and syringes. It should be emphasized that distinctions between hard and soft drugs or recreational and addictive drugs were often overlapping or vague. Nevertheless, it is noteworthy that the staff did discriminate, stressing that cannabis was useful in preventing inmates misusing other drugs. This contrasts sharply with recent initiatives in central policy, such as new urine testing programmes that identify safer drugs with a long half-life, such as cannabis, but which are less likely to identify harder drugs that remain in the bloodstream for a far shorter time.

As in the outside world, there is also confusion between the purposes of prescribing drugs for dependence and prescribing drugs for anxiety, depression and stress. If the respondents see drug misuse as harmless or self-medicating they will recommend its continuation in prison.

Officers discussed the effect of the presence of drugs in prison and the value of drug prevention strategies, together with the practical difficulties of balancing control, health and rehabilitative priorities. They stressed their concern at the extent of a black market, yet they were ambivalent about the possible influences of drug prevention strategies such as the recent withdrawal regimes and drug-free wings. Many saw the misuse of recreational drugs as far too widespread in the community to allow any rehabilitative effect in prison. In particular, they stressed the inherent conflict between the need for control and the rehabilitative philosophy of maintaining family links.

CONTROL ISSUES

Control of drug misuse is problematic in prison and simultaneously so is the effort to prevent it. Drug misuse can bring about problems by causing particular individuals to become volatile through intoxication or withdrawal and by providing the opportunity for a black market to develop with the violence, threats, bullying and intimidation that frequently accompany it. However, prevention or tightening up on drugs also causes control problems; it threatens the status quo and the maintenance of order and can lead to confrontation between staff and inmates. It also reduces the sedatives available to aggressive, unstable inmates and those who are anxious or depressed.

HEALTH ISSUES

As with control issues, there are health consequences of drug misuse and also of the control of drugs. The drug-related health problems include overdose, infected injecting sites, general health and transmission of HIV and hepatitis. These problems would of course disappear if it were possible to prevent drug misuse completely. In practice, they are exacerbated by driving drug misuse underground where it inevitably becomes more risky in health terms. Denying access to clean needles/syringes and decontaminants and to prescribed oral drugs is likely to increase the use of unsafe practices, including sharing infected needles and syringes. Recent research (Turnbull et al., 1992; Bloor et al., 1994) shows that HIV risk behaviours are restricted in the community to small groups of intimates: in prison the interaction with strangers is far greater and the opportunities for obtaining clean equipment extremely limited. Although fewer people use needles and syringes in prison, those that do are therefore at increased risk of contracting HIV or hepatitis and transferring it to their local communities on their release.

Health problems in prisons could be approached in the same way as in the community, with a regime designed to reduce the amount of drug-related health harm, rather than prevention. People are even less likely to stop misusing drugs in prison than in the community as stress, anxiety and boredom are greater in custody and therefore they may be prepared to take more health risks (Carvell and Hart, 1990; Turnbull, Dolan and Stimson, 1992; Power et al., 1992).

REHABILITATION ISSUES

As with control and health problems, the priorities of rehabilitation also conflict with a policy of drug prevention. It is impossible to rehabilitate inmates to fit better in the community when the outside world itself has the problem and it is of course unrealistic to try to change the community by making changes in prison. The practical consequences of drug prevention strategies in prisons involve tightening up visits, increasing surveillance cameras and searching visitors. Such measures engender bad relations with visitors and weaken links with the community for individual inmates and the prisons as a whole.

In summary, the effects of prison as a 'total institution' on drug misuse point up many of the issues that face professionals more generally. The problems related to drug misuse assume particular significance in the prison environment. Control is exercised at the cost of providing appropriate support in public health terms. The balance between them is, of course, exceptionally difficult to achieve.

Part Three
Practical Guidelines

INTRODUCTION

The misuse of drugs is now so widespread that most social and health care professionals are likely to have drug misusers on their caseload. Parts One and Two have given an understanding of the problems and an overview of the literature in the field. Part Three will address the practical implications of these issues for generic professionals.

In the Introduction to this book it was pointed out that social and health care professionals have reported a lack of confidence in their ability to deal with drug problems and an ambivalence about their role or responsibilities in this area. They often do not know where to go for help or where to refer clients. They need to know which of their own professional skills are appropriate, what specialist skills are necessary and what alternative support is available from other professional groups and agencies. In order to enable them to deal confidently and competently with drug problems these initial questions need to be addressed. The answers are outlined in Chapters 10 and 11, where the relevant health and social work skills are identified for each professional group.

It is also clear from Part Two that generic workers have knowledge and experience of the problems associated with drug misuse and the practical skills necessary for working with drug misusers. Nevertheless, drug users do present additional problems. Once professionals have decided to work with drug misusers, the following practical issues arise:

- How do I ensure the safety of my clients?
- How can I motivate my clients?
- How do I make an assessment?
- Should I use harm minimization and/or treatment methods?
- What about aftercare and preventing relapse?

These questions will be resolved through the development of a sequential helping process incorporating the necessary generic and specialist skills. The answer to each question will provide part of a structured programme incorporating safety precautions, motivational interviewing, assessment, harm minimization and/or treatment and aftercare maintenance, in that order.

- Safety is concerned with giving clients information and contacts.
- Motivational interviewing is a specialist skill.
- Assessment and care plans require generic skills and specialist knowledge.
- The aims and methods of harm minimization and treatment are discussed in Part Two.
- Aftercare and relapse prevention include both generic and specialist skills.

Chapter 10 will describe this sequential helping process for working with clients from the first contact through to aftercare and relapse prevention. It includes deciding which intervention is appropriate for which clients and when it is likely to be effective. The chapter is designed to inform readers how to use their own professional skills in conjunction with some specialist methods, within a sequential helping process designed specifically for work with drug misusers. It will outline the use of generic and specialist skills which form an integral part of this process.

Chapter 11 will then describe the different roles, responsibilities and skills available from a range of professional groups and identify where clients can get different forms of help and support. Professionals can use this chapter as a practical guide to working with other agencies, whether they decide to refer directly, co-work or simply seek advice and information.

A sequential helping process

10

INTRODUCTION: NEW BEGINNINGS AND NO END

Many of the methods discussed in previous chapters will be familiar to social and health care workers and generic counsellors. This chapter is concerned with encouraging generic workers to use the skills they already have in a specialist setting. To do this, a framework is outlined and new specialist skills introduced. Effective work with drug misusers requires the professional to work through a sequence of steps and tailor interventions at each stage. The stages at the beginning and the end of this sequential process assume particular importance in this field; at the outset these are health and safety precautions and motivation and, at the end, maintenance of change. Specialist methods are necessary only at these stages and these will be examined below in the context of the sequential helping process as a whole.

Although many of the methods and skills necessary for working with drug misusers are similar to those used in all health and social care work, the helping process itself is different for three reasons:

1. clients are often uncertain about reporting drug misuse;
2. clients are often ambivalent about changing their drug misuse;
3. clients often relapse within a short period.

For effective working, these differences must be taken into account.

The need for client motivation before assessment

This is done by changing the order and emphasis of the helping process. The initial stage is not assessment, as would be the case with many clients, but motivation and the early emphasis is on development of a working relationship with the client.

The need for health care maintenance support for those who do not want to change

This is followed by assessment and the negotiation of aims. In this field the client has a choice of two aims, either maintenance (harm minimization) or change (treatment). The choice must to a large extent remain with the client, though the professional may influence the client to move from harm minimization support to gradual changes in drug misuse and/or lifestyle.

The need for aftercare maintenance

Work with drug misusers is coloured by the high proportion of clients who relapse. It is therefore necessary with this client group to spend a good deal of time and effort on the aftercare period. This involves completing assessments and careplans for the post-treatment maintenance period as an integral part of the treatment process. These plans should contain clearcut procedures in the event of crisis or relapse.

THE RELEVANCE OF SPECIALIST THEORY AND PRACTICE FOR GENERIC PROFESSIONALS

Part Two described the theory and practice of prevention, harm minimization and treatment for drug misusers. The major conclusion was that the initial emphasis of any intervention should be health and safety. The first priority is to ensure that any drug-misusing client has access to information about safer drug misuse and access to the means to achieve it (see Chapters 6 and 7).

Once these basic health and safety precautions have been taken, the professional can then consider whether to work alone with the client's drug problems or to collaborate with other agencies. If the former, they can use the range of methods that specialist drug workers themselves employ to maintain and support clients and to help the client to change behaviours and lifestyle. These methods and the relevant skills are part of the basic training of most social and health care professionals.

In contrast, there is no pressing need for generic professionals to adhere to any one specialist theory or interpretation of drug dependence. As can be seen from the earlier chapters, many drug misusers are unlikely to be dependent and those who are may not want to stop anyway. If clients are dependent and want to stop being so, the controversy regarding abstinence may still be largely redundant, as it seems effective to offer help both before and after abstinence; either way, specialists often use the same methods to deal with similar problems. It is simply necessary to be flexible and adapt the sequence of events (and possibly the theoretical model) to fit the client's own beliefs and choices and so increase the likelihood of cooperation.

The status of a particular theory in clinical practice is, in effect, dependent on its usefulness. In this case several competing theories are all useful for therapeutic purposes. If clients believe they have an addictive disease it may be a useful belief in helping them give up drug misuse; if they believe they can control their dependence, it will also be useful for some clients. It should be remembered that the most respectable of scientists will use untested theories when they help make sense of a situation and even maintain several contradictory theories at the same time if they are practically useful (e.g. the wave and particle theories of light).

It is only when there is a need to refer a client to one or another type of agency that it becomes important to know about the theories and philosophies underlying specialist practice (e.g. a harm minimization outpatient service or a Twelve Step residential service), in order to determine if they are appropriate to the needs of each particular client. The theories of particular agencies may conflict with the beliefs of particular professionals, but this conflict should be of little consequence if the agency beliefs are in accordance with the beliefs and values of the client. It is the match of client to agency that is important, not the match of professional to professional. If a particular theoretical interpretation is useful to a client, it can be counterproductive for professionals to limit their client's choices to their own theoretical orientation.

Once armed with a basic understanding of drug problems, the professional can use specialist drug agencies as a source of specialist expertise and advice, rather than a place to refer all clients who use occasionally misuse drugs. Drug agencies can provide advice and support if the situation become problematic, together with the option of syringe exchange and health care.They will carry out initial assessments, offer detoxification facilities and long-term prescribing. It is up to both the generic professional and their client how many of these services will be taken up.

THE SEQUENTIAL HELPING PROCESS

The sequential helping process incorporates the core elements of work with drug misusers within a structured framework.

The essential first step for all clients is the provision of **information and health and safety precautions** (safety net). This is followed by the task of motivating some clients to misuse drugs more safely or to change their drug misuse in some way. Motivation is a primary task when working with drug misusers, as many may not wish to seek help or to change.This stage involves building a good relationship with the client before asking questions about possible drug misuse problems. In contrast to most work in the helping professions, motivation comes before assessment, as clients who are not motivated to deal with drug issues are likely to resist accurate assessment.

Following effective client motivation a comprehensive **assessment and care plan** will include the client's problems, skills and resources together with a review of appropriate assistance available through health and social care agencies. It should include assessment of the client's ability to change and also their ability to maintain any changes in behaviour, whether these be safer drug misuse or stopping dependence. The assessment will determine the intervention (**harm minimization and/or dependency treatment**) and it forms the basis for agreement of goals and the development of a careplan. The careplan will include the basic services and methods described in Chapters 7 and 8, together with planned **aftercare maintenance** including continuing support, relapse prevention skills training and a clearcut procedure for immediate intervention in the event of relapse.

NEW SKILLS AND METHODS

Whereas most social and health care professionals are familiar with counselling and helping skills, some practical guidance is necessary concerning the introduction of two new elements into general counselling practice: preintervention motivation and postintervention maintenance or relapse prevention. These elements are critical when working with drug misusers as both address the issue of high failure rates in this field. Clients are unlikely to respond in the first place if assessment and/or counselling takes place when they are unmotivated and they are likely to deteriorate or relapse if there is no maintenance or relapse prevention.

The research literature has demonstrated that initial relationship building and ongoing minimal aftercare maintenance are of particular relevance to drug misusers for harm minimization purposes. As we have seen, two new specialist methods have been developed to deal with dependent drug misusers before and after treatment – The motivational change model of Miller and Prochaska and DiClemente and the relapse prevention model of Marlatt and colleagues. Both have been developed specifically for work with substance misusers as a whole and can be used independently of any particular model of dependence.

THE SEQUENTIAL HELPING PROCESS: NEW SPECIALIST METHODS AND OLD GENERIC SKILLS

The majority of work with drug misusers is not concerned with changing their drug misuse but with providing safeguard and support until they choose to cut down or give up. This help consists of the distribution of information to the general population, followed in some cases by the identification of problems and the provision of social and health care resources to help clients cope and prevent deterioration (harm minimization). It should be emphasized that, although the interventions in the following framework form a sequence, there is no necessity for all drug misusers to progress along it, as the earlier steps can be used independently of those that follow. For instance, it may be decided that education or provision of pharmacist contacts is the most appropriate intervention. It is quite possible that this is all that is necessary or practicable.

If, however, the professional decides to embark on a working relationship to deal with drug problems, the total sequence becomes necessary.

If a client progresses through the motivation stage to assessment and harm minimization or treatment, these interventions should lead inevitably to continuing aftercare maintenance. In effect, harm minimization does not have a 'treatment outcome' as there is a continuing possibility that the client will take risks if services are removed. Likewise, with treatment, although there is an 'outcome', it is the following two years of aftercare support that are most critical in maintaining treatment change. If no such aftercare support is available, research indicates that clients will relapse to old behaviours. The answer is to develop a system of minimal maintenance support which not only guards against relapse but allows for rapid reestablishment of contact in the event of crisis or relapse.

ASSESSMENT AND HELP WILL VARY DEPENDING ON THE STAGE OF THE SEQUENTIAL HELPING PROCESS

No assessment is necessary for the safety net stage; it could indeed be counterproductive. The help given is largely information and contacts for easy access to services and resources if necessary.

For those clients seen as at risk, assessment of health care needs is necessary. Help may include the provision of needles and syringes prescribed drugs and health care support. Counselling for therapeutic change is not necessary at this stage unless specific psychological problems are identified independently of the drug misuse, or chaotic drug misuse points to a need for cognitive behavioural change.

The majority of clients will want to give up drugs at some time. It may be unproblematic for many, but some who are dependent will need help to change. For those clients assessed as dependent, further assessment is essential to determine whether change is likely (Does the client want to change?) and whether it is sensible (Is the client capable of changing and do they have the ability and resources to cope after treatment?). It is only when there is a likelihood of, or immediate necessity for, change and when clients are assessed as capable of maintaining change that a therapeutic change process should be undertaken. Even then, it is pointless to undertake this process without an aftercare programme. If a client relapses after treatment or decides to have treatment after maintenance for harm minimization, there is a need to progress through the sequence again to ensure the motivation and accurate assessment necessary for effective treatment and post-treatment aftercare.

The harm minimization sequence:

• Motivation to reduce risks and/or seek help;
• Assessment and assessment of needs and resources for safer drug misuse;
• Harm minimization;
• Aftercare maintenance support to avoid risk taking in the future.

The dependency treatment sequence:

• Motivation (motivational interviewing method);
• Reassessment of needs and resources for drug-free life, during and after treatment;
• Treatment intervention (cognitive behavioural and/or medical prescribing schedules to reduce or stop drug misuse);
• Aftercare and relapse prevention skills and development of social support systems.

GENERIC COUNSELLING SKILLS

Counselling skills of some form or another are part of the basic training and experience of most social and health care professionals. The term 'counselling' is rather vague and can mean anything from chatting to a client to a structured programme of behavioural change or intensive psychotherapy. Most structured counselling programmes use a mixture of psychodynamic client-centred methods with cognitive behavioural techniques. The skills developed from these programmes are outlined below, but for practical guidebooks and exercises the following are useful: Truax and Carhuff (1967), Kanfer and Goldstein (1986) and Egan (1990).

The important distinction between different kinds of counselling is the aim. Many are designed to achieve therapeutic change, such as the cognitive behavioural and therapeutic programmes for dependent drug misusers discussed in Chapter 8. However, only a small proportion of drug-misusing clients are likely to respond to such programmes. It is therefore necessary to distinguish between those counselling techniques aimed at change and those simply designed to build constructive helping relationships and to clarify problems.

Table 10.1 The sequential helping process

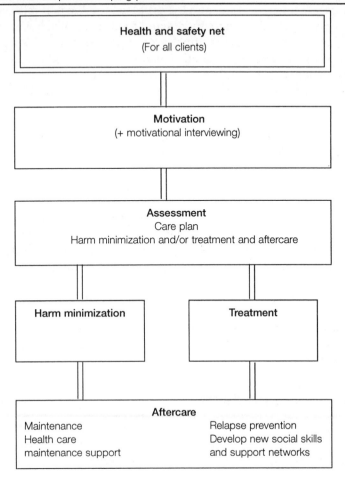

Much has been written about non-directive, client-centred counselling – how to listen and help the client to understand their problems. This is useful for building relationships before offering support and practical help. It can also result in changing the client's attitudes and in their desire to change their behaviour. If the aim is to bring about actual changes in behaviour through counselling techniques, it should be noted that these particular measures are not sufficient.

Much less has been written about translating changes in attitudes into changes in behaviour. The therapeutic guidelines in Chapter 8 addressed this problem by offering specific cognitive behavioural programmes (psychological approach) or straightforward step-by-step guidelines and social support networks (disease model). They had in common a clearcut definition of problems and practical solutions. For generic professionals this difficulty is best addressed by not only helping the client to understand the problem and the possible solutions, but also by assisting them to understand the ways of achieving the solutions. The essence of goal setting is to clearly define the sequence of steps necessary to achieve the aims; that is, to define exactly how a goal will be achieved.

Qualities of a good counsellor

Much research has been undertaken to identify which factors effective counselling sessions have in common. The most significant is the counsellor. The qualities of a good counsellor have been variously listed as warmth, genuineness, empathy, respect, caring, together with the use of feedback, confrontation, self-disclosure, commitment and careful listening. Carl Rogers has suggested that warmth, empathy and unconditional positive regard lead to a more genuine and authentic relationship; however, he was not convinced that these qualities could be learned. Gerard Egan (1986) points out that, though some personal qualities lead to a more effective working relationship, many can be seen as skills and can be learned and practised. Particular emphasis is placed on empathy and the ability to communicate understanding of the client's feelings and point of view, followed by a stress on genuineness, unconditional positive regard, respect and a non-judgemental attitude.

Skills of a good counsellor

The ability to demonstrate the qualities of empathy, respect and genuineness can be improved by good listening and feedback skills and the ability to confront clients and negotiate contracts.

Collecting information

Careful listening or attending is important. Body language is significant, together with appropriate eye contact and smiling, nodding, etc. It is important to concentrate, ask open-ended questions, identify significant issues, ask for specific detail and prevent too many deviations in the conversation. An integral part of collecting information is to remain silent for much of the time and allow the client to fill awkward pauses.

Feedback

An essential part of counselling is accurate empathic feedback. This involves the ability to paraphrase what the client has said (rephrasing or active listening) and to be specific and identify significant issues. The object of paraphrasing is to avoid misunderstandings and clarify issues. it also serves as clear indication to the client that you are listening and taking seriously what is said. Feedback also includes empathic responses in terms of attempting to understand what the client may have felt in certain situations and asking specific questions about emotional states. Positive comments and questions are an integral part of feedback but it is important to be brief. Summarizing is similar to paraphrasing but can be used specifically to clarify what has occurred so far and to determine what issue or direction should be pursued next.

Challenging and confrontation

Challenging a client's point of view can only be constructively attempted after the initial stages of listening and feedback. It involves introducing a new perspective on the client's problem or situation, usually taken from your own frame of reference or that of a significant other. It includes suggesting possible patterns or themes and pointing out discrepancies. Confrontation often requires self-disclosure or giving the client information. Information giving is a skill in itself which, at its most effective, uses simple words, short sentences and explicit categorizations. It is necessary to be concrete and specific, avoid jargon, summarize and ask for feedback from the client.

Clarifying the concrete steps necessary to achieve goals

The counsellor should clarify the practical means of achieving goals, step by step, so the client not only knows what they are going to do, but how they are going to do it. They need to know who will do what, what will happen when each stage is completed, what will happen if a goal is not achieved and what will happen in the event of relapses. The more specific and concrete each step can be, the better. So, for example, the main goal may be to stop sharing injecting equipment outside the home. The practical steps to achieving this goal may be to ensure that injecting equipment is always available at home, by developing links with local pharmacists and the local syringe exchange; by ensuring that drugs will be available at home when needed; and by the provision of safe disposal tubes.

Basic counselling within the sequential helping process

Counselling skills can be used at different stages of the process. First, for the motivation stage, when attempting to understand the client and develop a working relationship; second, at the assessment stage, when understanding problems and needs; third, at the care plan stage, when helping the client to work out what choices are available, make decisions about which to choose and develop strategies for achieving these goals, fourth, for therapeutic change in drug treatment.

Client-centred listening skills are the essence of motivation and assessment, whereas more directive and confrontational skills are an integral part of therapeutic change. Cognitive behavioural interventions are useful for both therapeutic change and relapse prevention.

In the sequential helping process, initial client-centred, reflective and non-directive counselling precedes the more task-orientated later stages. If there is a need for therapeutic change, the therapeutic component (whether cognitive behavioural or psychodynamic) follows initial non-directive listening. Finally, the emphasis on maintenance of change and relapse prevention requires further cognitive behavioural counselling skills.

SAFETY NET

The first intervention consists of education, providing a safety net of information about safer drug misuse, HIV and access to services such as syringe exchange. The aim is to prevent unnecessary harm.

This first phase establishes life-saving precautions for all clients and must be carried out for everyone, before anything else. Most drug misusers will not want or need anything more than these simple safety precautions, information and access to safety equipment such as needles and syringes. This phase can also be seen as the pre-care plan for those clients who do decide to seek help.

In metaphorical terms, clients can be seen as travellers in foreign parts. They do not want to tell you where they are going or how they will travel, so you offer life jackets, parachutes, safety belts, etc. and they make a choice of what they need privately without having to discuss the details with anyone else.

This should also be seen as a pre-primary, preassessment stage of damage limitation. While completely distinct and separate from health care and treatment itself, it is a necessary precondition of any intervention. Giving information leaflets and contact names is less intrusive and more efficient than other interventions. It involves giving general information to protect the individual and others; as for tuberculosis, it is a public health or health promotion measure. This information should be made generally available to all clients, as it is difficult to identify which are misusing drugs (though you can be fairly sure some of them are). It is better to assume that they do misuse drugs rather than not, to normalize drug misuse and talk about it as if it is a part of everyday life.

This information should include the range of services and products available, from syringe exchanges to health care and prescribed drugs. It should also provide contact points at different levels of confidentiality, from telephone to letter, from outreach to drop-in to formal appointment, so that the clients can decide how much they are prepared to reveal to whom.

It may well be you that the client chooses, in which case you must make it clear what your own limits are in terms of confidentiality and what services you can offer that drug misusers might find useful. Perhaps most important here is that you find out what your employer's policy is, so that you will know what support you have. Similarly, if you find out information about the services you and others provide, you are in a better position to offer a range of options. It is therefore necessary to be familiar with information about safer drug misuse and local services. It is also a good idea to practise giving life-saving information in difficult situations.

The safety net should be seen as a public health or health promotion measure designed for whole populations rather than aimed at individual clients. Many drug misusers will benefit from basic information about safer drug misuse and local service provision and some will ask you for further help. It is also useful to target more specific help at those with apparent drug problems.

Screening

Screening for alcohol problems is well advanced in many hospitals and primary health care settings internationally, the appropriate instruments often leading to brief interventions from basic literature to short counselling sessions (for example, the WHO screening questionnaire AUDIT – see Saunders and Aasland, 1987).

This form of intervention is not so far advanced for drug misuse, where screening programmes are generally introduced as a control procedure rather than as a health care intervention, for example in English prisons and the probation service. Although the utility of such schemes seems obvious (Weisner and Schmidt, 1993), the control issues make such programmes controversial (Keene, 1994a, 1995).

MOTIVATION

MOTIVATION TO SEEK HELP

The importance of motivation has long been understood by professionals working with drug misusers. Although there is little research in this area, as the phenomenon is difficult to measure, lack of motivation is often cited as a primary reason for treatment failure.

Motivating the client is the first step of the sequential helping process. it necessitates developing a working relationship or 'linking the client in' to the helping process. This should take place before assessment in order to ensure cooperation. Linking in and motivating the client means selling the service you are providing, saying to a client 'These things are available if you have a problem and want to take advantage of them'.

The main aim at this stage is to build understanding and trust, by listening to the client's view of their problems and being clearcut about your own professional position. It is particularly necessary to develop the relationship by clarifying what you can and cannot do. This stage should be designed to remove any blocks to client help-seeking by establishing rules of confidentiality and clarification of your professional responsibilities regarding children or criminal activities.

The first thing is to greet the client in a non-judgemental way and make them feel at ease. This is particularly important with drug misusers who will expect to be judged adversely, if not treated badly, because of their drug misuse.

MOTIVATION TO CHANGE

In recent years the use of motivational interviewing (Miller, 1983) and the change model (Prochaska and DiClemente, 1983) in the drugs and alcohol field has become widespread. The rather more traditional Twelve Step model also focuses on the importance of motivation and the change process, structuring therapeutic change through twelve separate stages. There has, however, been little research into motivational factors or the use of this type of approach. The exception is perhaps the work of Murphy and Bentall (1992) who carried out a factor analysis and revealed three main components of motivation to withdraw from heroin: private affairs, external constraint and negative effects of heroin misuse.

Using motivational change models for drug misusers

The best known authors on the importance of individual motivation in dependency treatment are Prochaska and DiClemente, who outline a process of individual change. Whilst a model of process can be used alongside a range of different theories of dependence, it is most closely aligned with a cognitive behavioural approach and the interventions suggested are concerned with cognitive and behavioural change.

The authors developed a model of change processes from their work with smokers, incorporating the notion of a sequential change process. They then modified it to develop a practical motivational counselling method for work with drug and alcohol users. The model consists of five stages (Prochaska and DiClemente, 1983):

1. pre-contemplation
2. contemplation
3. action
4. maintenance of change
5. relapse.

Precontemplation

The model predicts that people in the precontemplation stage of the change process will respond less well to information and advice than people in the contemplation stage. It is therefore suggested that counselling for clients in this stage will be more effective if it is designed simply to raise the client's awareness of the possible issues connected with drug misuse (rather than attempting to offer information and advice), in order to move them to the next stage. Intervention at the precontemplation stage is seen as less constructive than at the contemplation stage.

Contemplation

At the contemplation stage information and advice are more likely to be assimilated and utilized. The interventions suggested at this stage include cognitive techniques such as awareness raising, information, decisional balances, re-evaluation of life events and drug-related problems. Clients at this stage are more receptive to alternative observations, interpretations and direct confrontation. Behavioural assessments and analysis of the functions of drug misuse as a coping strategy can also be useful here.

Action

This stage is sometimes split into two, the decision stage, including a firm commitment to change, which can be of very short duration, followed rapidly by the action stage, where treatment interventions are most effective.

Maintenance of change

This stage follows treatment and indicates that achieving change involves ongoing maintenance.

Relapse

The authors have demonstrated that smokers progress through this whole cycle on average about three times. They have therefore incorporated a relapse phase into the overall model.

Although the later stages of maintenance and relapse are integral parts of the model the emphasis is on the precontemplation and contemplation stages and the later stages are less well developed. The model is therefore particularly useful for initial assessments and appropriately timed treatment interventions, but less helpful for working in the maintenance and relapse stages.

Uses of motivational interviewing

This work was paralleled by W.R. Miller's development of the concept of motivational interviewing. His work was published in the journal *Behavioural Psychotherapy* in 1983 and again (with Sovereign and Krege) in 1987.

It has the following key principles:

- the client must be accepted in a complete and unconditional way;
- the client is a responsible person;
- the client must be ready for change and not forced into it by the counsellor;
- the goals and the forms of treatment must be negotiated.

The stages are:

1. Assessment, to determine if a client wants to change or is 'ready' for change;
2. increasing motivation by encouraging the client to recognize that the drug misuse is a problem and/or causes other problems;
3. helping the client to understand the functions and effects of the drug misuse itself in order to weigh up the advantages and disadvantages;
4. monitoring progress through the cycle of change from precontemplation, contemplation, decision and action to maintenance of change.

Miller worked with clients with alcohol problems and argued that it was possible to increase motivation and reduce denial among them by placing emphasis on the 'decisional balance' or helping clients to weigh up the pros and cons of their drinking; by clarifying the range of choices open to them and by highlighting the discrepancies between their beliefs and behaviour. (There are similarities here between the decisional balance and the more confrontational 'breaking denial' of the Twelve Step approach, although Miller's work is not often compared with it.) The basic underlying premise is that change is a process with clearly definable sequential stages. The helper can be more effective if they identify which stage the client is at and make the appropriate intervention for that stage.

Limitations of the motivational change model for recreational drug misusers

Motivational interviewing is only for those clients who need to change: it selects from this group those who want to change. This model has been shown to be useful for specialists working with heavy drinkers (Davidson, Rollnick and MacEwan, 1991) and for smokers. However, both heavy drinkers and smokers are concerned with the long-term health consequences of heavy use. 'Readiness to change' is therefore a useful concept and change in drinking or smoking is the logical step. When these concepts are transferred to drug misusers this is not necessarily the case. Recreational drug misuse may be more like social drinking and risky drug misuse more like drunkenness than heavy drinking. The consequences of social drinking and drug misuse may not be sufficient to warrant change.

For example, the clients apparently at the precontemplation stage may not consider their drug misuse a problem, because it is not (see Chapter 2, where clients tested for 'readiness to change' did not understand the concept, as they did not see how their drug misuse could possibly be problematic). In contrast, the client whose drug misuse is problematic but who does not want to change cannot simply be equated with the precontemplation stage of Prochaska and DiClemente as many drug misusers have actually contemplated the risks and weighed up the pros and cons and already made the decision to carry on misusing drugs.

An individual process of change may therefore actually involve several options at any one stage, rather than simply progressing from one to another along a continuum. As Nick Heather (1991) remarks, 'one wonders whether the process of change is quite so smooth and unidirectional as it seems to suggest or whether change is often more fluctuating and inconsistent'.

Despite the limitations of using motivational interviewing and the change model for clients who do not need to change, they are very useful as assessment tools, allowing professionals to determine if the client is 'ready' and 'contemplating change' or not and determining which interventions are likely to be most effective at the precontemplation and contemplation stages. They also help make the crucial distinction between those who need harm minimization alone and those who are more likely to respond to treatment.

WHY A PROCESS MODEL IS USEFUL

A process model of change enables professionals to understand two important factors: that clients may not be ready to change and that if they do change, they will pass through a process (several times) and interventions should therefore be tailored to their current stage.

It is interesting to examine why a process model of motivation and change should achieve such popularity among professionals in the absence of research evidence of its effectiveness. Academics and other researchers have tended to focus on outcome studies of particular interventions, as a scientific methodology is designed for this purpose, rather than study the individual processes of change. The practical relevance of a model of change processes has highlighted the limitations of this approach.

However, as discussed in Chapter 8, clinical data and experience are based on a different kind of knowledge, that of individual cases over time. In other words, clinical experience resembles qualitative data; both provide the necessary indepth understanding of process, individual differences and subjective interpretation. Clinical experience indicates that a change model is useful for understanding an individual's progress over time and that motivational interviewing is a useful therapeutic tool.

The concepts used in the Twelve Step approach come from a different conceptual framework and as such are not accessible to some professionals and clients. However, the essence of this approach is that a person with experience of the therapeutic change process guides an inexperienced person through stages in the correct sequence. For example, the initial step, 'breaking denial', is designed to increase motivation in the same way as moving from precontemplation to contemplation by developing a greater awareness of the problems caused by drug misuse.

ASSESSMENT

Having established a working relationship, the most important task is to find out how urgent the client feels their problems are and if they need immediate help. Only after the client's own understanding of their problem has been clarified is it appropriate to start the structured assessment to establish the categories of the client's problems and consequent needs. It is important to stress the influence of comprehensive assessments prior to each intervention. The factor most likely to cause failure of drug treatment is the inadequacy of assessments prior to clients making demanding life changes. One of the most likely factors accounting for HIV transmission is the inability of professionals (and clients) to determine what the client actually needs to stay safe. The amount of support and help required is often underestimated as a consequence of incomplete assessments.

WHY DO DEPENDENT DRUG MISUSERS SEEK HELP?

Before listing the various problems drug misusers face, it is worthwhile to ask why they seek help. The respondents in Chapters 3 and 4 varied considerably in the kind of help they wanted, from clean needles and syringes to basic health care and a prescription. Similar questions were asked of dependent misusers by the Drug Indicators Project (ISDD, 1989). The answers resembled those of dependent drug misusers attending agencies and differed from those not dependent and/or not attending agencies. The researchers interviewed a group of 240 dependent opiate misusers (120 from agencies and 120 drug misusers not in contact with agencies), using the snowballing technique of contacting respondents through social networks of drug users. The criteria for inclusion were that they misused drugs such as heroin, cocaine, amphetamines or barbiturates at least 3–5 days a week for a minimum of two years. They found that the opiate-dependent agency attenders reported d _ _ dependence as the main reason for needing help (79%) and a third sought help for 'psychological and emotional state' and health.

The researchers also asked agency clients about their perceptions of the process leading up to seeking help. They found that a growing dissatisfaction with lifestyle and accumulating problems over time were the most important factors, followed by an acute crisis and/or a pattern of long-term recurring problems. These data highlight the relevance of a wide range of physical, psychological and social problems but perhaps more importantly, they demonstrate that the reasons why people seek help are related to how they perceive agencies and what type of help they think is available. For example, clients may be less likely to seek drug agency help for health problems or depression if they expect therapy for drug dependence. This was apparent when interviewing drug misusers in agencies. Many saw their problems in terms of dependence and therefore saw the solutions in therapeutic terms as this was the type of support available. In contrast, those drug misusers who did not attend agencies saw their drug problems as health related rather than psychological and their social problems as independent of their drug misuse. Drug misusers tended not only to see drug services as orientated towards physically dependent drug misusers, but also as designed specifically for heroin addicts and therefore no use for drug misusers with other types of drug problem (such as heavy chaotic amphetamine misusers). This perspective reflects the actual emphasis on methadone prescribing in many inner city drug agencies.

PURPOSES OF ASSESSMENT AND CARE PLANS

The overall purpose of assessment is simply to determine what help and which interventions are appropriate at which stages in the helping process. Information that does not serve this purpose is largely irrelevant.

- The motivation stage of the helping process can include a motivational interviewing assessment which enables the professional to determine whether or not the client wants to change and is ready for change.
- An initial assessment of client problems, needs and resources (skills and social support) allows the professional to identify whether or not a client has drug-related problems and if so, whether there is a need for support.
- An assessment of drug misuse category enables the professional to determine what is needed in terms of information, harm minimization and/or treatment.
- Assessment for aftercare informs the professional whether the client is capable of maintaining change, in terms of motivation, ability and social support networks and consequently what support they will need to do so.
- Ongoing assessment enables the professional to monitor client progress and intervene rapidly if necessary.
- Assessment as a whole also serves to inform the client of their needs, functioning as a feedback mechanism about the extent and nature of problems, necessity for change and progress.

Assessment can be a long and complex procedure. Each type of assessment should be limited to what is both necessary and appropriate to the demands of each stage of the helping process. Assessment should follow client motivation and choices. If clients choose help, then they should first be assessed for basic health care maintenance and minimization of harm. Following this minimal initial assessment, they can be assessed for drug dependence. If they are assessed as dependent and they choose to change, only then should they be assessed for treatment and aftercare support.

Assessment can therefore be seen as a continual process, particularly in drug treatment where clients are working towards changes in behaviour and lifestyle. If a client is changing behaviour or working towards practical short-term goals, monitoring of progress on a regular basis allows for the identification of client readiness for decisions and action and also the renegotiation of short-term goals on the basis of progress made or failure to achieve these goals.

Finally, assessment can also be used for clinical or 'therapeutic' purposes: first, to encourage the client to recognize that the drug misuse is a problem and/or that it causes other problems; second, to help the client to understand the functions and effects of the drug misuse itself in order to weigh up the advantages and disadvantages; and third, to monitor progress.

Care plans

Care plans use the initial assessment of individual needs together with an assessment of individual potential and resources to decide on what support is necessary. The care plan can be designed to ensure that the client's situation does not get worse, they do not take unnecessary risks and/or that the client can change and maintain change.

In the same way that knowing what drug a person misuses does not necessarily indicate whether or not they need help, knowing the individual and social difficulties of a client does not necessarily indicate that they need professional assistance. It is quite possible that clients also have the potential and resources to deal with their own problems. Assessment is therefore concerned with estimating both the extent of problems and the resources already available to the client.

DECIDING WHICH DRUG MISUSE CATEGORY PROVIDES THE BEST 'FIT'

In order for both professional and client to gain greater understanding of the client's problems it is necessary first to clarify the actual types and patterns of drug misuse and to explain the situations and circumstances that lead to this misuse and its consequences. It should be remembered that physical dependence is only one of a range of different problems that drug misusers may have. It is a serious mistake to focus only on signs of dependence for identification purposes and assessment procedures. Knowing that a person misuses drugs does not necessarily indicate whether or not they have problems and it is therefore necessary to ascertain first if the individual has any or is likely to have any. If so, it is then possible to move to the next stage of assessment.

While all problems associated with drugs can be categorized conveniently as 'problematic' drug misuse, this definition is of little practical use to the professional as it gives no information about the particular problems and needs associated with different types of drug misuse. Assessment is concerned with identifying different categories of drug misuse and drug misuser, in order to distinguish the problems and so clarify the needs associated with each category. There is a need to have a clearcut idea of the risks and problems associated with each form of drug misuse. This will lead logically to an understanding of the needs, which can then be assessed within the context of the individual and agency resources available.

As has been discussed throughout the book, there are many difficulties in determining which category is most useful for defining different types of drug misuse. The use of the categories defined by the earlier data in Part One and the research in Part Two allows the identification of characteristic problems and therefore the provision of appropriate services for the clients in each group.

Drug misuse has been classified into three categories:

1. recreational (including experimental),
2. risky,
3. dependent (including self-medication).

It is important to emphasize that these are not mutually exclusive nor permanent and neither are the categories clearcut. It is also difficult to distinguish between them at any one time.

Two categories at once

In practical terms it is important to emphasize that these categories are not mutually exclusive. It is quite possible that any one client's drug misuse will fall into more than one category, for example, both risky and dependent. It is then necessary to consider the problems and needs associated with both categories. In effect, a client can be several types of drug misuser at the same time; in the same way that a person can be both a schizophrenic and a depressive, a drug misuser can be both a risky misuser and a dependent misuser.

Different categories at different times

It is possible that clients may fall into two categories simultaneously and also that they are more than likely to fall into different categories at different periods in their lives. This is an extremely important concept as it emphasizes that drug problems change over time. In the long term this may mean that a small proportion of drug misusers move along a continuum from recreational to risky and eventually to dependent misuse. However, this is unusual and it is much more common for different categories of drug misuse to be temporary phenomena and to change at various times in a person's life.

Short periods of problematic use at different times in the life course

In the short term, these variations in patterns of drug misuse mean that one person will misuse drugs in chaotic, risky ways for short periods of time only, gaining and losing control in response to different circumstances. For example, a young mother may respond to the stress of a broken relationship by misusing drugs in a chaotic way but will quickly get her drug misuse back under control if she receives support and help at the right time.

Whilst it is clear that drug problems get worse as more drugs are misused in a more uncontrolled manner, it is probably more accurate to say that individuals will have short periods of problematic misuse at different times in their lives.

Recreational or risky?

The difficulty here is to determine when drug misuse becomes a health hazard and/or causes psychological and/or social problems. If there are few or no risks being taken there is no rationale for intervention, which can be worse than useless, stigmatizing clients unnecessarily. It also wastes professional time and resources.

There is some guidance on safe and unsafe use in the alcohol field, where health promotion authorities and the British government have set (variable) safe and unsafe limits on amounts of alcohol consumed where it becomes statistically more likely that health will suffer. However, there is little general agreement as to what is 'safe' or 'risky' recreational drinking in terms of psychological or social factors, nor what patterns of use are likely to cause greater risks.

As it is unlikely that any government would have the knowledge or inclination to define the 'safe health limits' for illicit substances, there is little guidance for determining criteria for safer forms of drug misuse, in terms of either health or psychological and social risks. The obvious difficulty with assessing illicit substance misuse is the lack of available, accurate information on the long-term effects of different drugs. Health promotion authorities occasionally attempt to develop more 'realistic' drug education campaigns (such as that launched in the UK in the summer of 1995), but are hindered by opposition from those who feel that all drug misuse is unequivocally bad.

In the absence of consistent research and formal guidelines, as for alcohol and cigarettes, professionals are thrown back on their own resources to a large extent and there is a greater need to carry out a detailed assessment of the long- and short-term risks taken by each individual.

Risky or dependent?

Similar difficulties apply in distinguishing between risky and dependent drug misuse. We cannot say simply that drug misusers' risky behaviour is illogical and therefore must be a consequence of dependence. It is clear from the earlier research data that many drug misusers think it only natural to take risks. It is necessary to draw a distinction between these two types of misuse as the interventions for each are different. Risky misuse can be made safer and dependence can be dealt with through withdrawal regimes and cognitive behavioural interventions.

Assessment of the dependency syndrome

This clinically useful concept was developed by Griffith Edwards (Edwards and Gross, 1976) and has since been adapted for drug misuse as a whole (Drummond, 1991). The following seven criteria are used to identify dependence: the first three are established medical criteria for physical dependence, the last four give an indication of the complexity of social and psychological factors associated with it:

1. increased tolerance to the drug;
2. repeated withdrawal symptoms;
3. subjective awareness of compulsion to take the drug (craving);
4. salience of drug-seeking behaviour;
5. relief or avoidance of withdrawal symptoms;
6. narrowing of the repertoire of drug taking;
7. reinstatement following a period of abstinence.

Salience of drug-seeking behaviour describes the growing priority of drug misuse over other areas in an individual's life, the narrowing of drug repertoire indicates the increasingly regular and rigid patterns of drug misuse and reinstatement of drug-misusing behaviour highlights the speed with which drug misuse escalates when an individual relapses after a period of abstinence.

This syndrome is very useful for identification and assessment of dependence on physically addictive drugs, particularly depressants such as opiates and benzodiazepines. It may also be useful for identifying a growing problem of psychological dependence among stimulant misusers.

Dependent or self-medicating?

It is useful to identify causal direction in the development of any problem as this too will often determine the type of intervention. A recurring question in the drugs field is: which comes first, the drug or the problem? For each of the drug misuse categories (recreational, risky and dependent), drug misuse can be functional, providing a solution to certain needs or problems, it can cause problems itself or, more commonly, it can be both. Problems are often dealt with by misusing drugs as a remedy and it is also true that drugs cause problems of their own. Therefore it is useful to discover if the client is self-medicating for serious underlying psychological problems such as clinical depression. If so, they will need to be dealt with separately from the drug misuse. This situation can call for two distinct assessments and two different interventions. This area of 'dual diagnosis' is a source of some controversy, particularly in the USA, where the abstinence aims of drug treatment may conflict with the necessary prescription of drugs for underlying problems (see Chapter 8).

Whilst it is necessary to determine the existence of underlying problems, it is often less than constructive to spend time consistently trying to disentangle cause and effect. Drugs initially misused to solve one set of difficulties can cause others, which remain long after the initial problem has been resolved. In contrast, problems arising from drug misuse, such as unemployment, can remain long after the drug misuse has been controlled.

The question of cause and effect is often a source of controversy between clients and helpers. Clients may see their drug misuse as a solution or remedy for various matters ranging from boredom and depression to social deprivation, whereas professionals tend to see drug misuse as the cause of these problems. This argument can become a serious block to communication and in actuality may be as unimportant as the proverbial chicken and egg; the phenomenon is perhaps most usefully understood as a circular self-perpetuating process.

CARRYING OUT AN ASSESSMENT

Patterns of drug misuse over time

A useful assessment attempts to understand the patterns of individual drug misuse over time within the social context of other life events. This helps clarify what type of problems the misuser has now, what precipitated these problems and what might lead to relapse after they have been dealt with.

The following areas are important:

- The client's own view of their drug misuse, definition of problem, pros and cons of misuse, resources and possible solutions.
- History of drug misuse and age at first misuse. Previous drug-misusing behaviour; types, amount, frequency and patterns of misuse;

changing patterns of misuse over time, including periods of abstinence. This enables determination of which categories best fit the client's present drug misuse patterns.

- The functions of misuse or problems leading to misuse. This can include social and individual reasons why drugs are misused; availability and group culture/norms, physiological effects and psychological reasons. This information helps determine whether drug misuse is recreational, risky, dependent or self-medicating.
- The problems arising from drug misuse.
- Medical history.
- Mental health history, including psychological problems, such as anxiety and depression.
- Social environment, problems and resources. Present social situation, relationships, dependants, accommodation, employment, legal status.
- Relationships; amount of emotional support available, including the level of family support and involvement and support from partner or spouse.
- Inter-relationships between drug misuse patterns and life circumstances and events in life history (particularly focusing on periods of abstinence and relapse).
- Present inter-relationships between daily and weekly patterns of drug misuse and other social/emotional life patterns (i.e. when do they misuse and why?).
- History of help seeking, interventions used and outcomes.
- History of previous withdrawal episodes (if applicable).

By using this information the client's present problems can be placed in one or more categories. This will then provide a structure for clarifying problems, risks, needs and possible solutions.

The types of drugs being misused and pattern of misuse can be verified by knowing the effects and withdrawal characteristics of a range of drugs and the problems associated with different drugs. But the most important aspect of assessment is the understanding of the inter-relationships between the events associated with drug misuse, as it is the antecedents and consequences of drug misuse that need to be understood in assessing and modifying problematic misuse.

Problems leading to and arising from drug misuse

As explained earlier, drug misuse is often a solution to particular underlying problems. It is therefore useful, where possible, to distinguish between problems that lead to and problems that arise from drug misuse.

Problems contributing to drug misuse

Significant here are social factors such as stress, unemployment, poverty and housing and psychological factors such as depression and anxiety. If drugs are misused as a solution to these problems, then clearly alternative solutions are needed if the drugs are removed.

Problems arising from drug misuse

Physical problems include the risks of chaotic, unplanned drug misuse, HIV, hepatitis and other health problems. Social problems include the risk to dependants and others and drug-related crime. Psychological problems include depression and paranoia and, finally, drug dependence itself.

It is important to identify the functions of drug misuse to a client. The main functions can be seen as negative reinforcement (to escape unpleasant feelings) and positive reinforcement, ranging from mood alteration to social facilitation and enhancement of perceived control.

Physical, psychological and social problems

A basic assessment grid

The grid in Table 10.2 gives a structure for assessing the different types of categories and problems. It can also be useful for identifying signs and symptoms of drug misuse in the first instance.

Table 10.2 Assessment grid

Signs and symptoms	Recreational	Risky	Dependent
Physical			
Psychological			
Social			

The physical, psychological and social problems associated with drug misuse have been discussed in Parts One and Two. For a detailed up-to-date account of specific drugs and their effects, the *ISDD Drug Abuse Briefing* is published each year.

PHYSICAL, PSYCHOLOGICAL AND SOCIAL EFFECTS OF SHORT- AND LONG-TERM DRUG MISUSE

The research data in Part One and Chapters 7 and 8 indicate that physical, psychological and social factors such as hedonistic and functional effects, dependence and social support affect both the continued misuse of drugs and the likelihood of relapse. Different drugs can have very different physical, psychological and social effects. It is therefore necessary to be able to distinguish between them.

The short-term effects of intoxication and the effects of prolonged drug misuse can influence mood, perception, memory and psychomotor ability. The effects of the drug leaving the bloodstream may likewise cause changes in normal physical and psychological states (whether withdrawals or simply 'comedowns'). It becomes difficult for professionals and for many drug misusers themselves to distinguish between the effects of a drug, the effects of withdrawal and associated problems. It is more than likely that the actual variation in levels of drug in the bloodstream is far more problematic than any effects of a constant supply.

This is complicated by the particular health problems associated with the effects of different drugs. For example, general signs of illness may be misinterpreted as side-effects of drug misuse or withdrawal. It is therefore important that professionals have a rough idea of the short- and long-term effects of different drugs and the basic health problems associated with each (see also Chapters 7 and 8). General health problems are related to infections such as HIV and hepatitis (particularly hepatitis C), to problems arising from injection such as skin infections and damage to arteries and veins, and problems such as respiratory difficulty, vitamin deficiency and anaemia.

Changes in levels of sensitivity and tolerance to different drugs give a rough indication of the extent of drug misuse. Recent research on the phenomenon of sensitivity suggests that individuals may physically become more sensitive to the effects of drugs before developing tolerance (see Chapter 8). Many drug misusers will modify the amount and frequency of drugs misused over time. It is likely that drug misusers are in effect regulating their physiological sensitivity and tolerance by misusing on an irregular basis to increase sensitivity and decrease tolerance. It is when they lose their control of the balance between them that problems are likely to occur. Physical dependence on drugs is associated with withdrawals, tolerance and craving. The physiological aspects of these three are also dealt with in Chapter 8.

Depressants

Depressant drug misuse can be fatal, as these drugs depress the respiratory system. Illicit drugs may be adulterated and there is a chance of overdose, particularly if tolerance is reduced or more than one depressant is misused. While it is possible to avoid physical dependence simply by not misusing on a regular daily basis, drugs such as opiates can lead to physical addiction after several weeks of continuous use.

The positive effects of depressants are the relief of stress, anxiety and pain. They depress the nervous system and consequently cause the user to feel relaxed. The sedative effects of depressants are associated with feelings of well-being, contentment and relaxation, particularly for those who suffer from anxiety, stress or depression.

The negative effects are the loss of motor coordination and cognitive efficiency. They reduce the inhibitions and lead to loss of control over behaviour. These drugs also induce sleep and can lead to unconscious-ness if misused in large enough quantities. The opiates, barbiturates and alcohol are physically addictive if used regularly on a daily basis.

The misuse of depressants has significant social effects, reducing inhibitions, causing a deterioration in perceptual and physical ability and leading to dependence. The misuse of benzodiazepines and alcohol can lead to aggression and violence and inability to function at work or when driving, as can abrupt withdrawals, particularly from benzo-diazepines. The main social effects are a product of the illicit nature of drug misuse, the heavy penalties associated with it and the black market. The illicit trade can result in violence and intimidation, the stigma of a drug offence and possibly prison. The stigma of drug misuse generally means there is a heavy price to be paid socially, whether this means being physically assaulted because one cannot pay a debt or being made unemployed because of a police arrest.

The importance of social factors in drug misuse has been pinpointed in Part One and in Chapters 7 and 8. It has become clear from research on both treatment and relapse that the social support networks are important in maintaining treatment gains. The effect of continued drug misuse is often to cause the breakdown of the social supports available, by placing stress on immediate family and loss of employment. The following social problems are associated with continued drug misuse: social isolation from non-drug-using peers and family; financial problems and debts; family and marital problems; homelessness; occupational problems; and criminal activity.

Opiates depress the respiratory system and can result in overdose. Risk is increased if methadone and other opiates are mixed with other depressants such as alcohol. Opiates are the most physically addictive of the depressants and the withdrawal syndrome includes serious flu-like symptoms and cramps and occasionally seizures. The probability of relapse is so high that it is unlikely that the withdrawal symptoms of physical dependence are the significant factor in preventing abstinence.

The psychological and physical effects of these drugs are on the whole positive; if the strong sedative effects are avoided, misuse does not affect the perception or motor coordination. Opiate misusers would consider these the most effective of the depressant drugs for inducing a sense of warmth and well-being where anxiety, stress and depression are eliminated. The problems are often connected to injection and unknown quality and quantity of illicit drugs. For this reason it is argued that a legitimate prescription allows relatively safe use of these drugs.

Barbiturates, though now very uncommon, are probably the most dangerous of the depressants, as the level of drug required for an effect is close to that of overdose. It is also dangerous if mixed with other depressants. Withdrawals can be fatal, as fitting is not uncommon.

Benzodiazepines are safer, inducing similar feelings of contentment and well-being but to a much lesser extent, perhaps appreciated most by those who suffer from stress and anxiety and therefore value the relief from these feelings. The effects last no more than three months before tolerance develops and dependence follows within a year or two. Withdrawals include anxiety, depression and sleeplessness, the very symptoms that lead to initial usage in many cases. It is therefore not uncommon for withdrawal symptoms to be misinterpreted as a deterioration of the initial disorders. Particular benzodiazepines such as temazepam can be dangerous if used in large quantities or in conjunction with other drugs. These drugs are also commonly injected; if the gel from temazepam capsules is injected, it can block veins and if tablets of any kind are crushed and injected, this can lead to a range of physical complications.

Finally, cannabis, whilst having a sensitizing effect on perception and the other senses, also has depressant effects including a reduction of inhibitions and relaxation. It serves to reduce anxiety but, in common with the stimulants, often induces a sense of paranoia. Continued use can lead to respiratory problems, but this is probably the least harmful of all drugs.

Stimulants

Stimulants, apart from nicotine, are less physically addictive than the depressants. They are also intrinsically less dangerous as they do not depress the respiratory system and it is more difficult to overdose. Stimulants lower blood pressure and whilst this in itself can lead to unconsciousness, it is less common.

The positive effects are an increase in energy, stamina and alertness, allowing misusers to stay awake for long periods or concentrate longer. The effects are pleasurable, inducing a happy mood and increased self-confidence.

The negative effects are largely concerned with the after-effects rather than the risks of the drug itself. Large amounts can cause increasing paranoia and anxiety. (For amphetamine there is a possibility that this can result in paranoid psychosis.) After-effects include extreme tiredness and depression. In contrast to physical withdrawal from depressants, these effects seldom have serious consequences. There can be increased irritability, anxiety and depression after long periods of misuse.

The social effects of stimulant misuse are similar to those of depressant misuse with the exception that, while depressant drugs (with the exception of opiates) impair performance, judicious use of amphetamines and cocaine can improve social performance at many tasks, at least in the short term before tolerance builds.

Amphetamines are the most common stimulant and far more common than opiates among drug misusers. The effects are entirely different from depressants, activating rather than depressing the central nervous system and causing a feeling of well-being, self-confidence, increasing energy and excitement, much like adrenaline. The after-effects are depression, anxiety and irritability. These take several days, sometimes weeks, to pass as the body's natural resources are depleted. Lack of sleep and food during long periods of amphetamine misuse result in ill health. Paranoia is common; this will pass as the drug wears off but it may result in a psychotic state in some instances.

Cocaine and its derivative crack induce more pleasant states than amphetamine, with a greater feeling of confidence, contentment and well-being, though the effects will last for a shorter period. There are similar effects from long-term use, such as anxiety, paranoia and nervous agitation, and whilst there is no physical withdrawal syndrome, giving up can result in prolonged periods of depression.

Ecstasy has stimulant and hallucinogenic effects and it produces feelings of relaxation, well-being, confidence and revitalizing energy. There is a possibility of occasional fatalities, apparently through heat stroke and occasionally heart attacks.

Steroid misuse results in increasing muscle development but may also cause liver and heart damage and may possibly lead to aggression. 'Poppers' or alkyl nitrates lead to a pleasant 'rush' but can lead to sickness and headaches.

Hallucinogens

LSD and the psilocybin mushroom affect sensory perception and sometimes heighten mood. Misuse can result in hallucinations and/or pseudo-hallucinations. Feelings of mystical enlightenment, insight and improvement of mood are counterbalanced by the possibility of panic and 'bad trips'. These drugs do not lead to physical dependence but tolerance develops very quickly, within a period of several days. Therefore prolonged periods of continual misuse are unusual as it is necessary to stop misusing in order to increase sensitivity.

This section has provided a brief overview of the possible effects of drug misuse. Whilst this can be associated with a range of physical, psychological and social problems, it is important to emphasize that not all drug misusers have problems. Those who attend specialist agencies and inform much of the research literature are not representative of drug misusers as a whole. In the same way that a probation officer's view of criminals is limited to those who get caught, the specialist drug worker or clinician sees only those drug misusers who have problems. Although they will also see these clients, the generic professional is much more likely to encounter those whose drug misuse is largely unproblematic.

CARE PLANS

The care plan uses the assessment to categorize the client's problems and determine appropriate solutions (see Table 10.2). It includes a review of the client's health, psychological and social needs and resources, together with other resources available to the practitioner. These might include, for example, the practitioner's counselling skills, health care facilities and welfare rights information, together with more specialist input from other agencies where necessary.

The care plan should follow the sequential process, providing the necessary help at both the intervention and postintervention stages and including a provisional plan of service provision in the event of a relapse or crisis. In order to do this, the plans require several distinct assessments in order to determine the ability of the client to change, obstacles to change and the client's own resources for change and maintenance of change. The care plan and aftercare plan should identify what support the client will need in order to reduce harm or to change in the first instance and what they will need in order to prevent relapse.

The plan prioritizes needs and sets practical goals within the resources available. It often includes providing equipment to ensure that clients reduce sharing of needles and syringes and use condoms and may include prescribing drugs in order to cut down the amount of illicit drug misuse and subsequent involvement in crime and debt.

Care plans include the following information:

- client's needs in the short term;
- client's needs for after-care maintenance;

- client's resources in the short term (personal coping strategies and skills);
- client's resources in the long term (skills and social support networks);
- agency resources for intervention and aftercare.

When the care plan is drawn up and a service is offered to a client, the terms and conditions should be made clear, including the extent of support available and penalties for non-compliance. Professional agreements with clients should include the negotiation of short- and long-term goals. Whilst all practitioners have a responsibility to help clients make informed decisions, there are often situations where the client and professional may have different priorities: it is necessary to clarify them when agreeing and defining goals for treatment.

Table 10.3 Completed care plan

	Physical resources	Psychological resources	Social resources
Safety net	Needles and syringes. First aid and health care equipment	N/A	Information and lists of available services
Harm minimization care plan	As above + prescribed drugs	N/A With exception of cognitive behavioural skills for some chaotic clients only	Developing social support networks and crisis intervention strategies
Treatment change care plan	Prescribed drugs	Counselling, psychotherapy cognitive behavioural methods	Ensure social support during change period
Aftercare maintenance and maintenance of change care plans	Prescribing if necessary	Improving cognitive behavioural coping skills; relapse prevention, relaxation + self-control techniques	Developing supportive social networks, new social activities + constructive occupations

NB. It should be emphasized that similar interventions may be given for different reasons. This has caused much confusion in the drugs field. For example, a prescription may be given as an alternative to illegal drugs, as part of a long- or short-term withdrawal strategy or as medication for the problems underlying drug misuse.

INTERVENTION: HARM MINIMIZATION AND/OR TREATMENT

Assessment is designed to determine which interventions are necessary. It also clarifies the physical, psychological and social problems which need to be addressed within each type of intervention. Practical guidelines for interventions are given in Chapters 7 and 8.

HARM MINIMIZATION (INTERVENTION FOR MAINTENANCE)

This intervention involves the stabilization, support and maintenance of drug misusers. The aim is to reduce the social and health damage of drug misuse. This intervention is enough for those whose drug misuse is temporary and/or non-addictive. Occasionally some clients may have short periods of chaotic drug misuse and at this time changes in behaviour or lifestyles are also necessary.

TREATMENT (INTERVENTION FOR CHANGE)

This intervention is used to facilitate therapeutic change; it includes different models of treatment, from psychological techniques, behaviour change and controlled drug misuse to medical prescribing regimes. The aim is to stop or curtail drug misuse.

AFTERCARE: MAINTENANCE OF HARM MINIMIZATION AND TREATMENT

Chapters 7 and 8 highlighted the need for continuing support and help over long periods of time. The problem with both harm minimization and treatment interventions is that they do not last forever. Research indicates that when the client stops attending the services they start to take risks again or they relapse. The most important task for generic professionals is to support and maintain clients after interventions.

Methods of relapse prevention were initially developed for preventing relapse after alcohol treatment but will be appropriated here for aftercare purposes for both harm minimization and drug treatment. The essence of relapse prevention is to equip the client to recognize risky situations and cope without reverting to old behaviours (whether these be drug misuse itself or associated risk behaviours).

This stage in the sequential helping process should be seen as distinct from the initial interventions; it requires a separate assessment, care plan and methods. Aftercare plans should be designed to maintain clients over long periods of time with the minimal necessary professional input and to enable the immediate reestablishment of more intensive contact if and when this is necessary.

The foundation of good aftercare is a comprehensive assessment of the reasons why a client might take risks or relapse and preparation to avoid or cope with these situations in the future. Initial preparation may involve the client developing a less stressful or dangerous lifestyle and learning the necessary relapse prevention and coping skills. Aftercare support itself involves the development of a programme of regular aftercare contacts, and procedures to re-establish contact in the event of a lapse or relapse.

Continuing maintenance and relapse prevention for harm minimization

Aftercare maintenance following harm minimization simply means maintaining a support structure to ensure that clients do not take unnecessary risks in the future. This can consist of the establishment of regular access to needles and syringes and the provision of crisis health care support if necessary. It may also involve teaching cognitive behavioural methods to control chaotic drug misuse or to help prevent relapse to old patterns of behaviour.

Post-treatment relapse prevention

Post-treatment aftercare involves a more structured approach to the maintenance of treatment change. Research indicates that social support networks and individual coping skills can be critical in preventing treatment relapse. Cognitive behavioural methods have been designed to prevent relapse. These methods can be used irrespective of the treatment theory itself, either to maintain abstinence or control drug misuse.

Whereas assessment for harm minimization aftercare should occur after initial provision of harm minimization services, assessment for drug treatment aftercare should take place before drug treatment starts. This is because there may be extra risks involved in drug treatment when clients make reductions in their drug misuse and it may not be possible for the client to maintain these changes over time. Reducing the amount of drugs misused can lead to more problems in the short term, rather than less. This can take the form of short-term withdrawals and mood swings and depression. In the long term, if clients start to misuse drugs again, reductions in dose or abstinence will have lowered tolerance levels and so the risks of overdose increase as clients try to give up, then start again and give up in turn. In effect, varying the amount of drug misused is likely to be more dangerous than maintaining a steady level of misuse.

The professional may therefore need actively to dissuade a client from engaging in drug treatment, reducing or stopping drug misuse until they are sure that the client has the necessary skills and support to cope in the long term. Assessment for aftercare maintenance provides the information to develop essential aftercare plans. Only when the future ground has been thoroughly prepared for changes in drug misuse should changes take place. If no account is taken of conditions contributing to drug misuse in the client's everyday life then relapse is more likely.

It has already been emphasized that motivation is an essential precursor to work with drug misusers. It is ineffective to try to make clients change their behaviour before they are ready. However, it may well be effective for professionals to suggest that clients postpone making drastic reductions in their drug misuse until they are fully prepared. Having motivated the client, the professional's job is then to help the client prepare for change. Part of making any change is preparing the ground; assessment and aftercare plans are therefore an integral part of drug treatment itself.

ASSESSMENT AND CAREPLANS FOR HARM MINIMIZATION
MAINTENANCE AND RELAPSE PREVENTION

Aftercare assessment is in effect a secondary assessment to establish what the client's needs will be after the intervention, to determine whether the client is capable of maintaining changes in the future and, if so, what individual resources/skills and professional support systems will be necessary.

Assessment will need to be reviewed and updated at the close of the treatment stage. (It is likely that these stages will in effect form a continuum, if not actually overlap.) This also includes assessing progress made in treatment to ensure that clients have resolved any underlying problems contributing to their drug misuse and have a clear understanding of their vulnerabilities to drug misuse in the future.

- Identification of potentially dangerous situations for risky behaviour and/or relapse.
- Assessment of social support and individual resources.
- Examination of functional purposes of individual drug misuse.

Following the assessment, the aftercare plan consists of preparing the client's social environment and developing their personal skills. A strategy can be planned to develop the client's own skills and resources with minimal professional support. This plan should out-line the extent of professional support and contact necessary and clarify a procedure for dealing with crisis situations, lapses and long-term relapses.

The aftercare plan should be seen as a continuation of the care plan. It involves:

- ongoing support on a regular basis;
- clear guidelines for intervention if the client lapses or relapses;
- the development and maintenance of family, social and community support systems and alternative occupations;
- the use of self-help groups and other community support services.

The input during this stage involves simple maintenance care and a reg-ular series of contacts and appointments to support and monitor the client's progress.

RELAPSE PREVENTION SKILLS FOR MAINTENANCE OF HARM
MINIMIZATION AND TREATMENT GAINS

In Chapter 8 the reasons for relapse or loss of treatment gains were identified as negative emotional states, interpersonal conflict and social pressure (Marlatt and Gordon, 1985; Marlatt, 1985; Wilson, 1992). The cognitions that are associated with these behavioural lapses can be placed in three categories; perceived inability to cope, decreased per-ception of control in high-risk situations and positive expectancies of outcome (Donovan and Marlatt, 1980). For example, if a client sees a situation as stressful and feels that they do not have the necessary skills to cope, there is a decrease in perceived control; if this is accompanied by the perception of drug misuse as an effective means of decreasing stress and increasing personal sense of control, then the client is likely to misuse drugs.

Marlatt and Gordon (1980) suggest that the emphasis of relapse pre-vention should therefore be on the situational demands and stresses placed on an individual and the beliefs individuals hold of their own ability to deal with these stresses. They propose a series of cognitive and behavioural exercises. The aims of these techniques are to increase the client's perceived and actual self-control, self-confidence and coping ability, by altering cognitions and teaching new behavioural skills and coping strategies.

These researchers carried out much of their work in the alcohol treatment field, but their conclusions are equally useful for both drug dependency treatment and for maintaining safer drug misuse. In other words, cognitive behavioural techniques can be used to prevent 'relapses' in dependent behaviour as a whole and also to prevent relapse to unsafe drug misuse practices.

The following harm minimization and drug treatment goals can all be maintained by using relapse prevention techniques.

- Giving up drugs completely.
- Misusing less regularly to avoid dependence.
- Misusing in a more controlled way and less chaotically.
- Misusing only with certain people in certain places.
- Misusing fewer drugs and/or less harmful drugs.
- Stopping sharing needles and syringes.
- Stopping injecting.

In order to maintain any of these behaviours the client can learn cognitive and behavioural skills designed specifically to prevent relapse to old behaviour patterns.

Self-monitoring

The aim of self-monitoring is for clients to gain insight into their drug misuse and learn to recognize patterns in their behaviour. Having done this, they can find ways of avoiding that behaviour in the future. Self-monitoring varies from the use of diaries to complex behavioural assessments.

Seemingly automatic, habitual behaviours can result in relapse because of a lack of self-awareness.The client simply falls back on old familiar patterns of behaviour without realizing that they have done so. Becoming self-aware is seen as an integral part of preventing relapse because habitual cognitions and behaviours can only be controlled through an increased understanding of the client's own particular arousal states and patterns of behaviour. Behavioural assessment allows the client to assess the relative risks of particular moods and situations (Marlatt and Gordon, 1985).

Descriptions of past relapses and rehearsals of possible future lapses

This information can be supplemented by learning from previous relapses and rehearsals of possible future relapses. Role play, desensitization and self-control methods are useful here.

Information about the immediate and delayed effects of substances

Basic practical information about the effects of short- and long-term drug misuse (including tolerance, withdrawals and craving) can help contradict less accurate and/or maladaptive beliefs. For example, information about length of time and the expected effects of withdrawals or the effects of decreased tolerance after periods of abstinence may help dispel inappropriate beliefs about personal pathologies and helplessness. Similarly, learning to estimate the period of time an intense feeling of craving may last can help clients to 'surf the urge'.

Teaching clients to assess risks

In the same way that clients can learn self-awareness, they can become more adept at assessing risks in any given situation and so withdraw before they lose control of their behaviour. For example, well-remembered cues may stimulate a lapse, but if an individual avoids places where these cues are likely to occur the danger of relapse is also avoided. Clients can learn to identify and respond to situational, interpersonal and intrapersonal cues as early warning signals.

Planning specific strategies for high-risk situations

It is possible to decrease the probability of relapse by helping clients to plan step-by-step coping strategies for particular high-risk situations involving negative emotional states, interpersonal conflicts and social pressure.

RELAPSE PREVENTION AFTER DEPENDENCY TREATMENT: SHORT-TERM AND LONG-TERM COPING STRATEGIES

A functional analysis of the purposes that drugs serve for each individual in lessening the effects of social and psychological stress can help identify individual deficits in coping skills which may cause relapse if drugs are given up after treatment. Social skills training has been shown to be effective in preventing relapse and a wide range of self-control or self-management programmes (Marlatt and Gordon, 1980) are designed to include the development of new skills and coping strategies.

Short-term coping skills

Skills training and coping strategies can include a wide range of generic techniques such as relaxation training, stress management, anxiety management and efficacy-enhancing imagery. In effect, the client is taught skills to help them cope and this in turn leads to increased confidence and perception of self-control, so relapse is less likely. Skills training programmes include problem-solving skills such as orientation, definition of problems, generation of alternatives and decision making. Teaching can be in the form of basic instruction, modelling or cognitive and behavioural rehearsal.

- Assertiveness training gives the ability to express aspects of covert hostility and resentment.
- Anger control teaches the client to recognize anger and its source in the environment and then communicate the anger in a less threatening way.
- Self-control skills lead to a decrease in levels of arousal and irritability and so reduce anger and prevent aggression.
- Cognitive behavioural stress management programmes teach the client to gain control over emotional and arousal states.
- Relaxation and meditation techniques teach the client how to relax physically and mentally.

It is clear that not all clients need social or coping skills training. However, many who stop misusing drugs find that these skills provide alternative coping mechanisms.

Long-term coping strategies

It is not enough to teach clients specific individual skills appropriate for personally defined high-risk situations. As Marlatt (1985) states, it is also necessary to develop the individual's ability to deal with a broader range of situations and to exercise general self-control strategies to reduce the likelihood of relapse in any situation.

The overall influence of negative affect (e.g. anger, frustration and depression) on relapse suggests that more general mood control strategies may also be effective; for example, learning self-management skills, increasing involvement in pleasant activities, learning to relax, becoming more socially skilful in interpersonal relationships, controlling negative or self-defeating thoughts, increasing positive self-reinforcing thoughts, providing appropriate coping self-statements and gaining constructive problem-solving skills.

IMPROVING THE ENVIRONMENT: DEVELOPING NEW SAFER, SUPPORTIVE ENVIRONMENTS

The alternative to changing individual behaviour directly is to change the environment itself. In Chapter 8 it became clear that maintenance of treatment gains was associated with psychological and social factors rather than treatment variables themselves (Moos et al., 1990; Lindstrom, 1992). In effect, the concept of 'treatment' focuses attention on individual rather than more general social factors that may precipitate relapse. In contrast, it may make more sense to reduce environmental stress, either by changing the environment or simply by helping clients to adopt a less stressful lifestyle and less stressful relationships. Marlatt and Gordon

(1980) argue that a 'lifestyle' balance is an essential part of a relapse prevention programme.

Although research is undeveloped in this area, it is possible that a more rewarding and less stressful lifestyle may provide a solution for certain types of vulnerable people. This can involve changes to reduce negative effects from stress at work or in relationships on a permanent basis. Clients can change jobs or relationships, start new activities and/or build new social support networks. Professional support can range from providing housing and occupational training to setting up support groups and working with families.

Moos *et al.* (1990) have demonstrated that individual variables interact with environmental factors to cause relapse. They suggest that support from family and friends and the existence of a relatively stable social and economic situation are important factors in the maintenance of behavioural change.

Work with drinkers indicates that pressure at work and lack of external supervision or structure leads to heavy drinking, especially if this is a socially acceptable means of releasing tension (Roman and Trice, 1970). Kaufman and Kaufman (1979) consider the positive and negative implications of family relationships for drug and alcohol misusers and Steinglass (1979) applies a family systems theory to drinking, in order to highlight the influence of family dynamic on an individual, interpreting excessive drinking as a means of reducing stress in the system itself. Therefore, within this model of understanding, changes in behaviour can be seen to pose a threat to the stability of the family system as a whole and would need to be monitored as part of relapse prevention. Family-orientated techniques are not common in this area, though the reactions of significant others to lapses and previous relapses can be influential, suggesting that it is important that relatives and family be involved in any relapse prevention strategies and that conjoint counselling to reduce more generalized stresses within a relationship may be useful.

The most influential research on environmental stressors and social supports is that of Rudolph Moos and his colleagues – see Cronkite and Moos (1980), Moos, Finney and Chan (1981), Billings and Moos (1983) and Moos, Finney and Cronkite (1990). This work has been discussed in more detail in Chapter 8, but it is relevant to add here that Moos and his colleagues found that relapsed alcoholics had more negative experiences and fewer positive life events that those who had not relapsed. They also discovered that those who retained treatment gains had apparently created benign conditions for themselves, developed social support networks and cohesion and revealed low conflict in family relationships and low time urgency and pressure in work settings. Hunt and Azrin (1973) designed and implemented a community-orientated programme to increase the quality of life. To do this they systematically programmed reinforcement schedules into areas of social, family and vocational activities by employing social reinforcers, such as marital and family counselling, social clubs, driver's licence, telephone service and newspaper subscription. They found that those who remained abstinent spent more time gainfully employed and with their families and more time on socially accepted recreational activities. Unfortunately to date little work has directly involved helping drug misusers systematically modify their post-treatment environment.

LAPSE AND RELAPSE AS AN INTEGRAL PART OF DRUG TREATMENT: RELAPSE INTERVENTION

Relapse **intervention**, as distinct from relapse **prevention**, refers to the techniques used by both client and professional helper to prevent initial lapses from becoming full-blown relapses.

Whereas relapse intervention serves a purpose for those attempting to maintain safer drug misuse behaviours, its main function is to prevent rapidly escalating loss of control among those who have been drug dependent, when they lapse after a period of abstinence or controlled misuse.

It will probably be necessary to continue to work with clients through several periods of dangerous drug misuse and/or relapse. The old-fashioned idea of meeting clients, helping them get off drugs and then ending the connection is not useful and it is now generally accepted that the relationship will last a long time and have no immediate outcome. Instead, workers support clients through difficult or dangerous periods of their drug misuse. They help them to control their drug misuse and to misuse in safer ways; to prevent relapse and to learn to cope after relapse. Drug misuse and drug problems vary during the client's life. There will be times when they misuse fewer drugs or misuse non-problematically and there will be occasions when they give up altogether for a varying length of time.

Professional aftercare procedures in the event of relapse

When clients go through drug treatment and stop misusing, the professional will help them to develop new skills and perhaps a new lifestyle to prevent relapse. The worker should then monitor the client's post-treatment progress in order to offer ongoing support and be available quickly in case the client lapses or relapses back to old dangerous patterns of drug misuse.

This continued support (regular phone calls, meetings and/or group support) and immediate availability at relapse should be built into the aftercare phase. At present it is not uncommon for client and worker to see the end of a treatment phase as the final 'outcome' and therefore clients may well feel any relapse is a failure and should be disguised from the worker. This mistake stems from the medical concept of treatment as a distinct time-limited programme with a clear outcome, rather than the notion of a recurring disorder or disability, where there is no 'outcome' as such, only a continued monitoring of progress and availability of support when necessary.

Aftercare support should include a clear procedure to be followed in the event of a lapse and/or relapse. The immediate priority is to be available to the client and offer harm minimization services if necessary. This can then be followed by the opportunity to re-enter treatment. A lapse can mean anything from a single mistake to a complete return to old lifestyles and patterns of behaviour. It is therefore important to determine its extent before acting further. A reassessment of the client's behaviour and circumstances will be necessary.

If potential relapse is presented to clients not as failure but as an expected part of the change process, they will then be more likely to return if they need help. Monitoring of progress is essential at this stage, to enable prompt intervention in the event of a lapse to prevent full relapse.

Relapse intervention: client action after a single lapse

The process of moving from lapse to relapse forms a continuum. The events that cause the initial lapse may be different from those which maintain it and from those which contribute to the re-establishment of the social and behavioural routines associated with a full-blown relapse. The practical implications of this are that relapse intervention as distinct from relapse prevention should become an essential part of an aftercare programme.

Different strategies are needed at different stages of relapse, in the same way as different methods are tailored to different stages earlier in the sequential helping process. These strategies are designed to deal with temporary loss of control. The situation where many men and women feel most helpless is after the first lapse back into drug misuse. The concept of an 'AVE' or 'abstinence violation effect' is helpful here as a way of understanding the intense craving that dependent drug misusers can feel after a lapse. Whilst there is undoubtedly a physiological component to this effect for physically dependent misusers, the cognitive interpretation of this craving may also influence the feeling of loss of control.

It can be helpful for the initial lapse to be seen as controllable. A relapse can be defined in many ways. If it is seen as a 'slip' or 'lapse' but not as a complete remission, it may make it easier to deal with. Marlatt and Gordon (1980) suggest that the interpretation placed on a single lapse may determine whether or not it develops into a full relapse. So if a client expects the lapse to precipitate a serious relapse this actually becomes more likely. This argument is often used as a criticism of the disease model and other abstinence-orientated approaches. By the same token, the client may be less likely to try a drug initially if they expect a full relapse. Marlett and Gordon suggest behavioural interventions to reduce the likelihood of transition from lapse to relapse. This involves developing a set of rules or guidelines that govern behaviour after the initial drink or drug use. Clients may be given cards to remind them what to do if they have a lapse in order to prevent a full relapse.

Inability to cope in high-risk situations and personal beliefs about the usefulness of drugs in helping to cope are likely to precipitate relapse. Once potentially dangerous relapse situations have been identified, the client and professional can decide on coping or avoidance strategies. These can involve individual skills such as learning to recognize and avoid high-risk situations, increasing personal confidence and developing coping skills, together with changes in the environment, from improving stressful relationships to developing new supportive circumstances.

- Provide clients with a cognitive behavioural framework, in which they can understand their drug misuse behaviour patterns.
- Help clients to identify high-risk situations (including the situational, relationship, cognitive, behavioural and emotional antecedents of drug misuse).
- Help clients to modify their own cognitions and behaviours that may lead them closer to relapse.
- Teach clients about relapse and relapse management.
- Teach new cognitions and behaviours to develop new confidence, coping skills and self-control.
- Teach clients to modify their own environmental antecedents, to improve relationships and make work and social environments less stressful.

SUMMARY

It can be seen that the old-fashioned belief that dependency happens once and can be treated like an illness and cured for good is non-sensical. Dependency, like other forms of problematic drug use, comes and goes. Those who have become physiologically or psychologically vulnerable are likely to have repeat episodes at intervals in their lives.

MONITORING AND EVALUATION

Monitoring is an essential part of the sequential helping process. It is important to keep notes to monitor progress and reassess where necessary. The clinical aim is to maintain regular assessments which will determine the type of intervention appropriate at each new stage.

Monitoring can involve:

- self-monitoring and self-assessment by the client. Clients can be asked to complete diaries, record feelings and behaviours and identify the antecedents and consequences of misusing drugs;
- professional monitoring of the client's progress can be used to modify interventions, change goals and increase support where appropriate (this will include reassessments for treatment and aftercare maintenance where necessary);
- monitoring of aftercare maintenance is also necessary to enable identification of danger signs and to facilitate early intervention if the client's situation deteriorates;
- professional monitoring and evaluation of their own input (with individual clients, with the agency population and with the targeted population as a whole).

Evaluation

Professional monitoring and evaluation of work must take account of the aims and purposes of each intervention. If the aim is prevention or education, outcome studies are inappropriate as the effects are often only visible in total populations over long periods of time. Epidemiological measures give an indication of outcome in terms of changes of prevalence and incidence of problems such as HIV and drug-related overdoses in total populations. Longitudinal studies of changes in attitudes and behaviour over time can be useful. Process measures at least give an indication of what has been done, if not what has been achieved.

If the aim is harm minimization, outcome measures become almost totally irrelevant. Ongoing maintenance precautions can only be evaluated as an absence of harm over periods of time. Many professionals and drug agencies will feel that any client ending contact is in effect a failure. If the aim is controlled drug misuse or abstinence, outcome measures have more significance, but here again it is the two or three years following treatment which are crucial indicators of 'success' or 'failure'.

It is therefore important to clarify the aims and objectives of any intervention before deciding what to measure and how to measure it. The understanding of the problem and its solution will influence what is counted as success. Events such as short lapses can be seen as failures or as part of the process of change itself. Abstinence at treatment outcome can be seen as success or the long-term maintenance of treatment gains can be assessed in follow-up several years later.

Professional practice: skills, knowledge and responsibilities

11

INTRODUCTION

This chapter explains how the knowledge and skills of different professions can be used for working with drug misusers.

The increase in drug misuse in recent years has left many professional groups unprepared for the numbers of drug misusers among their client groups. It is clear from the previous chapters that not only is drug misuse becoming fairly widespread among certain groups but that problematic drug misusers are more likely to present to the social and health care professions. This could be because their existing clientele may misuse illicit drugs to help them cope, in effect compounding the initial problems. Alternatively, drug misuse may be the primary reason that the client is drawn to the attention of service providers. Research (Rush *et al.*, 1987) has shown that there are greater proportions of drug misusers among social and health care clientele than in the general population. Although that study refers mainly to minor tranquillizers, it highlights the fact that social and health problems are often remedied by drug misuse and that drug misuse causes social and health problems.

Professionals may choose to refer clients, work jointly with drug agencies, co-work with other professionals or work on their own. These decisions are particularly relevant with reference to Community Care and new Home Office initiatives. This chapter helps clarify the options, outlining what support is available (specialist and generic) and what different agencies have to offer. It outlines the areas of expertise and responsibility of different professions and gives guidelines for practice for each area. It will be useful for those professionals working in each area and at the same time it also provides a chart of possible referral points for different types of drug-related problems.

The following professionals have skills and resources which can be mobilized for helping drug misusers. Most groups, including the specialist drug agencies, will now work in conjunction with each other. So, for example, a health visitor could refer to a specialist drug agency for an assessment and continue to co-work with a general practitioner.

- Specialist drug services – community drug teams, drug agencies, residential care.
- Prescribing practice – general practitioners and other doctors.
- Health care – doctors, nurses and others.
- Psychological problems – psychiatrists, psychologists, psychiatric nurses.
- Public health and HIV – specialist drug agencies, health and social care agencies.
- Drug education – health promotion and teachers.
- Pregnancy – midwives, health visitors, doctors and social workers.
- Child care – social workers.
- Crime and custody – probation, the police and prison.

SPECIALIST DRUG SERVICES – COMMUNITY DRUG TEAMS, DRUG AGENCIES, RESIDENTIAL CARE

Drug agencies vary from the informal 'low threshold', non-statutory service to more structured, formal, community drug teams. These services include psychiatrists, doctors, nurses, psychologists, social workers, probation officers and unqualified specialists. They may also use volunteer counsellors and are increasingly providing outreach workers to contact clients outside the agency. There are also drug specialists within social services departments, employed to coordinate care resources for residential rehabilitation and professionals within probation and health care, particularly in psychiatric and health promotion units.

Community drug teams

The form and function of a community drug team are described in the case study in Chapter 8. Traditionally, statutory providers operated a referral-only service. This is changing as they open syringe exchanges and develop a more user-friendly approach in accord with harm minimization. It is not unusual for statutory drug teams to provide drop-in advice and information, though this is not as large a part of the service as for non-statutory provision.

Services available include assessment, needle and syringe exchange, health care, social and welfare advice, individual counselling, psychotherapeutic interventions (client-centred) and cognitive behavioral interventions (task-centred), relapse prevention, and prescribing for withdrawal and/or maintenance.

Non-statutory drug agencies

The case study in Chapter 7 describes a typical drop-in syringe exchange service within a non-statutory drug agency. These services are often very informal and the service offered far more confidential than the statutory provision. 'Drop-in' clients, assessment and crisis work often form the bulk of the workload. These agencies also operate as low-threshold entry points into other services.

Voluntary agencies are often isolated and financially insecure. Ettore (1988) suggests that there should be a overall strategy for voluntary drug agencies and that there is a need to create a balance between the different functions of working with misusers who continue to take drugs and those who need residential care to become drug free. There is a worry within British voluntary services that, as they are required to work towards the outcomes required by purchasers, they may change their initial aims and tasks (Mason and Marsden, 1994).

Services available include drop-in advice and information, needle and syringe exchange, counselling, social help on housing, information and aid, welfare rights and diversionary activities.

Residential services

There are nearly 50 residential treatment facilities for drug misusers in Britain, which vary from those providing intensely therapeutic programmes to those who provide more educational approaches or simply hostel accommodation.

Because of the wide range of different types of service provided, it is necessary to accumulate specific information about what is available. Detailed information about each of these projects is available in the SCODA Residential Rehabilitation Guide (for information contact the Standing Conference on Drug Abuse, Waterbridge House, 32–36 Loman Street, London SE1 0EE).

Referrals can be made direct to an agency or via social services. Either way, it is necessary to secure funding through social services assessment procedures, unless the facilities are not registered. Some local authorities have drawn up contracts with particular residential services, but others are more flexible.

A large proportion of residential services are either based on the Twelve Step or Minnesota model or an alternative programme of intense group psychotherapy and counselling. Both types offer highly structured programmes of therapeutic and practical activities and these are interspersed with educational input such as lectures and videos. The more therapeutic environments may present a regimented and intrusive regime with little privacy which does not suit all clients. Some services have a strong Christian element and others are less intense and therapeutic. It is therefore very important to be aware of the type of programme and the content of each service before deciding to refer a particular client.

As there can be a high drop-out rate from these services, it is helpful (for both the client and the service providers) to inform the client what to expect before they go. The length of stay varies from a few weeks to several months and in some cases a year or more. Some projects insist on abstinence, requiring that clients are detoxified before arrival, while some offer detoxification as part of their service. One or two places will take clients receiving a prescription. Some places will accept couples and mothers and children, some will not. The issue of resettlement is important to many, who will offer half-way houses post-treatment and/or continued aftercare support.

The staffing of these services varies from consultants, psychologists, qualified nurses and social workers to ex-drug misusers and volunteers. Most will have access to medical support. As with all residential facilities, levels of staffing reflect levels of funding.

Non-professional staff and volunteers

Easthope and Lynch (1992) examined the development of voluntary drug agencies in the UK and Tasmania. They found that services had expanded greatly in the past few years, were generally underfunded and relied on unqualified staff who saw their main role as counselling.

Similarly, the non-professional drug counsellor in the USA has traditionally played a large role within the framework of the Twelve Step, Anonymous Fellowship (the Narcotics Anonymous programme is described in Chapter 8). There is much controversy about the working relationship between professionals and non-professionals in non-statutory agencies and in AA/NA residential treatment. Kostyk *et al.* (1993) found that peer co-leadership was effective and that the peer co-leader provided a role model by offering hope and optimism. This role for non-professionals has also extended outside the Anonymous framework, as outreach and education become primary issues in terms of HIV. Brown (1993) points out that in some areas of America the traditional role of ex-addict, or 'professionals of experience', as counsellors and advisors has been phased out as health educators become more numerous, but that a new task for non-professionals is that of outreach educators.

PRESCRIBING PRACTICE – GENERAL PRACTITIONERS, CONSULTANTS AND OTHER DOCTORS

This book does not cover the medical management of drug misuse. The reader is instead recommended to follow the guidelines laid out in *Drug Misuse and Dependence: Guidelines on Clinical Management* (Department of Health, 1991). However, the following information gives a useful overview.

Prescribing can be a means of achieving basic objectives such as establishing and maintaining contact with drug misusers. It can serve the function of harm minimization or treatment, as a response to chaotic, dangerous drug misuse and/or injecting drug misuse and also as a response to a physical withdrawal syndrome. Some drugs are safer than others. Drugs in an injectable form carry more of a risk, though occasionally they can serve as the only means of preventing misusers buying illicit injectable drugs. It is not always necessary to prescribe a direct substitute or a controlled drug. The prescribing of amphetamine substitutes is discussed in Chapter 7 and methadone in Chapter 8.

It is important to emphasize that polydrug misuse is the norm rather than the exception among drug misusers. It is therefore sometimes difficult to determine which is the most problematic so it may be necessary to treat more than one drug. This is important as, if a drug misuser is withdrawn from one substance, they may simply replace it with another. The issue of dual diagnosis, particularly use of alcohol and opiates, was discussed in Chapter 8. It should be emphasized again here that illicit use of benzodiazepines and sedatives is commonplace among all drug misusers. Research findings discussed earlier indicate that, although a particular individual may misuse stimulants and depressants in sequence in order to benefit from both types of effect, many drug misusers have a preference for one or the other.

Methadone prescribing

As discussed in Chapter 8, methadone maintenance over long periods of time has been shown to be safer than long-term heroin misuse (Novick *et al.*, 1993) and reduces the chance of HIV infection (Serpelloni *et al.*, 1994). There is no doubt that methadone prescribing is useful for both harm minimization and treatment of opiate users. However, it is sometimes difficult to determine whether maintenance or withdrawal programmes are most likely to be constructive.

- Whichever purpose the prescription serves, the first stage is to assess what is necessary as a stabilizing dose.

- The practitioner may then decide to continue prescribing at this level as a maintenance strategy.
- Alternatively the prescriber may arrange a withdrawal schedule to reduce or stop the prescription completely.
- If the initial prescription regime has served to change or reduce drug taking, this must then be followed by a prescribing strategy to maintain this change.

The drug is usually prescribed as methadone mixture 1 mg/1 ml or methadone mixture DFT 1 mg/1 ml. Methadone mixture can be made without sugar and/or colourings (methadone linctus can refer to the mixture). Methadone can also be prescribed as tablets (5 ml) or ampoules (10 ml), often under the trade name of Physeptone. Most GPs will not prescribe Physeptone as it is easily injectable in either form. This is not to say that some misusers will not inject methadone mixture or linctus, but it is less common.

Urine testing

Urine testing facilities are now available from some health authorities and pharmacology departments. It is important to label samples clearly as urine may be positive for several infections, such as hepatitis B. It is important to stress that although urine testing can be a useful adjunct to treatment, it is less appropriate for harm minimization purposes (see Chapter 7).

Managing withdrawal

Depressant withdrawals can lead to a mild pyrexia, increased blood pressure, tachycardia and vasoconstriction. The physiological effects of withdrawal from stimulants are less clear (see Chapter 8) but depression, tiredness and lethargy are common. Convulsions can occur in withdrawal from barbiturates and can lead to brain damage. It is unwise to attempt detoxification from barbiturates on an out-patient basis.

The reader is again referred to the Department of Health *Guidelines on Clinical Management* (1991) for detailed information. It is important to emphasize the need for a period of stabilization prior to a withdrawal schedule. This is particularly the case if a substitute drug is being used instead of that normally taken.

The basic aim of a withdrawal regime is to provide a therapeutic dose that prevents withdrawals without causing intoxication. This dose can then be reduced at a pace that causes least distress. Substitute drugs can be used to medicate for withdrawal symptoms. For example, problems such as sleeplessness, anxiety and physiological symptoms can often be dealt with through the use of benzodiazepines such as diazepam.

HEALTH CARE – GPS, NURSES AND OTHERS

Practical guidelines for health problems and health care are given in Chapter 7.

Health care, of course, remains the province of the health care professions. Non-specialists may nevertheless become involved in referral to health care services as a basic element in the care of drug misusers is a general health check. Accident and emergency departments provide health care in crisis. Roche and Richards (1991) completed a survey of GPs and found that the majority were willing to work with drug misusers. The authors found, in common with Alcohol Concern, that knowledge of drugs and confidence in clinical skills were necessary precursors to intervention. Offering support and co-working with community drug teams is often the answer for many GPs (Greenwood, 1992).

Glass-Crome (1994) points out that in the UK, the Government's *Health of the Nation* strategy includes as a target the reduction of drug-related harm by lowering the percentage of drug misusers sharing needles and syringes to 5% by the year 2000. The strategy highlights the need for professionals adequately trained in health and social work. Hopkins (1991) argues that health workers should no longer try to get drug misusers to stop using, as this is unrealistic, but instead they should aim to reduce the risks. In common with Hopkins, Robertson (1989) also states that, as it is not possible to cure drug misuse, GPs should concern themselves with the reduction of drug-related harm among their patients, both drug misusers and their social and sexual contacts. He gives guidelines and recommends working with non-statutory drug workers.

Espeland (1993) suggests that nurses take the lead in assessing clients for solvent abuse. Preston (1992) states that nurses should take a role in informing young people about Ecstasy misuse. The Children Act in the UK gives children the right to refuse drug treatment. Harding-Price (1993) argues that this means that nurses have a greater responsibility to work to convince drug-misusing children to enter treatment, giving particular thought to issues of confidentiality. Nurses also have a role in working with drug-misusing mothers under stress (Reider, 1990) and the role of nurses in working with HIV and AIDS has implied increasing contact with drug misusers (Faugier, 1988).

Fridinger and Dehart (1993) suggest that the psychological benefits of exercise warrant the development of health and fitness programmes for substance misusers. They also emphasize the usefulness of stress management and other aspects of health promotion such as nutrition.

PSYCHOLOGICAL PROBLEMS AND MENTAL HEALTH – PSYCHIATRISTS, PSYCHOLOGISTS AND COMMUNITY PSYCHIATRIC NURSES

Psychiatrists can offer psychiatric assessment and treatment. Drug specialist psychiatric consultants are often available in mental health services and attached to community drug teams. The psychiatric consultant who specializes in drug problems provides a useful referral point, as several areas of expertise are available from one professional: these include a general health check, psychiatric assessment and review of drug problems. Psychiatrists can also provide treatment, including prescribed drugs and/or detoxification and withdrawal programmes.

Psychologists can provide assessment of psychological problems, cognitive behavioural therapy (usually brief, task-centred) and group therapies. They particularly focus on problems such as depression and anxiety, as these can lead to self-medication and also occur as a consequence of the misuse of drugs and withdrawal (see Chapter 7 for details). Psychologists may also design cognitive behavioural programmes for relapse prevention.

Community psychiatric nurses and other qualified nurses and health visitors work in hospital and the community to provide health care. Some can give prescriptions to clients at home and a CPN is therefore useful for visiting clients attempting a home detoxification. (This service is now often available through community drug teams, rather than health centres.) These professionals often have extensive knowledge of the family and circumstances of clients.

Mental health workers

Acute psychiatric disturbance can occur as a result of drug misuse, for example, amphetamine psychosis. Affective disorders can occur as a result of drug misuse or withdrawal; for example, depression is often associated with depressant drug misuse and with withdrawal from stimulants. It is also clear from Parts One and Two that many drug misusers are attempting to self-medicate for anxiety, depression and other disorders. The difficulty is in determining which came first: this is important as the causal direction will influence the type of intervention but the complexity of the interaction between drug misuse and mental health disorders often precludes a simple answer.

In the USA combined substance misuse or dual diagnosis is increasingly recognized as a problem (Anonymous, 1994). Fine and Miller

(1993) state that trained mental health professionals are traditionally prone to deny or minimize addiction when other problems are present. Kadden and Kranzler (1992) describe a multidisciplinary team approach to treatment of drug dependence and co-morbid disorders. The focus of treatment is on the identification of high-risk and other problem situations, training in coping skills to handle those situations, developing insight and enhancing motivation. Buckley and Bigelow (1992) give an example of a multidisciplinary service incorporating mental health, probation and social and housing services, together with drug and alcohol treatment agencies, suggesting that this comprehensive type of provision is more cost effective than dealing with an individual's problems separately.

PUBLIC HEALTH AND HIV – SPECIALIST DRUG AGENCIES, HEALTH AND SOCIAL CARE AGENCIES

As discussed in Part Two, there is a difference between the aims of HIV prevention implicit within harm minimization policies and drug treatment itself. That is, between promoting public health generally and treating individual cases of drug misuse and dependence. For health care agencies the emphasis is often quite clearly on public health priorities; other agencies may be less clear.

HIV prevention

The main health problem associated with drug misuse is HIV and risk of HIV infection. This is now a far greater risk, in terms of public health, than the misuse of drugs itself (Advisory Council on Misuse of Drugs, 1988/1989). The increased focus on HIV has pinpointed not only the basic health care needs of those who have the infection but also those drug misusers who do not inject or have not (yet) contracted the virus. HIV specialist workers are now attached to health trusts, genitourinary medicine clinics and social services departments. These provide confidential assessment and pre/post-test counselling. It is HIV and the threat of infection that has led to such radical changes in drug service funding and type of service provision in the past decade. HIV is now the major source of concern and main priority in all work with drug misusers.

As Farrell (1991) points out, as a consequence of HIV there has been a growing concern with the physical problems associated with drug misuse. 'As HIV infected individuals become progressively more immunocompromised they will experience increased rates of infective complications such as pneumonia, septicaemia, endocarditis and tuberculosis'. While it is clear that HIV depresses the immune system, it is not clear whether or not drug misuse itself does so.

There is not space here to deal with the care of people who are HIV positive or have AIDS: the reader is therefore referred to the following texts: Advisory Council on Misuse of Drugs 1988/1989, Gaitley and Seed (1989), Bould and Peacock (1989) and Brettle, Farrer and Strang (1990).

There are particular issues for treating drug dependence in people with HIV and AIDS. Brettle *et al.* (1994) describe how a multidisciplinary outpatient medical team functions to deal with the medical, social and public health problems of working with this group. This issue is sharpened when the aims of methadone treatment concur with those of HIV prevention. As Summerhill (1990) points out, there is a public health element to prescribing for clients with HIV, as this can serve to prevent its spread.

DRUG EDUCATION – HEALTH PROMOTION AND TEACHERS

Health promotion units will provide literature regarding drug misuse and HIV, staff will give lectures to the general public and organize training for professional groups. There is often a specialist in substance misuse and/or HIV within each health authority and specialist teachers within local education authorities with specific responsibility for drug education.

PREGNANCY AND THE NEWBORN CHILD – MIDWIVES, HEALTH VISITORS, DOCTORS, SOCIAL WORKERS

Detailed information can be found in *Drugs, Pregnancy and Childcare: A Guide for Professionals* (ISDD, 1995).

In practical terms it is useful to offer information to women drug misusers before they become pregnant concerning contraception, health and the implications of drug misuse for pregnancy. Women using opiates, for example, may think themselves less fertile if menstrual periods stop. This is not necessarily the case and causes added problems in that a woman may not realize she is pregnant until later than usual. For this reason many pregnant drug misusers do not attend for antenatal care until very late in pregnancy. An added complication is the fear of professional intervention. Many women are afraid of the involvement of social services. The reaction from social services departments and paediatricians varies; certain areas may have more flexible policies than others. Although most policies make clear that drug misuse is not a sufficient condition to require social work intervention, there is a delicate balance between ensuring the safety of the child and preserving a cooperative relationship with the mother.

It is difficult to give accurate information about the effect of drugs on the foetus, as drug misuse is often correlated with poor health and the quantities and combinations are difficult to monitor. It appears that low birth weight, susceptibility to infection and feeding problems are possible and that there may be an increase in perinatal mortality. There is some suggestion that for the most serious 2–3% of pregnant drug misusers, congenital abnormalities are possible. The main risk is in the first three months, the time when the heavy drug misuser may not realize that she is pregnant. There is some evidence that drugs such as amphetamine and benzodiazepine may very occasionally cause cleft palate in babies (ISDD, 1995).

It is important to stress that drug withdrawal can be dangerous for the foetus, so the risks of drug misuse need to be weighed against the risks of withdrawal. Drug withdrawal in pregnancy is least stressful if carried out slowly in the middle trimester. It is also probable that if the amount can be reduced greatly or terminated 4–6 weeks prior to birth, withdrawal symptoms are unlikely in the newborn child. Sudden withdrawal is dangerous at any stage in pregnancy because of the risk of intrauterine death.

If the dependent mother has not greatly reduced drug misuse, it is possible that babies will go through a neonatal withdrawal syndrome starting up to a week after delivery and continuing for several weeks. Withdrawals can range in severity from slight irritability to convulsions (it is therefore necessary to inform the mother of the need for neonatal paediatric care). There may also be a risk of HIV transmission from mother to baby, though this is by no means inevitable. It is wise to advise the mother to avoid breastfeeding if she is HIV positive.

It is extremely important to provide aftercare support and this can involve a range of services from the health visitor and GP to social services if there are worries about the safety of the child. Drug misuse is a particular concern in the treatment of mothers and babies. There is a lack of programmes targeted specifically at mothers or pregnant women (Malpas, 1990). This may be linked to the lack of emphasis on services for women generally. Mumme (1991) considers the lack of research and resources for women and stresses the importance of aftercare facilities. Copeland et al. (1993) compared a woman's drug/alcohol service with mixed sex services and found that replicating mixed treatment service provision on a woman-only basis was not enough. They suggest that the treatment content itself needs to be changed. Finkelstein (1993) gives a review of a range of guidelines and protocols and Dawe, Gerada and Strang (1992) describe a pregnancy liaison service attached to a community drug team. It would be an effective and efficient use of resources to focus on the prenatal care for women who misuse drugs. If services could be made accessible and user-friendly for this group, much later damage might be avoided.

CHILD CARE – SOCIAL WORKERS

There is a growing awareness of the need to resolve conflicts between help for drug-misusing parents and child care issues. This is largely concerned with training both drug workers and child care social workers to collaborate and consider both the welfare of the parent and child in tandem. Kearney and Norman-Bruce (1993) state that drug workers need training in child protection. These authors emphasize that the Children Act of 1989 gives opportunities for agencies to gain support for clients with children. Tracy and Farkas (1994) in America also argue that specialists need to be trained for working with children in families with substance misuse. In contrast, Cohen (1990) discusses the problems for social workers trying to deal with drug misuse problems.

The task of social workers is the completion of assessment and care plans for drug-misusing clients with children who may be at risk. Social workers can be involved in the coordination of other professionals to fulfil care plans; they can mobilize resources, offer counselling and, perhaps more importantly, can offer advice regarding welfare rights, housing, child care and financial problems.

There are no grounds for supposing that children are at risk simply because the parents misuse drugs and social workers would be unwilling to be involved unless there were other reasons to suspect that the children are not cared for properly.

However, social workers have a particular responsibility for the needs of children of drug-misusing parents. When receiving a referral or identifying drug misuse within known families, the procedure is not necessarily to focus on the drug misuse *per se*. It is preferable to assess the family on a general basis including health, lifestyle and social support before considering the effects of drug misuse within this context. The relevance of drug misuse to the welfare of children is determined to a large extent by the stability or chaotic nature of parental lifestyle, where responsibilities may be more difficult to meet, and the presence or absence of social support to provide alternative means of meeting these responsibilities. Periods of chaotic drug misuse may lead to intoxication and neglect of responsibilities; these periods may be transitory and alternative support may be available.

The Children Act

The Children Act 1989 greatly influenced social services response to children and families. For information on the changes in social services see

An Introduction to the Children Act 1989 (Department of Health, 1989) and *The Care of Children: Principles and Practices in Regulations and Guidance* (Department of Health, 1990). See also Cobley (1995) for a thorough review from the legal perspective.

The Act is concerned with providing support to parents and families to enable children to remain in the home. An emphasis has been placed on the parental rights and responsibilities and the need for professionals to provide good reason why removal into care or 'being looked after by the local authority' is a preferable alternative to the child remaining at home. The emphasis is instead of providing support and prevention services to maintain the child in the home environment. This has implications for drug misusers.

The Act allows some flexibility to local authorities in defining the term 'in need'. First, 'S/he is likely to achieve or maintain, or have the opportunity of achieving or maintaining a reasonable standard of health or development with the provision of services by a Local Authority'. Second, 'Health or development is likely to be significantly impaired or further impaired without the provision of such services' (The Children Act 1989, 17.10 (a) and (b)).

The resources available for support of drug-misusing parents include contractual child minding, family centres, nurseries, day care facilities, local play groups and nurseries.

Guidelines for the social work assessment of drug misusers

In 1986, the Standing Conference on Drug Abuse, UK (SCODA) drew up guidelines for the social work assessment of drug misusers. They are aimed at enabling the social worker to establish in what ways, if any, drug misuse is putting children at risk. These guidelines are published by the Institute for the Study of Drug Dependence (Standing Conference on Drug Abuse, 1986). Although they were designed in 1986, they share the underlying philosophy of the Children Act 1989.

The guidelines emphasize that it is extremely difficult to assess emotional neglect and there is no evidence to lead us to suppose that parents misusing drugs are less emotionally available than those who are not.\The main point raised is that chaotic drug misuse is a far greater danger to children than dependent drug misuse. The patterns and type of drug misuse are therefore of more importance in an assessment than drug dependency itself. \Drug misuse is seen as problematic if it leads to erratic behaviour, either for a short period or over a longer period.

The guidelines therefore stress that it is not necessary for parents to be abstinent to be able to care for their children. However, as changes in amount of drug used leads to intoxication and withdrawal which may lead to erratic behaviour, it may be necessary to stabilize and regulate drug misuse rather than encourage attempts at withdrawal and abstinence. The reasons underlying drug misuse need to be carefully assessed, as chaotic misuse may be precipitated by certain needs or events, which when resolved return the drug misuse to a safer stable pattern offering less threat to the safety of a child.

As with any social work assessment, it is necessary to look beyond the immediate nuclear family for evidence of supportive environments, as often extended families and friends offer emotional and practical support networks for children. Women particularly will often make alternative child care arrangements if they expect to be away from the home or intoxicated for any period of time.

The social environment of the child may be more risky if it includes drug dealing, violence and/or criminality. It is, of course, not necessarily the case that a drug misuser is closely involved with criminality or violence. Most will acquire their drugs from friends outside the home and are unlikely to deal to any large extent. However, it is possible that the child may be living in environments where their health is at risk from the presence or drugs and/or needles and syringes.

The most important point made in these guidelines is that drug misuse in itself is not a justification for automatic Child Protection Registration. It may well, however, be grounds for a thorough assessment procedure. It is important to consider what help and support can be offered, both in terms of child care provision and harm minimization and drug treatment. Drug-misusing parents may well be able to overcome periods of chaotic drug misuse if given temporary support with drug problems and child care and ongoing aftercare maintenance.

Working with children under 16

The needs of young people range from advice and information to prescribed drugs and clean needles and syringes. If children under 16 request these services it presents serious dilemmas to many different types of professional and the response will vary depending on the responsibilities and priorities of each. It may be appropriate for professionals to follow the guidance given during the landmark case *Gillick v West Norfolk and Wisbech Area Health Authority* which involved the provision of contraceptive advice and treatment of under-16s without parental consent. In the House of Lords judgement some guidelines were given concerning responsible practice which can be generalized to drug misusers.

This guidance stressed that to deny young people under the age of 16 the confidentiality of professional relationships:

... might cause some not to seek professional advice at all. They could then be exposed to the immediate risks of pregnancy and of sexually transmitted diseases, as well as long-term physical, psychological and emotional consequences.

Jane Goodsir, in a Release pamphlet (n.d.), points to the similarities between the harm reduction objectives of intervention with drug misusers and the guidance in the Gillick case. She emphasizes that the Lords held that 'In general the rights of parents applied to duties over the child's welfare'. However, circumstances might arise when a doctor (and other professionals) could use their discretion to act in the child's best interests without parental consent. Lord Frasier outlined five main conditions where treatment without parental consent might be justified. The following adaptions of these recommendations are given by Goodsir, who has simply substituted 'drugs' for 'contraceptive' and 'young person' for 'girl'.

First, that the young person, although under 16 years of age, will understand the professional's advice; second, that the professional cannot persuade the young person to inform parents or to allow him/her to inform the parents that the young person is seeking drugs advice; third, that the young person is very likely to begin or to continue misusing drugs with or without drug treatment; fourth, that unless the young person received drugs advice or treatment the young person's physical or mental health or both are likely to suffer; and finally, that the young person's best interests require the professional to give the young person drugs advice or treatment or both without parental consent.

As Goodsir points out, the judges had effectively delegated responsibility to professionals (in this case, doctors) better qualified to deal with cases on an individual basis, implying that professionals must themselves resolve the tension between parental rights and confidentiality, judging each case on its merits.

CRIME AND CUSTODY – PROBATION, THE POLICE AND PRISON OFFICERS

As with social work, the aims of probation and the police are not concerned primarily with drug misuse itself. Instead, these professions are concerned with the relationship between drug misuse and crime. The focus for these professionals must therefore be on the crime rather than the drugs.

The population of drug misusers attending agencies is not representative of the general population, but instead resembles the probation/prison population in terms of age, employment and criminality (Jarvis and Parker, 1989). It is possible that drug misusers and criminals constitute a large part of the same social groups, though each can of course exist independently (for example, the majority of recreational drug misusers in Chapter 2 did not have a criminal record). Studies have consistently demonstrated a high correlation between problem drug misuse and crime, but researchers have been unable to determine a causal relationship between them (Bean and Wilkinson, 1988; Hammersley, Forsyth and Lavelle, 1990).

It is of course possible that some types of drug misuse, such as regular heroin misuse, may lead to crime to finance the habit, but also that crime can enable people to develop contacts and buy drugs. Therefore, although there is a strong correlation between drugs and crime, there is at present little evidence to suggest that drug misuse necessarily leads to crime. If anything, the causal relationship may be in the other direction, *viz.* that criminal activities precede drug misuse and that termination of criminal activity happens before drug misuse stops. In other words, if you stop people committing crimes, they will stop misusing drugs, as they no longer have the necessary finances or the contacts to carry on. It is, however, more likely that there is no direct causal relationship between crime and drug misuse, but rather that similar underlying conditions facilitate both activities.

Both the police and probation departments have recently developed partnership with drug agencies – the police by arranging cautioning schemes, whereby drug misusers are referred to drug agencies rather than charged with offences, probation by developing referral schemes that involve probation and drug agencies working in partnership to assess, refer and treat drug misusers.

Police

The police are increasingly involved in multidisciplinary projects to prevent drug misuse and to treat drug misusers. The participation of the police in drug prevention was discussed in Chapter 6. They are also active in the referral of drug misusers to drug agencies. Arrest referral schemes are a recent phenomenon and as a consequence there is little national literature on the subject. Wright discusses the increased use of police caution and diversion schemes (*Probation Journal*, **37**, 1990) and examines Northamptonshire's initiative for diversion of adults. He argues that, although the bulk of energy has been targeted at offenders at risk of custody, there is also a potential to increase police cautioning generally. Savage (*Justice of the Peace*, **152**, 1988) acknowledges the value of increased diversion from custody for juvenile offenders, but questions why they should be diverted from a court hearing. A range of issues has arisen from the early schemes, for example, disparity in practice across the country, whether the juvenile has admitted the offence before being cautioned and diversion without cautioning. The most common problem is a reflection of the differing aims and priorities of both partners. The great majority of those cautioned by the police are cannabis misusers, who by definition have committed a criminal offence. These people do not usually have drug problems and therefore drug agency intervention is unnecessary.

Drug misusers present a more immediate problem when in police custody. It is important that a detained drug misuser is closely observed and assessed by a doctor as soon as possible. There can be a danger of overdose and need for resuscitation. It may be appropriate for the doctor to prescribe to prevent withdrawals and render the drug misuser capable of responding appropriately in an interview.

Probation

Probation officers can co-work with specialist agencies. They are responsible for court reports and may have resources available for clients, such as drug education groups and diversionary activities. Probation–drug agency partnerships also (in common with police arrest referral programmes) have to contend with differing aims and priorities. Whereas the probation task is to prevent reconviction, many drug agencies focus on preventing drug-related harm and providing help and support for drug misusers. These control and health-orientated priorities may well be in conflict.

Probation officers and drug agency staff have had to compromise their own professional priorities in order to deal with these conflicts. For example, probation staff have adapted to the health needs of drug misusers by providing needle and syringe disposal points in probation offices, despite the legal implications of injecting drugs for their clients. Similarly, many drug agencies have tightened up contractual arrangements with clients in order to increase the control function.

This conflict between the priorities of crime prevention and health care is not limited to police referral schemes and probation partnerships.

Prison

This area has been dealt with in some detail in Parts One and Two. We need therefore only note that on reception to prison a medical questionnaire and examination may reveal drug misuse or withdrawals. A doctor should provide detoxification programmes over time where regular drug misuse is identified. Where the patient has been receiving prescriptions or injecting equipment in the community, it is useful to contact the drug agency or previous doctor for information.

If the patient has injected drugs, special care should be taken to provide information concerning the dangers of HIV in the prison environment, particularly the dangers of sharing equipment. It is particularly important to explain where disinfecting tablets are available and how these should be used. All drug misusers should be informed of the dangers of reduced tolerance and overdose when leaving prison. It may also be useful to arrange support from a GP or drug agency in the community.

MULTIDISCIPLINARY WORKING

The expansion of drug misuse generally has resulted in a need for social and health care workers to develop a new area of expertise regarding the extent, seriousness and implications of drug misuse. It can be seen from the information in this chapter that a range of different professionals provide a variety of services of value to drug misusers. General practitioners, community nurses, social workers and health visitors may be best placed to identify and work with drug problems, as they visit people in their own homes, have regular contact and have the relevant skills and expertise. This work can be facilitated through specialist support, if generic professionals work in conjunction with each other and with specialist drug agencies. Social and health care workers can provide support facilities from child care, transport and housing to health care and mental health treatment. Specialist agencies can provide assessment, harm minimization, treatment and aftercare services (Gustavsson, 1991).

There are problems inherent in multidisciplinary working as a consequence of differing beliefs and priorities. Each separate discipline will approach problems of drug and alcohol misuse in different ways depending on the objectives and priorities of their particular profession. For example, a general practitioner would focus on the health care aspects, whereas a probation officer would focus on the criminal implications. These can be best negotiated through prior knowledge of the roles and responsibilities of each.

In essence, social and health care professionals provide for the social and health care needs of drug misusers: in this respect drug misusers should be treated no differently from other clients. The moral issues of child care and crime are the specific remit of social workers and probation.

Unfortunately, the moral and control issues often become confused with social and health care tasks. Before referring drug misusers to specialist or generic agencies it is as well to consider the reasons why they may not want to go. In the classic report of drug misuser views (Drug Indicators Project, ISDD, 1989), the respondents listed what they considered to be the barriers to help-seeking. These included Home

Office notification, lack of confidentiality and 'expressed fear regarding the consequences for their children should they seek help'. The majority of drug misusers, in common with those in Chapters 2 and 3, do not see their drug misuse as the most significant problem, whereas they can be fairly sure that a professional would.

Appendix – A brief glossary of drugs

Drugs	Colloquial and trade names
Stimulants	
Amphetamines	Speed, whiz, sulphate
Dexamphetamine	Dexedrine
Cocaine	Coke, charlie
Crack (freebased cocaine)	Ice, crack, base, rock
Depressants	
Opiates:	
Heroin (diamorphine)	H, junk, skag, smack
Methadone	Meth, linctus
Buprenorphine	Temgesic
Dextromoramide	Palfium
Dextropropoxyphene	Distalgesic
Barbiturates:	
Quinalbarbitone	Seconal, Tuinal
Pentobarbitone	Nembutal
Minor tranquillizers:	
Diazepam	Valium
Lorazepam	Ativan
Temazepam	Normison, 'tems', 'eggs', 'jellies'
Nitrazepam	Mogadon
Cannabis	Dope, blow, grass
Hallucinogens	
Lysergic acid	LSD, acid
Psilocybin	Magic mushrooms

Others

Amyl nitrate	Poppers
Methylenedioxyamphetamine	Ecstasy, E
MDA, MDEA	

References

Adger, H. (1991) Problems of alcohol and other drug use and abuse in adolescents. *Journal of Adolescent Health*, **12**, 606–13.

Advisory Council on the Misuse of Drugs (1984) *Prevention*, HMSO, London.

Advisory Council on the Misuse of Drugs (1988/1989) *AIDS and Drug Misuse*, Parts 1 and 2, HMSO, London.

Advisory Council on the Misuse of Drugs (1991) *Drug Users and the Criminal Justice System*, HMSO, London.

Advisory Council on the Misuse of Drugs (1993) *Drug Education in Schools: The Need for New Impetus*, HMSO, London.

Alberts, J.K., Hecht, M.L., Miller-Rassulo, M. and Krizek, R.L. (1992) The communicative process of drug resistance among high school students. *Adolescence*, **27**, 203–26.

Alcoholics Anonymous (1981) *Survey of AA in Great Britain*, General Service Board of Alcoholics Anonymous Ltd.

Anderson, D. (1981) *The Minnesota Experience*, Hazelden Foundation, Minnesota.

Andrews, J.A., Hops, H., Ary D., Lichenstein, E. and Tildesley, E. (1991) The construction, validation and use of a Guttman scale of adolescent substance use: an investigation of family relationships. *Journal of Drug Issues*, **21**, 557–72.

Anonymous (1970) *The Little Red Book*, Hazelden Foundation, Minnesota.

Anonymous (1976) *Alcoholics Anonymous*, 3rd edn, Watson and Witney.

Anonymous (1994) Position statement on the need for improved training for treatment of patients with combined substance use and other psychiatric disorders. *American Journal of Psychiatry*, **151**(5), 795–6.

Antze, P. (1979) The role of ideologies in peer psychotherapy groups, in *Self Help Groups for Coping with Crisis*, (eds M.A. Lieberman, L.D. Borman and P. Antze), Jossey-Bass, San Francisco.

Azrin, N.H., Sisson, R.W., Meyers, R. and Godley, M. (1982) Alcoholism treatment by Disulfiram and community reinforcement therapy. *Journal of Behaviour Therapy and Experimental Psychiatry*, **13**, 105–12.

Babor, T.F., Ritson, E. and Hodgson, R. (1986) Alcohol related problems in the primary health care setting: a review of early intervention strategies. *British Journal of Addiction*, **81**, 23–46.

Babor, T.F., Dolinsky, Z., Rounsaville, B. and Jaffe, J. (1988) Unitary versus multidimensional models of alcoholism treatment outcome: an empirical study. *Journal of Studies on Alcohol*, **49**, 167–77.

Baekland, F., Lundwall, L. and Kissen, B. (1975) Methods for the treatment of chronic alcoholism: a critical appraisal, in *Research Advances in Alcohol and Drug Problems*, **2**, (eds R.J. Gibbens, Y. Israel, H. Kalant, R.E. Popham, W. Schmidt and R.G. Smart), Wiley, New York, pp. 247–327.

Bailey, S.L. (1992) Adolescents' multisubstance use patterns: the role of heavy alcohol and cigarette use. *American Journal of Public Health*, **82**(9), 1220–4.

Balfour, D.J.K. (1990) *Psychotropic Drugs of Abuse*, Pergamon, New York.

Balfour, D.J.K. (1994) Neural mechanisms underlying nicotine dependence. *Addiction*, **89**(11), 1419–23.

Ballard, R. (1988) *Teacher Training for Effective School-based Drug Education*, Queensland Government Printer, Brisbane.

Bandura, A. (1977) *Social Learning Theory*, Prentice-Hall, Englewood Cliffs.

Banks, A. and Waller, T. (1988) *Drug Misuse, A Practical Handbook for GPs*, Blackwell Scientific Publications, Oxford.

Barbee, J.G., Clark, P.D. and Crapanzano, M.S. (1989) Alcohol and substance abuse among schizophrenic patients presenting to an emergency psychiatric service. *Journal of Nervous and Mental Disease*, **177**, 400–7.

Barnea, Z., Rahav, G. and Teichman, M. (1987) The reliability and consistency of self reports on substance use in a longitudinal study. *British Journal of Addiction*, **82**, 891–8.

Battjes, R.J., Pickens, R. and Amsel, Z. (1991) *Trends in HIV Infection and AIDS Risk Behaviours among Intravenous Drug Users in Selected US Cities*. Seventh International Conference on AIDS, Florence, Abstract THC46.

Bean, P. and Wilkinson, C. (1988) Drug taking, crime and the illicit supply system. *British Journal of Addiction*, **83**, 533–9.

Beck, A.T. (1989) *Cognitive Therapy and the Emotional Disorders*, International Universities Press, Inc., New York.

Becker, H.S. (1964) *The Other Side*, Free Press, New York.

Becker, M.H. (ed.) (1974) The health belief model and personal health behaviour. *Health Education Monographs*, **2**, 324–508.

Becker, M.H. and Maiman, L.A. (1975) Sociobehavioural determinants of compliance with health and medical care recommendations. *Medical Care*, **13**, 10–24.

Bekir, P., McLellan, T., Childress, A.R. and Gariti, P. (1993) Role reversals in families of substance misusers. *International Journal of the Addictions*, **28**(7), 613–30.

Bell, D.S. (1990) The irrelevance of governmental policies on drugs. *Drug and Alcohol Dependence*, **25**, 221–4.

Bell, J., Digiusto, E. and Byth, K. (1992) Who should receive methadone maintenance? *British Journal of Addiction*, **87**(5), 689–94.

Berberian, R.M., Gross, C., Lovejoy, J. and Parerella, C. (1976) The effectiveness of drug education programs: a critical review. *Health Education Monograph*, **4**, 377–98.

Berg, I.K. and Hopwood, L. (1991) Doing with very little: treatment of homeless substance abusers. *Journal of Independent Social Work*, **5**(3/4), 109–19.

Bergin, A. and Garfield, S. (eds) (1978) *Psychotherapy and Behaviour Change*, Wiley, New York.

Billings, A.G. and Moos, R.H. (1983) Psychosocial processes of recovery among alcoholics and their families: implications for clinicians and programme evaluators. *Addictive Behaviours*, **8**(3), 205–18.

Birke, S.A., Edelmann, R.J. and Davis, P.E. (1990) An analysis of the abstinence violation effect in a sample of illicit drug users. *British Journal of Addiction*, **85**, 1299–307.

Blankertz, L.E., Chaan, R.A., White, K., Fox, J. and Messinger, K. (1990) Outreach efforts with dually diagnosed homeless persons. *Families in Society*, **71**, 387–95.

Blaze-Temple, D. and Kai Lo, S. (1992) Stages of drug use: a community survey of Perth teenagers. *British Journal of Addiction*, **87**, 215–25.

Bloor, M. (1995) *The Sociology of HIV Transmission*, Sage, London.

Bloor, M., Frischer, M., Taylor, A. *et al.* (1994) Tideline and turn: possible reasons for the continuing low HIV prevalence among Glasgow's injecting drug users. *Sociological Review*, **42**(4), 738–57.

Bolles, R.C. (1979) *Learning Theory*, Holt, Rinehart and Winston, New York.

Bolton, K. and Sellick, S. (1991) Out on your own: making solo outreach work. *Druglink*, **6**, 9–10.

Botvin, G.J. (1990) *Substance Abuse Prevention: Theory, Practice and Effectiveness*, University of Chicago Press, Chicago.

Bould, M. and Peacock, G. (1989) *AIDS Models of Care*, King's Fund Centre, London.

Bowman, W.C. and Rand, M.J. (1980) *Textbook of Pharmacology*, Blackwell, Cambridge.

Bratt, I. (1953) *Alcoholism, a Disease?*, Bonniers, Stockholm.

Brettle, R.P., Farrell, M. and Strang, J. (1990) The clinical manifestations of HIV in drug users, in *AIDS and Drug Misuse*, (eds S. Strang and V.G. Stimson), Routledge, London.

Brettle, R.P., Willocks, L., Hamilton, B.A. and Shaw, L. (1994) Out-patient medical care in Edinburgh for IDU-related HIV. *AIDS Care*, **6**(1), 49–58.

Brinker, R.P., Baxter, A. and Butler, L.S. (1994) An ordinal pattern analysis of four hypotheses describing interactions between drug addicted, chronically disadvantaged and middle-class mother–infant dyads. *Child Development*, **65**(2), 361–72.

Brown, B.S. (1979) *Addicts and Aftercare: Community Integration of the Former Drug User*, Sage, Beverly Hills.

Brown, B.S. (1993) Observations on the recent history of drug user counselling (review). *International Journal of the Addictions*, **28**(12), 1243–55.

Brown, B.S. and Ashery, R.S. (1979) Aftercare in drug abuse programming, in *Handbook on Drug Abuse*, (eds R.I. DuPont, A. Goldstein and J. O'Donnell), National Institute of Drug Abuse, Washington DC.

Brown, L.S., Mitchel, J.L., DeVorre, S.L. and Primm, B.J. (1989) Female intravenous drug users and perinatal HIV transmission. *New England Journal of Medicine*, **320**(22), 1493–4.

Brown, S.A., Goldman, M.S. and Christiansen, B.A. (1985) Do alcohol expectancies mediate drinking patterns of adults? *Journal of Consulting and Clinical Psychology*, **53**(4), 419–26.

Buckley, R. and Bigelow, D.A. (1992) The multiservice network: reaching the unserved multiproblem individual. *Community Mental Health Journal*, **28**(1), 43–50.

Buning, E. (1991) The role of harm reduction programmes in curbing the spread of HIV by drug injectors, in *AIDS and Drug Misuse*, (eds J. Strang and G.V. Stimson), Routledge, London.

Burke, A.C. (1992) Between entitlement and control: dimensions of U.S. drug policy. *Social Service Review*, **66**(4), 571–81.

Burt, J. and Stimson, G.V. (1990) *Strategies for Protection: Drug Injecting and the Prevention of HIV Infection*. Monitoring Research Group, The Centre for Research on Drugs and Health Behaviour, Charing Cross and Westminster Medical School, London.

Caddy, G.R. and Block, T. (1985) Individual differences in response to treatment, in *Determinants of Substance Abuse: Biological, Psychological and Environmental Determinants*, (eds M. Galizio and S.A. Maisto), Plenum Press, New York.

Caldwell, C. (1976) Physiological aspects of cocaine usage, in *Cocaine; Chemical, Biological, Clinical, Social and Treatment Aspects*, (ed. S.J. Mule), CRC Press, Cleveland, Ohio.

Caplehorn, J.R.M., Bell, J., Kleinbaum, D.G. and Gebski, V.J. (1993) Methadone dose and heroin use during maintenance treatment. *Addiction*, **88**(1), 119–24.

Carroll, K.M., Rounsaville, B.J. and Bryant, K.J. (1993) Alcoholism in treatment-seeking cocaine abusers: clinical and prognostic significance. *Journal of Studies on Alcohol*, **54**(2) 199–208.

Cartwright, D. (1979) Contemporary social psychology in historical context. *Social Psychology Quarterly*, **42**, 82–93.

Carvell, A.L.M. and Hart, G.T. (1990) Risk behaviours for HIV infection among drug users in prison. *British Medical Journal*, **300**(6736), 1383–4.

Catalano, R.F. and Hawkins, J.D. (1985) Project skills: preliminary results from a theoretically based aftercare experiment, in *Progress in the Development of Cost-Effective Treatment for Drug Abuse*, (ed. R.S. Ashery), National Institute on Drug Abuse, Rockville, MD.

Catalano, R.F., Hawkins, J.D., Wells, E.A. and Miller, J. (1990/91) Evaluation of effectiveness of adolescent drug abuse treatment, assessment of risks for relapse and promising approaches for relapse prevention. *International Journal of Addictions*, **25** (9A and 10A), 1085–140.

Catalano, R.F., Hawkins, J.D., Krenz, C. and Gilmore, M. (1993) Special populations: using research to guide culturally appropriate drug abuse prevention. *Journal of Consulting and Clinical Psychology*, **61**(5), 804–11.

Catania, C., and Harnad, S. (eds) (1988) *The Selection of Behaviour*, CUP, Cambridge.

Chalmers, J.W.T. (1990) Edinburgh's community drug problem service, pilot evaluation of methadone substitution. *Health Bulletin*, **48**, 62–72.

Chasnof, I. (1989) Drug use and women: establishing a standard of care, in *Prenatal Abuse of Licit and Illicit Drugs*, (ed. D. Hutchings), New York Academy of Sciences, New York, pp. 208–10.

Chavkin, W. (1990) Drug addiction and pregnancy: policy crossroads. *American Journal of Public Health*, **80**(4), 483–7.

Cheung, Y.W. (1993) Approaches to ethnicity: clearing roadblocks in the study of ethnicity and substance use. *International Journal of the Addictions*, **28**(12), 1209–26.

Clayton, R. and Ritter, C. (1985) The epidemiology of alcohol and drug abuse among adolescents. *Advances in Alcohol and Substance Abuse*, **4**, 67–97.

Cobley, C. (1995) *Child Abuse and the Law*, Cavendish Publishing, London.

Coggans, H., Shewan, D., Henderson, M., Davies, J.D. and O'Hagen, F.J. (1990) *National Evaluation of Drug Education in Scotland*, Scottish Education Department.

Cohen, P. (1989) *Cocaine Use in Amsterdam in Nondeviant Subcultures*, Institute voor Social Geografie, Universiteit van Amsterdam, Amsterdam.

Cohen, P. (1990) Balancing act. *Social Work Today*, **22**, 18–19.

Conrad, K.M., Flay, B.R. and Hill, D. (1992) Why children start smoking. Predictors of onset. *British Journal of Addiction*, **87**(12), 1711–24.

Cook, C. (1988) The Minnesota Model in the management of drug and alcohol dependency: Part 1, philosophy and programme; Part 2, guidance and conclusions. *British Journal of Addiction*, **83**, 626–34 and 735–43.

Copeland, J., Hall, W., Didcott, P. and Biggs, V. (1993) A comparison of a specialist women's alcohol and other drug treatment service with two traditional mixed-sex services: client characteristics and treatment outcome. *Drug and Alcohol Dependence*, **32**(1), 81–92.

Craig, R.J. and Olson, R.E. (1990) MCMI comparisons of cocaine abusers and heroin addicts. *Journal of Clinical Psychology*, **46**(2), 230–7.

Cronkite, R.C. and Moos, R.H. (1980) Determinants of post treatment functioning of alcoholic patients: a conceptual framework. *Journal of Consulting and Clinical Psychology*, **48**, 305–16.

Czarnecki, D.M., Russell, M., Cooper, M.L. and Salter, D. (1990) Five-year reliability of self-reported alcohol consumption. *Journal of Studies on Alcohol*, **51**, 68–76.

Dackis, C.A. and Marks, S.G. (1983) Opiate addiction and depression: cause or effect. *Drug and Alcohol Dependence*, **11**, 105–9.

Dai, B. (1937) *Opium Addiction in Chicago*, Commercial Press, Shanghai.

Darke, S., Swift, W., Hall, W. and Ross, M. (1994) Predictors of injecting and injecting risk-taking behaviour among methadone maintenance clients. *Addiction*, **89**(3), 311–16.

Davidson, R., Rollnick, S. and MacEwan, I. (eds) (1991) *Counselling Problem Drinkers*, Routledge, London.

Davies, J.B. (1993) *The Myth of Addiction*, Harwood Academic Publishers, Switzerland.

Davies, J.B. and Coggans, N. (1991) *The Facts about Adolescent Drug Use*, Cassel, London.

Davies, M. (1985) *The Essential Social Worker: A Guide to Positive Practice*, Community Care Practice Handbooks. Ashgate, Hants.

Dawe, S., Gerada, C. and Strang, J. (1992) Establishment of a liaison service for pregnant opiate-dependent women. *British Journal of Addiction*, **87**, 867–71.

Deakin, E.Y., Levy, J.C. and Wells, V.W. (1987) Adolescent depression, alcohol and drug abuse. *American Journal of Public Health*, **77**, 178–82.

De Haes, W. and Schurman, J. (1975) Results of an evaluation study on three drug education models. *International Journal of Health Education*, **18**, supplement.

Dembo, R., Williams, L. and Getrev, A. (1991) A longitudinal study of the relationship among marijuana/hashish use, cocaine use and delinquency in a cohort of high risk youths. *Journal of Drug Issues*, **21**, 271–312.

Denzin, N.K. (1986) *The Recovering Alcoholic*, Sage, Beverly Hills.

Denzin, N.K. (1987a) *Treating Alcoholism: An Alcoholics Anonymous Approach*, Sage, Beverly Hills.

Denzin, N.K. (1987b) *The Alcoholic Self*, Sage, Beverly Hills.

Department of Health (1989) *An Introduction to the Children Act*, HMSO, London.

Department of Health (1990) *The Care of Children: Principles and Practices in Regulations and Guidance*, HMSO, London.

Department of Health (1994) *Drug Misuse Statistics – Six months ending 31 March 1993*, Government Statistical Service, HMSO, London.

Department of Health (1995) *Tackling Drugs Together*, HMSO, London.

Department of Health, Scottish Home and Health Department and Welsh office (1991) *Drug Misuse and Dependence: Guidelines on Clinical Management*, HMSO, London.

Dermott, F. and Pyett, P. (1994) Co-existent psychiatric illness and drug abuse: a community study. *Psychiatric and Psychological Evidence* **22**, 45–52.

Des Jarlais, D.C. (1990) Stages in the response of the drug abuse treatment system to the AIDS epidemic in New York City. *Journal of Drug Issues*, **20**, 335–47.

Des Jarlais, D. and Friedman, S.R. (1988) HIV and intravenous drug use. *AIDS*, **2** (suppl 1), S65–S69.

Des Jarlais, D.C., Friedman, S.R. and Casriel, C. (1990) Target groups for preventing AIDS among intravenous drug users: 2 The 'hard' data studies. *Journal of Consulting and Clinical Psychology*, **58**(1), 50–6.

Deykin, E.Y., Buka, S.L. and Zeena, T.H. (1992) Depressive illness among chemically dependent adolescents. *American Journal of Psychiatry*, **149**(10), 1341–7.

Dickerson, D.P. (1991) Teacher education and its role in urban drug prevention. *Early Child Development and Care*, **73**, 133–8.

DiClemente, C.C. and Prochaska, J.O. (1985) Processes and stages of change: coping and competence in smoking behaviour change, in *Coping and Substance Abuse*, (eds S. Shipman and W. Wills), Academic Press, New York.

Dolan, K.A., Donoghoe, M.C., Jones, S. and Stimson, G.V. (1991) *A Cohort Study at Four Syringe Exchange Schemes and Comparison Groups of Drug Injectors*, Report to the Department of Health, The Centre for Research on Drugs and Health Behaviour.

Donoghoe, M.C., Stimson, G.V. and Dolan, K. (1989) Sexual behaviour of injecting drug users and associated risks of HIV infection for non-injecting sexual partners. *AIDS Care*, **1**, 51–8.

Donoghoe, M.C., Dolan, K.A. and Stimson, G.V. (1991) *The 1989–1990 National Syringe-Exchange Monitoring Study*, Monitoring Research Group, The Centre for Research on

Drugs and Health Behaviour, Charing Cross and Westminster Medical School, London.

Donovan, D.M. and Marlatt, G.A. (1980) Assessment of expectancies and behaviours associated with alcohol consumption: a cognitive behavioural approach. *Journal of Studies on Alcohol*, **41**, 1156–85.

Dorn, N. and Murji, K. (1992) *Drug Prevention: A Review of the English Language Literature*, ISDD Research Monograph 5, Institute for the Study of Drug Dependence, London.

Drummond, C. (1991) Dependence on psychoactive drugs: finding a common language, in *Addiction Behaviour*, (ed. I.B. Glass), Routledge, London.

Drummond, D.C., Cooper, T. and Glautier, S.P. (1990) Conditioned learning in alcohol dependence: implications for cue exposure treatment. *British Journal of Addiction*, **85**(6), 725–43.

Duran, H. and Brooklyn, J. (1988) Inherent problems in substance abuse education on university campuses: student perspectives. *Journal of College Student Psychotherapy*, **3** and **4**, 63–87.

Easthope, G. (1993) Perceptions of the causes of drug use, in a series of articles in the *International Journal of the Addictions*, **28**(6), 559–69.

Easthope, G. and Lynch, P.P. (1992) Voluntary agencies dealing with drug and alcohol misusers: a comparison of three surveys conducted in London, Scotland and Tasmania. *International Journal of the Addictions*, **27**(12), 1401–11.

Edwards, G. (1986) The Alcohol dependence syndrome. *British Journal of Addiction*, **81**(2), 171.

Edwards, G. (1988) Long term outcome for patients with drinking problems: the search for predictors. *British Journal of Addiction*, **83**(8) 917–27.

Edwards, G. (1989a) As the years go rolling by: drinking problems in the time dimension. *British Journal of Psychiatry*, **154**, 18–26.

Edwards, G. (1989b) What drives British drug policies? *British Journal of Addiction*, **84**(2), 219–26.

Edwards, G. and Grant, M. (eds) (1976) *Alcoholism: New Knowledge and Responses*, Croom Helm, London.

Edwards, G. and Grant, M. (eds) (1977) *Alcoholism*, Oxford University Press, Oxford.

Edwards, G. and Gross, M.M. (1976) Alcohol dependence: provisional description of a clinical syndrome. *British Medical Journal*, **1**, 1058–61.

Edwards, G. and Orford, J. (1977) Alcoholism: a controlled trial of treatment and advice. *Journal of Studies on Alcohol*, **38**, 1004–31.

Edwards, G., Brown, D., Duckitt, A., Oppenheimer, E., Sheehan, M. and Taylor, C. (1987) Outcome of alcoholism: the structure of patient attributions as to what causes change. *British Journal of Addiction*, **82**(5) 533–45.

Egan, G. (1990) *The Skilled Helper: A Systematic Approach to Effective Helping*, Brooks Cole, California.

Eggert, L.L. and Herting, J.R. (1991) Preventing teenage drug abuse: exploratory effects of network social support. *Youth and Society*, **22**(4), 482–524.

Eggert, L.L. and Herting, J.R. (1993) Drug involvement among potential drop-outs and 'typical' youth. *Journal of Drug Education*, **23**(1), 31–55.

Eggert, L.L., Thompson, E.A., Herting, J.R. and Nicholas, L.J. (1994) Preventing adolescent drug abuse and high school dropout through an intensive school-based social network development programme. *American Journal of Health Promotion*, **8**(3), 202–15.

Eiser, J.R. (1982) Addiction as attribution, cognitive processes in giving up smoking, in *Social Psychology and Behavioural Medicine*, (ed. J.R. Eiser), Wiley, Chichester.

Eiser, J.R. and Gossop, M. (1979) 'Hooked' or 'sick': addicts' perceptions of their addiction. *Addictive Behaviours*, **4**, 185–91.

Eiser, J.R. and Sutton, S.R. (1977) Smoking as a subjectively rational choice. *Addictive Behaviours*, **2**, 129–34.

Eiser, C., Eiser, J.R. and Pritchard, M. (1988) Reactions to drug education: a comparison of two videos. *British Journal of Addiction*, **83**(8), 955–63.

Ellen, R.F. (ed). (1984) *Ethnographic Research: A Guide to General Conduct*, Academic Press, London.

Ellickson, P.L., Bell, R.M. and McGuigan, K. (1993) Preventing adolescent drug use: long term results of a junior high program. *American Journal of Public Health*, **83**(6), 856–61.

Ellis, A. (1962) *Reason and Emotion in Psychotherapy*, Lyle Stuart, New York.

Ellis, A. (1987) The evolution of rationale-emotive therapy (RET) and cognitive-behaviour therapy (CBT), in *The Evolution of Psychotherapy*, (ed. J. Zeig), Brunner/Mazel, New York.

Ellis, R. and Stephens, R.C. (1976) The arrest history of narcotic addicts prior to admission. A methodological note. *Drug Forum*, **5**, 211–24.

Elmquist, D.L., Morgan, D.P. and Bolds, P.K. (1992) Alcohol and drug use among adolescents with disabilities. *International Journal of the Addictions*, **27**(12), 1475–83.

Engelsman, E.L. (1991) Drug misuse and the Dutch. *British Medical Journal*, **302**, 484–5.

Ennett, S.T., Tobler, N.S., Ringwalt, C.L. and Flewelling, R.L. (1994) How effective is drug abuse resistence education? A meta-analysis of Project DARE outcome evaluations. *American Journal of Public Health*, **84**(9), 1394–401.

Erickson, P.G. (1990) A public health approach to demand reduction. *Journal of Drug Issues*, **20**, 563–75.

Erikson, L., Bjornstad, S. and Gotestam, K.G. (1986) Social skills training in groups for alcoholics: one year treatment outcome for groups and individuals. *Addictive Behaviours*, **11**, 309–29.

Espeland, K. (1993) Inhalant abuse: assessment guidelines. *Journal of Psychosocial Nursing & Mental Health Services*, **31**(3), 11–14.

Ettore, B. (1988) London's voluntary drug agencies. III. A 'snapshot' view of residential and non residential clients. *International Journal of the Addictions*, **23**(12), 1255–69.

Farrell, M. (1990) Beyond platitudes: problem drug use, a review of training. *British Journal of Addiction*, **85**(12), 1559–62.

Faugier, J. (1988) AIDS and drug abuse. *Senior Nurse*, **8**, 9–10.

Feigelman, B. and Jaquith, P. (1992) Adolescent drug treatment, a family affair: a community day center approach. *Social Work in Health Care*, **16**(3), 39–52.

Fine, J. and Miller, N.S. (1993) Evaluation and acute management of psychotic symptomatology in alcohol and drug addictions. *Journal of Addictive Diseases*, **12**(3), 59–72.

Fingarette, H. (1988) *Heavy Drinking: The Myth of Alcoholism as a Disease*, University of California Press, Berkeley.

Fink, E.B., Longabaugh, R., McCrady, B.S. *et al.* (1987) Effectiveness of alcoholism treatment in partial versus impatient settings: twenty four month outcomes. *Addictive Behaviours*, **10**, 235–48.

Finkelstein, N. (1993) Treatment programming for alcohol and drug-dependent pregnant women. *International Journal of the Addictions*, **28**(13), 1275–309.

Finney, J.W., Moos, R.H. and Mewborn, C.R. (1980) Post treatment experiences and treatment outcome of alcoholic patients six months and two years after hospitalisation. *Journal of Consulting and Clinical Psychology*, **48**, 17–29.

Fishbein, M. and Ajzen, I. (1985) *Belief, Attitude, Intention and Behaviour: An Introduction to Theory and Research*, Addison-Wesley, Reading, Mass.

Fitzpatrick, J.L. and Gerard, K. (1993) Community attitudes towards drug use: the need to assess community norms. *International Journal of the Addictions*, **28**(10), 947–57.

Fletcher, H. (1990) *Drug Users and Custody*, NAPO, London.

Frank, J. (1974) *Persuasion and Healing: A Comparative Study of Psychotherapy*, Johns Hopkins University Press, Baltimore.

Frank, J. (1981) Therapeutic Components Shared by All Psychotherapies, Master Lecture Series, 1, Psychological Association, London.

Frank, J., Marel, R. and Schmeidler, J. (1984) An overview of substance use among New York State's upper income householders. *Advances in Alcohol and Substance Misuse*, **4**, 14–26.

Fraser, A. (1991) Into the Pleasuredome. *Druglink*, **6**, 12–13.

Fraser, A. and George, M. (1988) Changing trends in drug use: an initial follow-up of a local heroin using community. *British Journal of Addiction*, **83**(6), 655–63.

Fridinger, F. and Dehart, B. (1993) A model for the inclusion of a physical fitness and health promotion. *Journal of Drug Education*, **23**(3), 215–22.

Frischer, M., Shaw, S., Bloor, M. and Goldberg, D. (1993) Modeling the behaviour and attributes of injecting drug users: a new approach to identifying HIV risk practices. *International Journal of Addictions*, **28**(2), 129–52.

Funkhouser, A.W., Butz, A.M., Feng, T.I. and McCaul, M.E. (1993) Prenatal care and drug use in pregnant women. *Drug and Alcohol Dependence*, **33**(1), 1–9.

Gaitley, R. and Seed, P. (1989) *A Social Network Approach*, Jessica Kingsley, London.

Galanter, M. (1993) Network therapy for addiction: a model for office practice. *American Journal of Psychiatry*, **150**(12), 28–36.

Galanter, M., Kaufman, E., Taintor, Z., Robinowitz, C.B., Meyer, R.E. and Halikas, J. (1989) The current status of psychiatric education in alcoholism and drug abuse. *American Journal of Psychiatry*, **146**(1), 35–9.

Galizio, M. and Maisto, S.A. (1985) Towards a biopsychosocial theory of substance abuse, in *Determinants of Substance Abuse: Biological, Psychological and Environmental*, (eds M. Galizio and S.A. Maisto), Plenum, New York.

Garrard, J. (1990) Evaluation in drug education: some issues and options. *Drug Education Journal of Australia*, **4**(1), 1–15.

Garrard, J. and Northfield, J. (1987) *Drug Education in Victorian Post-Primary Schools*, Monash University, Victoria.

Gawin, F.H. and Kleber, H.D. (1984) Cocaine abuse treatment: an open pilot trial with lithium and desipramine. *Archives of General Psychiatry*, **41**, 903–9.

George, S.L., Shanks, N.J. and Westlake, L. (1991) Census of single homeless people in Sheffield. *British Medical Journal*, **302**, 1387–9.

Gibson, D.R., Sorensen, J.L., Wermuth, L. and Bernal, G. (1992) Families are helped by drug treatment. *International Journal of the Addictions*, **27**(8), 961–78.

Gibson, G.S. and Manley, S. (1991) Alternative approaches to urinalysis in the detection of drugs. *Social Behavior and Personality*, **19**(3), 195–204.

Gilman, M. (1991) Beyond opiates … and into the '90s. *Druglink*, **6**, 16–18.

Gilman, M. (1992) *Outreach*, ISDD Drugs Work 2, Institute for the Study of Drug Dependence, London.

Ginzburg, HI.M. (1989) Syringe exchange programmes: a medical or policy dilemma? *American Journal of Public Health*, **79**(10), 1350–1.

Glanz, A., Byrne, C. and Jackson, P. (1990) *Prevention of AIDS Amongst Drug Users: The Role of the High Street Pharmacist*, Department of Health, London.

Glaser, B.G. and Strauss, A. (1970) *The Discovery of Grounded Theory*. Aldine Press, New York.

Glaser, F. (1980) Anybody got a match? Treatment research and the matching hypothesis, in *Alcoholism Treatment in Transition*, (eds G. Edwards and M. Grant), Baltimore University Press, Baltimore.

Glass-Crome, I.B. (1994) Alcohol misuse; a challenge to medical education: a belated remedy (review). *British Medical Bulletin*, **50**(1), 164–70.

Glatt, M. (1972) *The Alcoholic and the Help He Needs*, Priory Press, London.

Globetti, G. (1988) Alcohol education programmes and minority youth. *Journal of Drug Issues*, **18**, 115–29.

Goddard, E. (1990) *Why Children Start Smoking*, HMSO, London.

Goffman, E. (1968) *Asylums*, Penguin Books, Harmondsworth.

Goldfried, S.L. and Bergin, A.E. (eds) (1986) *Handbook of Psychotherapy and Behaviour Change*, Wiley, New York.

Gonzalez, G.M. (1988) Should alcohol and drug education be a part of comprehensive prevention policy? The evidence from the college campus. *Journal of Drug Issues*, **18**, 355–65.

Goodstadt, M. (1974) Myths and mythologies in drug education: a critical review of research evidence, in *Research on Methods and Programs of Drug Education*, (ed. M. Goodstadt), Addiction Research Foundation, Toronto.

Goodstadt, M. (1987) Prevention strategies for drug abuse. *Issues in Science and Technology*, **3**(2) 28–35.

Gorman, D.M. (1993) A theory-driven approach to the evaluation of professional training in alcohol abuse. *Addiction*, **88**(2), 229–36.

Gossop, M. (1992) Severity of dependence and route of administration of heroin, cocaine and amphetamines. *British Journal of Addiction*, **87**(11), 1527–36.

Gould, A. (1993) Opposition to syringe exchange schemes in the UK and Sweden. *Journal of European Social Policy*, **3**(2), 107–18.

Gould, L.C., Walker, A.C., Crane, L.E. and Lidz, C.W. (1974) Connections: Notes from the Heroin World, Yale University Press, New Haven.

Greenwood, J. (1992) Persuading general practitioners to prescribe – good husbandry or a recipe for chaos? *British Journal of Addiction*, **87**(4) 567–75.

Griffiths, P., Gossop, M. Powis, B. and Strang, J. (1992) Extent and nature of transitions of route among heroin addicts in treatment – preliminary data from the Drug Transitions Study. *British Journal of Addiction*, **87**, 485–91.

Griffiths, P., Gossop, M., Powis, B. and Strang, J. (1994) Transitions in patterns of heroin administration: a study of heroin chasers and heroin injectors. *Addiction*, **89**(3), 301–9.

Gross, J. and McCaul, M.E. (1992) An evaluation of a psychoeducational and substance abuse risk reduction intervention for children of substance abusers. *Journal of Community Psychology*, special issue, 75–87.

Grunberg, N. (1994) Overview: biological processes relevant to drugs of dependence, *Addiction*, **89**(11), 1443–6.

Gunne, L.M. (1990) Politicians and scientists in the combat against drug abuse. *Drug and Alcohol Dependence*, **25**(2), 241–4.

Gustavsson, N.S. (1991) Chemically exposed children: the child welfare response. *Child and Adolescent Social Work*, **8**(4), 297–307.

Guttman, L. (1950) The basis for scalogram analysis, in *Measurement and Prediction*, (ed. S.A. Stouffer), Wiley, New York, pp. 60–90.

Hall, S.M., Havassy, B.E. and Wasserman, D.A. (1990) Commitment to abstinence and acute stress in relapse to alcohol, opiates and nicotine. *Journal of Consulting and Clinical Psychology*, **58**(2), 175–81.

Hammer, T. and Vaglum, P. (1990) Use of alcohol and drugs in the transition from adolescence to young adulthood. *Journal of Adolescence*, **13**, 129–42.

Hammersley, R., Forsyth, A. and Lavelle, T. (1990) The criminality of drug users in Glasgow. *British Journal of Addiction*, **85**(12), 1583–94.

Hammersley, R., Lavelle, T.L. and Forsyth, A.J.M. (1992) Adolescent drug use, health and personalty. *Drug and Alcohol Dependence*, **31**(1), 91–9.

Hanslope, J. (1994) Healthy women. *Druglink*, **9**(2), 16–17.

Harding, G. (1988) *Opiate Addiction, Morality and Medicine*, Macmillan, London.

Harding-Price, D. (1993) A sensitive response without discrimination. Drug misuse in children and adolescents. *Professional Nurse*, 8(7), 419–22.

Harford, T.C., Wechsler, H. and Rohman, M. (1983) The structural context of college drinking. *Journal of Studies on Alcohol*, 44(6), 722–32.

Harris, J.A. (1988) Assessing patterns of drug use in local jurisdictions. *International Journal of the Addictions*, 23(10), 1071–81.

Harrison, L. (1992) Substance misuse and social work qualifying training in the British Isles: a survey of CQSW courses. *British Journal of Addiction*, 87(4), 635–41.

Harrison, L.D. (1992) Trends in illicit drug use in the United States: conflicting results from the National Surveys. National Institute on Drug Abuse (NIDA). *International Journal of the Addictions*, 27(7), 817–47.

Hawkins, J.D. and Catalano, R. (1985) Aftercare in drug abuse treatment. *International Journal of the Addictions*, 20(6–7), 917–45.

Hawkins, J.D., Catalano, R.F. and Miller, J.Y. (1992) Risk and protective factors for alcohol and other drug problems: implications for substance abuse prevention. *Psychological Bulletin*, 112(1), 64–105.

Heather, N. (1989) Psychology and brief interventions. *British Journal of Addiction*, 84(4), 357–70.

Heather, N. (1991) Foreword, in *Counselling Problem Drinkers*, (eds R. Davidson, S. Rollnick and I. MacEwan), Tavistock/Routledge, London.

Heather, N. and Robertson, I. (1981) *Controlled Drinking*, Methuen, New York.

Heather N. and Robertson, I. (1986) *Problem Drinking, The New Approach*, Penguin Books, Harmondsworth.

Henley, G.A. and Winters, K.C. (1989) Development of psychosocial scales for the assessment of adolescents involved in alcohol and drugs. *International Journal of the Addictions*, 24(10), 973–1001.

Henry, J.A. (1992) Ecstasy and the dance of death. *British Medical Journal*, 305(6856), 775.

Hester, R. and Miller, W.R. (1989) *Handbook of Alcoholism Treatment Approaches: Effective Approaches*, Pergamon, Oxford.

Hodgson, R. (1976) Modification of compulsive behaviour, in *Case Histories in Behaviour Therapy*, (ed. H. Eysenk), Routledge and Kegan Paul, London.

Hodgson, R. (1994) Treatment of alcohol problems; section 5, Treatment. *Addiction*, 89(11), 1529–34.

Hodgson, R.J. and Stockwell, T.R. (1977) Does alcohol reduce tension? in *Alcoholism* (ed. G. Edwards), Oxford University Press, Oxford.

Hoffman, J.P. (1993) Exploring the direct and indirect family effects of adolescent drug use. *Journal of Drug Issues*, 23(3).

Holder, H., Longabaugh, R., Miller, W.R. and Rubonis, A.V. (1991) The cost of effectiveness of treatment for alcohol problems: a first approximation. *Journal of Alcohol Studies*, 52(6), 517–40.

Home Office (1993) *Statistics of Drugs Seizures and Offenders Dealt With, United Kingdom 1992*. Government Statistical Service, London.

Home Office (1995) *Tackling Drugs Together: A Strategy for England 1995–1998*, Cmd 2846, HMSO, London.

Homel, P.J., Daniels, P.M., Reid, T.R. and Lawson, J.S. (1981) Results of an experimental school based health development program in Australia. *International Journal of Health Education*, 24(4), 263–70.

Hopkins, T. (1991) Safe from harm? *Nursing Times*, 87, 42–3.

Howitt, D. (1990) Britain's 'substance abuse policy': realities and regulation in the United Kingdom. *International Journal of the Addictions*, 25, 353–76.

Hsu, N. (1993) Drug use and the family, *World Health*, 46(6), 21–3.

Huebert, K. and James, D. (1992) High-risk behaviours for transmission of HIV among clients in treatment for substance misuse. *Journal of Drug Issues*, **22**(4), 885–901.

Hunt, G.H. and Azrin, N.H. (1973) A community-reinforcement approach to alcoholism. *Behaviour Research and Therapy*, **11**, 91–104.

Iguchi, M.Y., Handelsman, L., Bickel, W.K. and Griffiths, R.R. (1993) Benzodiazepine and sedative use/abuse by methadone maintenance clients. *Drug and Alcohol Dependence*, **32**(3), 257–66.

Institute for the Study of Drug Dependence (1989) *Drug Indicators Project; Study of Help Seeking and Service Utilisation by Problem Drug Takers*, ISDD London.

Institute for the Study of Drug Dependence (1991) *Annual National Audit of Drug Misuse*, ISDD, London.

Institute for the Study of Drug Dependence (1994a) *Drug Abuse Briefing*, ISDD, London.

Institute for the Study of Drug Dependence (1994b) *Drug Misuse in Britain*, ISDD, London.

Institute for the Study of Drug Dependence (1995) *Annual National Audit of Drug Misuse*, ISDD, London.

Institute for the Study of Drug Dependence (1995) *Drugs, Pregnancy and Childcare: a Guide for Professionals*, ISDD, London.

Irwin, R.P. (1990) *The School Development in Health Education Project Model: Drug Education with a Past, a Present, and a Future?* Proceedings of the Fifth National Drug Educators' Seminar, Alcohol and Drug Foundation, Melbourne.

Ives, R. (1988) Developing our understanding of young people and drugs. *Children and Society*, **2**, 35–42.

James, R. and Carruthers, S. (1991) The role of the school in the national campaign against drug abuse. *Drug Education Journal of Australia*, **5**(3), 185–98.

Jarvis, G. and Parker, H. (1989) Young heroin users and crime. How do the new users finance their habits? *British Journal of Criminology*, **29**(2), 175–89.

Jellinek, E.M. (1960) *The Disease Concept of Alcoholism*, Hillhouse Press, New Haven.

Jellinek, E.M. and Bowman, G. (1946) Alcohol addiction and its treatment. *Quarterly Journal of Studies on Alcohol*, **2**, 98–176.

Jensen, E.L., Gerber, J. and Babcock, G.M. (1991) The new war on drugs: grass roots movement or political construction? *Journal of Drug Issues*, **21**, 651–67.

Jessor, R., Collins, M.I. and Jessor, S.L. (1972) On becoming a drinker: social-psychological aspects of an adolescent transition, in *Nature and Nurture in Alcoholism*, (ed. F.A. Seixas), Annals of the New York Academy of Sciences.

Jonas, S. (1994) A public health approach to reducing harm from drug use. *American Journal of Health Promotion*, **8**(4), 247–51.

Kadden, R. and Kranzler, H. (1992) Alcohol and drug abuse treatment at the University of Connecticut Health Center. *British Journal of Addiction*, **87**(4), 521–6.

Kail, B.L. and Litwak, E. (1989) Family, friends and neighbours: the role of primary groups in preventing the misuse of drugs. *Journal of Drug Issues*, **19**, 261–81.

Kandel, D.B. (1982) Epidemiological and psychosocial perspectives on adolescent drug use. *Journal of the American Academy of Child Psychology*, **4**, 328–47.

Kandel, D. and Yamaguchi, K. (1993) From beer to crack: developmental patterns of drug involvement. *American Journal of Public Health*, **83**(6), 851–5.

Kanfer, F. and Goldstein, A. (eds) (1986) *Helping People Change: A Textbook of Methods*, Pergamon Press, Oxford.

Kaufman, E. and Kaufman, P.N. (eds) (1979) *Family Therapy and Drug and Alcohol Abuse*, Gardner Press, New York.

Kearney, P. and Norman-Bruce, G. (1993) The Children Act. *Druglink*, **8**(1), 10–12.

Keene, J. (1994a) High risk groups and prison policies. *International Journal of Drug Policy*, **5**(3), 142–6.

Keene, J. (1994b) *Alcohol Treatment: A Study of Therapists and Clients*, Avebury, Aldershot.

Keene, J. (1996) Drug use among prisoners before, during and after custody. *Addiction Research*, in press.

Keene, J. and Raynor, P. (1993) Addiction as soul sickness: the influence of client and therapist beliefs. *Addiction Research*, **1**(1), 77–87.

Keene, J. and Stimson, G.V. (1991) HIV prevention: the role of the community pharmacist. *Pharmaceutical Journal*, **252**, 408–9.

Keene, J. and Stimson, G.V. (1992) Rural HIV prevention, *Druglink*, **7**, 8–9.

Keene, J. and Stimson, G.V. (1996) Professional ideologies and the development of syringe exchange: Wales as a case study. *Medical Anthropology*, in press.

Keene, J. and Trinder, H. (1995) *Evaluation and Comparison of Different Treatment Approaches for Drug and Alcohol Problems in Wales*. Report to the Welsh Office.

Keene, J. and Williams, M. (1996) Drug education and the role of the police in Britain. *Druglink*, **11**, 2.

Keene, J., Parry-Langdon, N. and Stimson, G. (1991) HIV prevention amongst drug users; specialist and community based provision. *International Journal on Drug Policy*, **2**(6), 26–9.

Keene, J., Stimson, G.V., Jones, S. and Parry-Langdon, N. (1993) Evaluation of syringe-exchange for HIV prevention among injecting drug users in rural and urban areas of Wales. *Addiction*, 88(8), 1063–70.

Keene, J., Willner, P. and James, D. (1996) *A Study of Drug Misuse and Drug Related Problems in the Neath and Afan Valleys*. Centre for Substance Abuse Research, University of Wales, Swansea.

Khantzian, E.J. (1985) The self medication hypothesis of addictive disorders; focus on heroin and cocaine dependence. *American Journal of Psychiatry*, **142**, 1259–64.

Khantzian, E.J. and Mack, J.E. (1989) Alcoholics Anonymous and contemporary psychodynamic theory, in *Recent Developments in Alcoholism* **7**, (ed. M. Galanter), Plenum Press, New York, pp. 67–89.

Kidorf, M. and Stitzer, M.L. (1993) Descriptive analysis of cocaine use of methadone patients. *Drug and Alcohol Dependence*, **32**(3), 267–75.

Kim, S., McLeod, J.H. and Shantis, C. (1989) An outcome evaluation of refusal skills program as drug abuse prevention strategy. *Journal of Drug Education*, **19**(4), 363–71.

Kirsch, J.P. (1983) Preventive health behaviour: a review of research and issues. *Health Psychology*, **2**, 277–302.

Klitzner, M., Gruenewald, P.J. and Bamberger, E. (1990) The assessment of parent led prevention programs. *Journal of Drug Education*, **20**(1), 77–94.

Kofoed, L. (1993) Outpatient vs. inpatient treatment for the chronically mentally ill with substance use disorders. *Journal of Addictive Diseases*, **12**(3), 123–37.,

Kostyk, D., Fuchs, D., Tabisz, E. and Jacyk, W.R. (1993) Combining professional and self-help group intervention: collaboration in co-leadership. *Social Work with Groups*, **16**(3), 111–23.

Kuhns, J.B., Heide, K.M. and Silverman, I. (1992) Substance use/misuse among female prostitutes and female arrestees. *International Journal of the Addictions*. 27(11), 1283–92.

Kurtz, E. (n.d.) *Not God: A History of Alcoholics Anonymous*, Hazelden Foundation, Minnesota.

Lamarine, R.J. (1993) School drug education programming: in search of a new direction. *Journal of Drug Education*, **23**(4), 325–31.

Landry, M.J., Smith, D.E., McDuff, D.R. and Baughman, O.L. (1991) Anxiety and substance use disorders: the treatment of high-risk patients. *Journal of the American Board of Family Practice*, **4**(6), 447–56.

Lart, R. and Stimson, G.V. (1991) 'Not just a syringe exchange' – a study of the organisation, working practices and philosophy of three syringe exchanges in England. Unpublished paper.

Lefevre, R. (n.d.) *The Promis Handbook on Alcoholism, Addictions and Recovery*, Promis Publishing, London.

Lehman, W.E.K., Barrett, M.E. and Simpson, D.D. (1990) Alcohol use by heroin addicts 12 years after drug abuse treatment. *Journal of Studies on Alcohol*, **51**, 233–44.

Leitner, M., Shapland, J. and Wiles, P. (1993) *Drug Usage and Drugs Prevention: The Views and Habits of the General Public*, Home Office Report, HMSO, London.

Leukefeld, C.G. and Tims, F.M. (1989) Relapse and recovery in drug abuse: research and practice. *International Journal of the Addictions*, **24**(3), 189–201.

Leventhal, H. and Cameron, L. (1987) Behavioural theories and the problem of compliance. *Patient Education and Counselling*, **10**, 117–38.

Leventhal, H. and Nerenz, D. (1985) The assessment of illness cognition, in *Measurement Strategies in Health Psychology*, (ed. P. Karoly), Wiley, New York, pp. 517–54.

Lindstrom, L. (1992) *Managing Alcoholism. Matching Clients to Treatments*, Oxford University Press, Oxford.

Littleton, J. and Little, H. (1994) Current concepts of ethanol dependence. *Addiction*, **89**(11), 1397–412.

Lofland, J. (1971) *Analysing Social Settings; A Guide to Qualitative Observation and Analysis*, Belmont, California.

Longabaugh, R. and Beattie, M. (1985) Optimising the cost effectiveness of treatment for alcohol abusers, in *Future Directions in Alcohol Abuse Treatment Research*, (eds B.S. McCrady, N.E. Noel and T.D. Nirenberg), Research monograph no. 15, National Institute on Alcohol Abuse and Alcoholics, Rockville, pp. 104–36.

Longshore, D., Hsieh, S., Danila, B. and Anglin, M.D. (1993) Methadone maintenance and needle/syringe sharing. *International Journal of the Addictions*, **28**(10), 983–96.

Lopez, J.M.O., Miron Redondo, L. and Leungo Martin, A. (1989) Influence of family and peer group on the use of drugs by adolescents. *International Journal of the Addictions*, **24**(11), 1065–82.

Ludwig, A.M. (1972) On and off the wagon: reasons for drinking and abstaining by alcoholics. *Quarterly Journal of Studies on Alcohol*, **33**, 91–6.

Lungley, S. (1988) *Intravenous Drug Use in New Zealand: A Baseline Study of Intravenous Drug Users and Their Risk of AIDS*. Wellington Health Services Research and Development Unit, Wellington.

Mack, J.E. (1981) Alcoholism, AA and the governance of self, in *Dynamic Approaches to the Understanding and Treatment of Alcoholism*, (eds M.H. Bean and N.E. Zinberg), The Free Press, New York, pp. 128–62.

Madden, A., Swinton, M. and Gunn, J. (1992) A survey of pre-arrest drug use in sentenced prisoners. *British Journal of Addiction*, **87**, 27–33.

Magruder-Habib, K., Hubbard, R.L. and Ginzburg, H.M. (1992) Effects of drug misuse treatment on symptoms of depression and suicide. *International Journal of the Addictions*, **27**(9), 1035–65.

Maisto, S.A. and Carey, K.B. (1987) Treatment of alcohol abuse, in *Developments in the Assessment and Treatment of Addictive Behaviours*, (eds T.D. Nirenberg and S.A. Maiston), Abllex, Norwood, New Jersey, pp. 173–211.

Makkai, T., Moore, R. and McAllister, I. (1991) Health education campaigns and drug use. *Health Education Research*, **6**(1), 65–76.

Mallams, J.H., Godley, M., Hall, G.M. and Meyers, R. (1982) A social systems approach to resocialising alcoholics in the community. *Journal of Studies on Alcohol*, **43**, 1115–23.

Malpas, D. (1990) The addicted mother's conflict. *Community Care*, March, 18–29.

Maltzman, I. and Schweiger, A. (1991) Individual and family characteristics of middle class adolescents hospitalised for alcohol and other drug use. *British Journal of Addiction*, **86**(11), 1435–47.

Maremmani, I., Hardini, R., Zolesi, O. and Castrogiovanni, P. (1994) Methadone dosages and therapeutic compliance during a methadone maintenance programme. *Drug and Alcohol Dependence*, **34**(2), 163–6.

Marlatt, G.A. (1985) Cognitive assessment and intervention procedures for relapse prevention, in *Relapse Prevention: Maintenance Strategies in the Treatment of Addictive Behaviours*. (eds G.A. Marlatt and J.R. Gordon), Guilford Press, New York.

Marlatt, G.A. (1988) Matching clients to treatment: treatment models and stages of change, in *Assessment of Addictive Behaviours*, (eds D.M. Donovan and G.A. Marlatt), Guilford Press, New York.

Marlatt, G.A. and George, W.H. (1984) Relapse prevention: introduction and overview of the model. *British Journal of Addiction*, **79**(3), 261–73.

Marlatt, G.A. and Gordon, J.R. (1980) Determinants of relapse: implications for the maintenance of behaviour change, in *Behavioural Medicine: Changing Health Lifestyles*, (eds P. Davidson and S. Davidson), Plenum, New York, pp. 410–57.

Marlatt, G.A. and Gordon, J.R. (eds) (1985) *Relapse Prevention: Maintenance Strategies in the Treatment of Addictive Behaviours*, Guilford Press, New York.

Marlatt, G.A., Demming, B. and Reid, J. (1973) Loss of control drinking in alcoholics: an experimental analogue. *Journal of Abnormal Psychology*, **812**, 233–41.

Marston, A.R., Jacobs, D.F., Singer, R.D., Wideman, K.F. and Little, T.D. (1988) Adolescents who apparently are invulnerable to drug, alcohol and nicotine use. *Adolescence*, **23**, 593–8.

Mash, E.J. and Terdal, L.G. (eds) (1976) *Behaviour Therapy Assessment: Diagnosis, Design and Evaluation*, Springer, New York.

Mason, D.T., Lusk, M.W. and Gintzler, M. (1992) Beyond ideology in drug policy: the primary prevention model. *Journal of Drug Issues*, **22**(4), 959–76.

Mason, P. and Marsden, J. (1994) The State of the market. *Druglink*, **9**(2), 8–11.

McCarty, D., Argerious, M., Huebner, R.B. and Lubran, B. (1991) Alcoholism, drug abuse, and the homeless. *American Psychologist*, **46**(11), 1139–48.

McCaughrin, W.C. and Price, R.H. (1992) Effective outpatient drug treatment organisations: program features and selection effects. *International Journal of the Addictions*, **27**(11), 1335–58.

McCoy, C.B., Rivers, J.E. and Khoury, E.L. (1993) An emerging public health model for reducing AIDS related risk behaviour among injecting drug users and their sexual partners. *Drugs and Society*, **7**(3/4), 143–59.

McCrady, B.S. and Sher, K.J. (1983) Alcoholism treatment approaches: patient variables, treatment variables, in *Medical and Social Aspects of Alcohol Abuse*, (eds B. Tabakoff, P.B. Sutker and C.L. Randell), Plenum Press, New York.

McDermott, P. and McBride, W. (1993) Crew 2000: peer coalition in action. *Druglink*, **8**(6), 13–15.

McDermott, P., Matthews, A. and Bennett, A. (1992) Responding to recreational drug use. *Druglink*, **7**, 12–13.

McKeganey, M. and Barnard, M. (1992) AIDS, *Drugs and Sexual Risk: Lives in the Balance*, Open University Press, Buckingham.

McLachlan, C., Crofts, N., Wodak, A. and Crowe, S. (1993) The effects of methadone on immune function among injecting drug users: a review. *Addiction*, **88**(2), 257–63.

McLaughlin, R.J., Holcomb, J.D., Jibaja-Rusth, M.L. and Webb, J. (1993) Teacher ratings of student risk for substance use as a function of specialised training. *Journal of Drug Education*, **23**(1), 83–95.

McMurran, M., Hollin, C. and Bowen, A. (1990) Consistency of alcohol self report measures in a male young offender population. *British Journal of Addiction*, **85**, 205–8.

Measham, F., Newcombe, R. and Parker, H. (1993) *An Investigation into the Relationship between Drinking and Deviant Behaviour amongst Young People*. Interim report to the Alcohol, Education and Research Council.

Mensch, B.S. and Kandel, D.B. (1988) Dropping out of high school and drug involvement. *Sociology of Education*, **61**, 95–113.

Meyers, C. and Moss, I. (1992) Residential treatments: linkage with community drug treatment programs. *Child Welfare*, **71**(6), 537–45.

Miller, W.R. (1983) Motivational interviewing with problem drinkers. *Behavioural Psychotherapy*, **11**(2), 147–72.

Miller, W.R. and Hester, R.K. (1986a) The effectiveness of alcoholism treatment: what research reveals, in *Treating Addictive Behaviours: Processes of Change*, (eds W.R. Miller and N. Heather), Plenum, New York.

Miller, W.R. and Hester, R.K. (1986b) Matching problem drinkers with optimal treatments, in *Treating Addictive Behaviours: Processes of Change* (eds W.R. Miller and N. Heather), Plenum, New York.

Miller, W.R., Sovereign, R.G. and Krege, B. (1988) Motivational interviewing with problem drinkers: II The drinkers check-up as a preventive intervention. *Behavioural Psychotherapy*, **16**(4), 251–68.

Millman, R.B. (1988) Evaluation and clinical management of cocaine abusers. *Journal of Clinical Psychiatry*, **49**, 27–33.

Mishara, B.L. and McKim, W. (1993) Methodological issues in surveying older persons concerning drug use. *International Journal of the Addictions*, **28**(4), 305–26.

Moos, R.H., Finney, J.W and Chan, D.A. (1981) The recovery process from alcoholism: 1. Comparing alcoholic patients and matched community controls. *Journal of Studies on Alcohol*, **42**, 383–402.

Moos, R.H., Finney, J.W. and Cronkite, R.C. (199) *Alcoholism Treatment. Context, Process and Outcome*, Oxford University Press, Oxford.

MORI/News of the World (1989) *Modern Living*, MORI, London.

MORI (1992) *Tomorrow's Young Adults*, Health Education Authority.

Mumme, D. (1991) Aftercare: its role in primary and secondary recovery of women from alcohol and other drug dependence. *International Journal of the Addictions*, **26**(5), 549–64.

Mundal, L.D., Van der Weele, T., Berger, C. and Fitsimmons, J. (1991) Maternal–infant separation at birth among substance using pregnant women: implications for attachment. *Social Work in Health Care*, **16**(1), 133–43.

Murphy, P.N. and Bentall, R.P. (1992) Motivation to withdraw from heroin: a factor-analysis study. *British Journal of Addiction*, **87**(2), 245–50.

Murray, J.B. (1986) An overview of cocaine use and abuse. *Psychological Reports*, **59**, 243–64.

National Campaign Against Drug Abuse (1988) *Task Force Evaluation*, NCADA, Australia.

Newcombe, R. (1991) *Raving and Dance Drugs: House Music Clubs and Parties in North-West England*, Rave Research Bureau, Liverpool.

Newcombe, R. (1992) A researcher report from the rave. *Druglink*, **7**, 14–16.

NOP Market Research Ltd. (1982) *Survey of Drug Use in the 15–21 Age Group*, NOP, London.

Nordstrom, G. and Burgland, M. (1986) Successful adjustment in alcoholism; relationships between causes of improvement, personality and social factors. *Journal of Nervous and Mental Disease*, **174**, 664–8.

Novick, D.M., Richman, B.L., Friedman, J.M. and Friedman, J.E. (1993) The medical status of methadone maintenance patients in treatment for 11–18 years. *Drug and Alcohol Dependence*, **33**(3), 235–45.

Nurco, D.M., Wegner, N., Stephenson, P., Makofsky, A. and Shaffer, J.W. (1983) *Ex-Addicts Self-Help Groups*, Praeger, New York.

Nutbeam, D. (1988) Planning for a Smokefree Generation. *Smoke-free Europe 6*, World Health Organization, Copenhagen.

O'Callaghan, F.V. and Caalan, V.J. (1992) Young adult drinking behaviour: a comparison of diary and quantity-frequency measures. *British Journal of Addiction*, **87**(5), 723–32.

O'Malley, P. (1991) The demand for intoxicating commodities: implications for the war on drugs. *Social Justice*, **18**, 49–75.

OPCS (1969) *General Household Survey*, OPCS, London.

Orford, J. (1977) Alcoholism and what psychology offers, in *Alcoholism*, (eds G. Edwards and J. Grant), Oxford University Press, Oxford.

Orford, J. (1985) *Excessive Appetites, A Psychological View of the Addictions*, Wiley, London.

Orford, J. and Edwards, G. (1979) *Alcoholism: A Comparison of Treatment and Advice*, Maudsley Monograph No. 26. London University Press, London.

Orford, J. and Velleman, R. (1990) Offspring of parents with drinking problems: drinking and drug taking as young adults. *British Journal of Addiction*, **856**, 779–94.

Orford, J., Oppenheimer, E. and Edwards, G. (1979) Abstinence or control: the outcome for excessive drinkers two years after consultation. *Behaviour, Research and Therapy*, **14**, 409–18.

Parker, H. (1993) *Pick and Mix: Alcohol, Drugs and the 1990s adolescent*. Paper given at the British Criminology Conference, Cardiff, Wales, July 1993.

Parker, H., Baker, K. and Newcombe, R. (1988) *Living with Heroin*, Open University Press, Buckingham.

Parole Release Scheme (1989) *The Parole Release Scheme Report 1987–89*, HMSO, London.

Parssinen, P. (1983) *Secret Passions, Secret Remedies*, Manchester University Press, Manchester.

Paton, S.M., Kessler, R. and Kandel, D. (1977) Depressive mood and adolescent illegal drug use: a longitudinal analysis. *Journal of Genetic Psychology*, **31**, 267–89.

Pattison, E.M., Sobell, M.B. and Sobell, L.C. (eds) (1977) *Merging Concepts of Alcohol Dependence*, Springer Publishing Company, New York.

Pearson, G., Ditton, J., Newcombe, R. and Gilman, M. (1991) Everything starts with an 'E'. *Druglink*, **6**, 10–11.

Petursson, H. and Lader, M. (1984) *Dependence on Tranquillisers*, Oxford University Press, Oxford.

Poland, M.L., Combrowski, M.P., Ager, J.W. and Sokol, R.J. (1993) Punishing pregnant drug users: enhanci ₒ ʰe flight from care. *Drug and Alcohol Dependence*, **31**(3), 199–203.

Powell, J.E. and Taylor, D. (1992) Anger, depression, and anxiety following heroin withdrawal. *International Journal of the Addictions*, **27**(1), 25–35.

Power, K., Markova, I., Rowlands, A. *et al.* (1992). Intravenous drug use and HIV transmission amongst inmates in Scottish prisons. *British Journal of Addiction*, **87**, 35–45.

Preble, E. and Casey, J.J. (1969) Taking care of business. *International Journal of the Addictions*, **4**, 1–24.

Preston, A. (1992) Pointing out the risk. *Nursing Times*, **88**, 24–6.

Preston, A. (1993) *The Methadone Handbook*, ISDD, London.

Prochaska, J.O. (1979) *Systems of Psychotherapy: A Transtheoretical Perspective*, Dorsey Press, Homewood, Ill.

Prochaska, J.O. and DiClemente, C.C. (1982) Transtheoretical therapy: towards a more integrative model of change. *Psychotherapy Theory*, Research and Practice, **19**(3), 276–88.

Prochaska, J.O. and DiClemente, C.C. (1983) Stages and processes of self change of smoking: towards a more integrative model of change. *Journal of Consulting and Clinical Psychology*, **51**, 390–5.

Prochaska, J.O. and DiClemente, C.C. (1986) Towards a comprehensive model of change, in *Treating Addictive Behaviours: Processes of Change*, (eds W.R. Miller and N. Heather), Plenum, New York.

Prochaska, J.O. and DiClemente, C.C. (1994) *The Transtheoretical Approach: Crossing Traditional Boundaries of Therapy*, Daw-Jones, Irwin, New York.

Project MATCH Research Group (1993) Project MATCH:L rationale and methods for a multisite clinical trial matching patients to alcoholism treatment. *Alcoholism: Clinical and Experimental Research*, **17**(6).

Raffoul, P.R. and Haney, C.A. (1989). Interdisciplinary treatment for drug misuse among older people of color: ethnic considerations for social work practice. *Journal of Drug Issues*, **19**, 297–313.

Raistrick, D. (1994) Report of the Advisory Council on the Misuse of Drugs; AIDS and drug misuse update. *Addiction*, **89**(10), 1211–13.

Raistrick, D. and Davidson, R. (1985) *Alcoholism and Drug Addiction*. Churchill Livingstone, Edinburgh.

Raveis, V.H. and Kandel, D.B. (1987) Changes in drug behaviour from the middle to the late twenties: initiation, persistence, and cessation of use. *American Journal of Public Health*, **77**(5), 607–11.

Ravndal, E. and Vaglum, P. (1994) Treatment of female addicts: the importance of relationships to parents, partners, and peers for the outcome. *International Journal of the Addictions*, **29**(1), 115–25.

Raymond, C.A. (1988) Study of IV drug users and AIDS finds differing infection rate, risk behaviours, *Journal of the American Medical Association*, **260**(21), 3105.

Reijneveld, S.A. and Plomp, H.N. (1993) Methadone maintenance clients in Amsterdam after five years. *International Journal of the Addictions*, **28**(1), 63–72.

Rhodes, F. and Humfleet, G.L. (1993) Using goal-orientated counselling and peer support to reduce H IV/AIDs risk amongst drug users not in treatment. *Drugs and Society*, **7**(3/4), 189–204.

Rhodes, T. and Stimson, G.V. (1994) Sex, drugs intervention and research: from the individual to the social. *International Journal of the Addictions*, **29**.

Rhodes, T., Hartnoll, R. and Johnson, A. (1991) *Out of the Agency and onto the Streets: A Review of HIV Outreach Health Education in Europe and the US*, ISDD Research Monograph 2, Institute for the Study of Drug Dependence, London.

Rieder, B.A. (1990) Perinantal substance abuse and public health nursing intervention. *Children Today*, **19**, 33–5.

Riley, D.M., Sobell, L.C., Leo, G.I., Sobell, M.B. and Klajner, F. (1987) Behavioural treatment of alcohol problems: a review and a comparison of behavioural and non-behavioural studies, in *Treatment and Prevention of Alcohol Problems: A Resource Manual*, (ed. W.M. Cox), Academic Press, Orlando, pp. 73–115.

Ringwalt, C., Ennett, S.T. and Holt, K.D. (1991) An outcome evaluation of project DARE. *Health Education Research*, **6**(3), 327–37.

Robertson, I. and Heather, N. (1986) *Let's Drink to Your Health: A Self Help Guide to Sensible Drinking*, British Psychological Society, London.

Robertson, J.R. (1989) Treatment of drug misuse in the general practice setting. *British Journal of Addiction*, **84**(4), 377–80.

Robertson, M.J. (1991) Homeless women with children: the role of alcohol and other drug abuse. *American Psychologist*, **46**(11), 1198–1204.

Robinson, D. (1979) *Talking Out of Alcoholism: The Self Help Process of AA*, Croom-Helm, London.

Roche, A.M. and Richards, G.P. (1991) Doctors' willingness to intervene in patients' drug and alcohol problems. *Social Science & Medicine*, **33**(9), 1053–61.

Rollnick, S., Heather, N., Gold, R. and Hall, W. (1992) Developments of a short 'readiness to change' questionnaire for use in brief, opportunistic interventions with excessive drinkers. *British Journal of Addiction*, **87**(5), 743–54.

Roman, P.M. and Trice, H.M. (1970) The development of deviant drinking behaviour:

occupational risk factors, *Archives of Environmental Health*, **20**, 424–35.

Rosenbaum, D., Flewelling, R., Bailet, S., Ringwalt, C. and Wilkinson, D. (1994) Cops in the classroom: a longitudinal study of drug abuse resistance education. *Journal of Research in Crime and Delinquency*, **1**(31), 3–31.

Rosenthal, D., Moore, S. and Buzwell, S. (1994) Homeless youths: sexual and drug related behaviour, sexual beliefs and HIV/AIDS risk. *AIDS Care*, **6**(1), 3–94.

Rosenstock, I.M. (1966) Why people use health services, *Millbank Memorial Fund Quarterly*, **44**, 94.

Rounsaville, B.J. (1986) *Clinical Implications of Relapse Research*, National Institute of Drug Abuse Research Monograph Series 72, Rockville, MD.

Royal College of Physicians (1992) *Smoking and the Young*, RCP, London.

Rush, B., Simmons, M., Timney, C.B., Evans, J. and Finlay, R. (1987) A comparison of psychotropic drug use between the general population and clients of health and social service agencies. *International Journal of the Addictions*, **22**, 843–59.

San, L., Tato, J., Torens, M. and Castillo, C. (1993a) Flunietrazepam consumption among heroin addicts admitted for TH – patient detoxification. *Drug and Alcohol Dependence*, **32**(3), 281–6.

San, L., Torrens, M., Catillo, C. and Porta, M. (1993b) Consumption of buprenorphine and other drugs among heroin addicts under ambulatory treatment: results from cross-sectional studies in 1988 and 1990. *Addiction*, **88**(10), 1341–9.

Sanchez-Craig, M. (1990) Brief didactic treatment for alcohol and drug related problems: an approach based on client choice. *British Journal of Addiction*, **85**(2), 169–77.

Sarvela, P.D. and McClendon, E.J. (1988) Indicators of rural drug use. *Journal of Youth and Adolescence*, **17**, 335–47.

Saunders, J.B. and Aasland, O.G. (1987) *WHO Collaborative Project on the Identification and Treatment of Persons with Harmful Alcohol Consumption*, Report on Phase 1 Development of a Screening Instrument, Geneva, WHO.

Saunders, W.M. and Kershaw, P.W. (1979) Spontaneous remission from alcoholism: a community study. *British Journal of Addiction*, **74**, 251–65.

Saxon, A.J., Calsyn, D.A. and Jackson, T.R. (1994) Longitudinal changes in injection behaviour in a cohort of injecting drug users. *Addiction*, **89**(2), 191–202.

Schapps, E., Dibartolo, R., Moskowitz, J., Palley, C.S. and Churgin, S. (1981) A review of 127 drug abuse prevention program evaluations. *Journal of Drug Issues*, **11**, 17-23.

Scheider, L.M. Newcombe, M.D. and Skager, R. (1994) Risk, protection, and vulnerability to adolescent drug use: latent variable models of three age groups. *Journal of Drug Education*, **24**(1), 48–82.

Schoenbaum, E.E., Hartel, D. and Selwyn, P.A. (1989) Risk factors for human immuno-deficiency virus infection in intravenous drug users. *New England Journal of Medicine*, **321**, 874–79.

Schwartz, G.E. (1982) Testing the biopsychosocial model: the ultimate challenge facing behavioural medicine. *Journal of Clinical Psychology*, **50**, 1040–53.

Selwyn, P.A. (1991) Injection drug use, mortality, and the AIDS epidemic. *American Journal of Public Health*, **81**(10) 1247–9.

Serpelloni, G., Carrieri, M.P., Rezza, G. and Morganti, S. (1994) Methadone treatment as a determinant of HIV risk reduction among injecting drug users: a nested case-control study. *AIDS Care*, **6**(2), 215–20.

Shearer, S.L. (1990) Frequency and correlates of childhood sexual and physical abuse in adult female borderline patients. *American Journal of Psychiatry*, **147**(2), 214–16.

Shewan, D., Gemmel, M. and Davies, J.B. (1994) Drug Use and Scottish Prisons, *Occasional Paper 6*, Scottish Prisons Service, Edinburgh.

Siegal, H., Baumgartner, K. and Carlson, R. (1991) *HIV Infection and Risk Behaviours among Injectable Drug Users in Low Seroprevalence Areas in the Mid-west*. Seventh International Conference on AIDS, Florence, Abstract MC3213.

Silver, R.L. and Wortman, C.B. (1980) Coping with undesirable life events, in *Human Helplessness*, (eds J. Garber and M. Seligman), Academic Press, New York, pp. 271–341.

Simpson, D.D. and Sells, S.B. (1982) *Effectiveness of Treatment for Drug Abuse, Institute of Behavioural Research*, Report 81–1, Texas Christian University, Fort Worth.

Simson, D.D. and Marsh, K.L. (1986) Relapse and recovery among opoid addicts 12 years after treatment, in *Relapse and Recovery*, (eds F. Tims and C. Leukefeld), National Institute of Drug Abuse Research Monograph 72, DHHA publication No. (ADM) 86–1473, US Department of Health and Human Services, Rockville, MD.

Singh, H. and Mustapha, N. (1994) Some factors associated with substance misuse among secondary school students in Trinidad and Tobago. *Journal of Drug Education*, 24(1), 83–93.

Single, E, Kandel, D. and Johnson, B.D. (1975) The reliability and validity of drug use responses in a large scale longitudinal survey. *Journal of Drug Issues*, 5, 426–43.

Sisson, R.W. and Azrin, N.H. (1986) Family-member involvement, to initiate and promote treatment of problem drinkers. *Journal of Behaviour Therapy and Experimental Psychiatry*, 17(1), 15–21.

Skinner, H.A. (1974) Alcoholic personality types: identification and correlates. *Journal of Abnormal Psychology*, 83, 653–66.

Skog, O. (1992) The validity of self reported drug use. *British Journal of Addiction*, 87(4), 539–48.

Smith, D.G. (1990) Thailand: AIDS Crisis Looms. *Lancet*, 335, 781–2.

Smith, J.P. (1991) Research, public policy and drug abuse: current approaches and new directions. *International Journal of the Addictions*, 25(2A), 181–99.

Solivetti, L.M. (1994) Drug diffusion and social change: the illusion about a formal social control. *Howard Journal of Criminal Justice*, 33(1), 41–61.

Solomon, L., Frank, R., Vlahov, D. and Astemborski, J. (1991) Utilisation of health services in a cohort of intravenous drug users with known HIV-1 serostatus. *American Journal of Public Health*, 81(10), 1285–90.

Solwiji, N., Hall, W. and Lee, N. (1992) Recreational MDMA use in Sydney: a profile of 'Ecstasy' users and their experiences with the drug. *British Journal of Addiction*, 87(8),1161–72.

Somers, NM. and Marlatt, A. (1992) Alcohol problems, in *Principles and Practice of Relapse Prevention* (ed. P. Wilson), Guilford Press, New York.

Sorenson, J.L., Costantini, M.F., Wall, T.L. and Bison, D.R. (1993) Coupons attract high-risk untreated users into detoxification. *Drug and Alcohol Dependence*, 31(3), 247-52.

Spear, S.F. and Mason, M. (1991) Impact of chemical dependency on family health status. *International Journal of the Addictions*, 26(2), 179–87.

Spencer, L. and Dale, A. (1979) Integration and Regulation in organisations: a contextual approach. *Sociological Review*, 27, 679–702.

Spradely, J.P. (1980) *Participant Observation*, Holt, Rhinehart and Winston, New York.

Stall, R. and Leigh, B. (1994) Understanding the relationship between drug or alcohol use and high risk sexual activity for HIV transmission: where do we go from here? *Addiction*, 89(2), 131–4.

Standing Conference on Drug Abuse (1986) *Guidelines for the Social Work Assessment of Drug Misusers*, Institute for the Study of Drug Dependence, London.

Stastny, D. and Potter, M. (1991) Alcohol abuse by patients undergoing methadone treatment programmes. *British Journal of Addiction*, 86(3), 307-10.

Steinberg, L., Fletcher, A. and Darling, N. (1994) Parental monitoring and peer influences on adolescent substance misuse. *Paediatrics*, **93**(6), pt. 2, 1060–4.

Steinglass, P. (1979) Family therapy and alcoholics: a review, in *Family Therapy and Drug and Alcohol Abuse* (eds E. Kaufman and P.N. Kaufman), Gardner Press, New York.

Stimson, G.V. (1973) *Heroin and Behaviour*, Wiley, New York.

Stimson, G.V. (1989) AIDS and HIV: The Challenge for British Drug Services, The Fourth Thomas James Okey Memorial Lecture, reprinted in *AIDS and Drug Misuse: Understanding and Responding to the Drug User in the Wake of HIV*, (J. Strang and G.V. Stimson), Routledge, London.

Stimson, G.V. (1990) AIDS and HIV: the challenge for British drug services. *British Journal of Addiction*, **85**(3), 329–39.

Stimson, G.V. (1995) AIDS and injecting drug use in the United Kingdom, 1988–1993: the policy response and the prevention of the epidemic. *Social Science and Medicine*, **41**(5), 699–716.

Stimson, G.V. and Keene, J. (1991) *Development of Syringe Exchange Services in Wales*. Welsh Office, Cardiff.

Stimson, G.V. and Oppenheimer, E. (1973) *Heroin Addiction*, Tavistock, London.

Stimson, G.V., Dolan, K. and Donoghoe, M.C. (1990) The future of UK syringe exchange. *International Journal of Drug Policy*, **2**(2), 14–17.

Stimson, G.V., Alldritt, L.J. Doland, K.A., Donoghoe, M.C. and Lart, R.A. (1988) *Injecting Equipment Exchange Schemes: Final Report*, Monitoring Research Group, Goldsmiths' College, London.

Stimson, G., Keene, J., Parry-Langdon, N. and Jones, S. (1992) *Evaluation of the Syringe Exchange Programme in Wales*. Welsh Office Report, Cardiff.

Stockwell, T. (1994) alcohol withdrawal: an adaption to heavy drinking of no practical significance? *Addiction*, **89**(11), 1397–412.

Strain, E.C., Stitzer, M.L., Liebson, I.A. and Bigelow, G.E. (1993) Methadone dose and treatment outcome. *Drug and Alcohol Dependence*, **33**(2), 105–17.

Strang, J. and Stimson, G.V. (1990) *AIDS and Drug Misuse*, Routledge, London.

Strunin, L. and Hingson, R. (1992) Alcohol, drugs and adolescent sexual behaviour. *International Journal of the Addictions*, **27**(2), 129–46.

Summerhill, D. (1990) Some reflections on dynamics and dilemmas in a DDU. *British Journal of Addiction*, **85**(5), 589–92.

Sutton, S.R. (1992) Commentaries. *British Journal of Addiction*, **87**, 24.

Swadi, H. and Zeitlin, H. (1988) Peer influence and adolescent substance abuse: a promising side? *British Journal of Addiction*, **83**(2), 153–7.

Swisher, J.D. (1993) Review of process and outcome evaluations of team training. *Journal of Alcohol and Drug Education*, **39**(1), 66–77.

Taylor, A. (1994) *Women Drug Users: An Ethnography of a Female Injecting Community*, Clarendon Press, New York.

Thompson, R. (1988) Action research applied to drug education – the DAPPS Study. *Drug Education Journal of Australia*, **2**(1), 7–14.

Tracy, E.M. and Farkas, K.J. (1994) Preparing practitioners for a child welfare practice with substance-abusing families. *Child Welfare*, **73**(1), 57–68.

Trinder, H. and Keene, J. (1996) Comparing substance misuse agencies: different substances, clients and models, but are they using the same methods? *Journal of Substance Misuse*, in press.

Trower, P., Casey, A. and Dryden, W. (1991) *Cognitive–Behavioural Counselling in Action*, Sage Publications, London.

Truax, C. and Carhuff, R. (1967) *Towards Effective Counselling and Psychotherapy*, Aldine, Chicago.

Tumim, J. (1992–95) *Chief Inspector of Prisons Report*, Home Office, London.

Turk, D.C., Rudy, T.E. and Salovey, P. (1986) Implicit models of illness. *Journal of Behavioural Medicine*, **9**(5), 453–74.

Turnbull, P.J., Dolan, K. and Stimson, G.V. (1991) Prisons, *HIV and AIDS: Risks and Experiences in Custodial Care*, The Centre for Research on Drugs and Health Behaviour, London.

Turnbull, P.J., Dolan, K.A. and Stimson, G.V. (1992) Prevalence of HIV infection among ex-prisoners in England. *British Medical Journal*, **304**, 90–1.

United States Department for Health and Human Services (1989) *Reducing Health Consequences of Smoking: Twenty-five Years of Progress*, A Report of the Surgeon General, DHHS, Washington DC.

Valliant, G.E. (1983) *The Natural History of Alcoholism: Causes, Patterns and Paths to Recovery*, Harvard University Press, Harvard.

Van Baar, A. (1990) Development of infants of drug dependent mothers. *Journal of Child Psychology and Psychiatry and Allied Disciplines*, **31**, 911–20.

Van den Hoek, J.A.R., Van Haastrecht, H.J.A. and Coutinho, R.A. (1989) Risk reduction among intravenous drug users in Amsterdam under the influence of AIDS. *American Journal of Public Health*, **79**(10), 1355–7.

Waldron, S. (1969) *Statement to Select Committee on Crime*, House of Representatives, 91st Congress, July 29.

Wallace, J.M. and Bachman, J.G. (1991) Explaining racial and ethnic differences in adolescent drug use. *Social Problems*, **38**, 333–57.

Warhiet, G.J. and Biafore, F. (1991) Mental health and substance abuse patterns among a sample of homeless post-adolescents. *International Journal of Adolescence and Youth*, **3**(1/2), 9–27.

Wechsberg, W.M., Dennis, M.L. Cavanagh, E. and Rachel, J.V. (1993) A comparison of injecting drug users reached through outreach and methadone treatment. *Journal of Drug Issues*, **23**(4), 668–87.

Weisner, C. and Schmidt, L. (1993) Alcohol and drug problems among diverse health and social service populations. *American Journal of Public Health*, **83**(6), 824–9.

Welsh Office (1993) *Welsh Drug Misuse Database: Report 5* (Oct 92–Mar 93), Health Statistics and Analysis Unit, Welsh Office, Cardiff.

Werch, C.E. and DiClemente, C.C. (1994) A multi-stage model for matching drug prevention strategies and messages to youth stage of use. *Health Education Research*, **9**(1), 37–46.

West, R. and Gossop, M. (1994) Overview: a comparison of withdrawal symptoms from different drug classes. *Addiction*, **89**(11), 1483–9.

Whelan, S. and Moody, D. (1994) *DARE, Mansfield. An Evaluation of a Drug Prevention Programme for Children attending a Middle School in Mansfield Notts*, North Nottingham Health Promotion and Nottingham Drug Prevention Team.

Williams, A.B., McNelly, E.A., Williams, A.E. and D'Aquila, R.T. (1992) Methadone maintenance treatment and HIV type 1 seroconversion among injecting drug users. *AIDS Care*, **4**(1), 34–41.

Williams, G.D., Aitken, S. and Malin, H. (1985) Reliability of self-reported alcohol consumption in a general population survey. *Journal of Studies on Alcohol*, **46**(3), 223–7.

Williams, M. and Keene, J. (1995) *Drug Education and the Police: A Review of Recent Research Studies*, West Glamorgan Drug Prevention Team, Swansea.

Wilson, P. (1992) Relapse prevention: conceptual and methodological issues, in *Principles and Practice of Relapse Prevention*, (ed. P. Wilson), Guilford Press, New York.

Winick, C. (1991) Social behaviour, public policy and non-harmful drug use. *The Millbank Quarterly*, **6**(3), 437–59.

Wodak, A. (1994) How do communities achieve reductions in alcohol and drug-related harm? *Addiction*, **89**(2), 147–50.

Woogh, C.M. (1990) Patients with multiple admissions in psychiatric record linkage system. *Canadian Journal of Psychiatry*, **35**(5), 401–6.

World Health Organization (1989) *International Classification of Diseases*, WHO, Geneva.

World Health Organization Expert Committee on Drug Dependence (1994) *28th Report*, WHO, Geneva.

Yin, R.E. (1989) *Case Study Research, Design and Methods*, Applied Social Research Methods Series 5, Sage Publications, London.

Young, J. (1971) *The Drug Takers*, MacGibbon and Kee, London.

Yu, J. and Williford, W.R. (1992) Drug and alcohol use. *International Journal of the Addictions*, **27**(11), 1313–23.

Zinberg, N.E. (1984) *Drug, Set and Setting*, Yale University Press, New Haven.

Index

Agencies *cont'd*
referrals to 239, 257
service provided 239
community drug team 222–9
Twelve Step programme centre 229–35
syringe exchange service 163–4, 165, 169
advantages 158
case study 155–8
disadvantages 158–60
use by clients 161
treatment 165, 166, 238–9
Agency attenders 4–5, 9, 10, 20, 46, 49, 149
age of 19
distance travelled 158
ex-prisoners 84–5
injecting behaviour 29
older misusers 140–1
prisoners 84
using different agencies 238
views about dependency 57–8
women 139
see also Dependent drug misusers
AIDS, *see* HIV
Aims
brief interventions 188
community drug team 223
counselling 260
defining, drug prevention 117–18, 120
harm minimization 124–7, 141, 167
outreach work 153
Twelve Step programme centre 230
Alcohol
dependence syndrome 179–80, 189, 208
with methadone 71, 281
misuse 29, 128, 140, 176
brief interventions 188–9
cognitive psychology 184–5
controlled 206–9
craving 209–10
early onset 96, 97, 106
family relationships 293
fashions in theory change 221–2
gender studies 98
matching clients to treatment 213–14

screening for 265
social context of 29
social treatment programmes 197, 201–3
treatment outcomes 199, 203–5, 293–4
opiate addicts 212
Alcoholics Anonymous 189, 189–90, 195, 196, 218
Alkyl nitrates 283
America
aftercare 220
drug education 101, 102, 106, 122–3
dual diagnosis 306
epidemiological studies 13
fall in smoking 107
harm minimization services 136, 136–7
Narcotics Anonymous 190, 220
non-professional counsellors 302
prescribed drugs 182
Project MATCH 200
statistics 11–12
syringe sharing 155
treatment of dependency 178–9, 195, 196
Amphetamines 7, 8, 10, 20, 182, 183
addict notifications 9
effects of 283
on foetus 309
high-risk misusers 34–5, 36, 40
injected 29
non-agency attenders 135
police seizures 8
prescribing for harm minimization 147–8
prison inmates 84, 245, 249
recreational use 19, 29
Amsterdam Methadone maintenance programme 183
Anger 78, 149
management 211, 227
Anonymous 157, 163, 164
outreach work 153
Anonymous Fellowship 180, 190, 191, 196, 203, 220, 230
drop-out rate 195
length of attendance 218
self-help subculture 219
Anxiety 149, 150, 210–11, 227

Arrest referral schemes 10, 314
Arterial occlusion 143
Assessment 258, 259–60, 270–86
  aftercare 288
  behavioural interventions 185–6
  carrying out 277–9
  community drug team 225
  drug-misusing parents 310, 311–12
  purposes of 272–3
  Twelve Step programme day centre
    232
Asthma 144
Attribution theory 187–8
Australia
  drug education 101
  treatment 178

Background of misuse xi–xii
Barbiturates 144, 281, 282
  withdrawal 304
Barriers to help-seeking 316–17
Behaviour
  changes, changing attitudes in
    261–2
  cognitions and action 100
  control, see Cognitive behaviour
    techniques
  violent 211, 248–9, 281
Behavioural interventions 185–6
Beliefs, clients'
  about effects of drugs 290–1
  about misuse behaviour 187–8
Benefits of misuse
  dependent misusers 69–70
  recreational drugs 24–6
Benzodiazepines 9, 19, 34, 303, 305
  dependent misusers 57
  effects of 144, 281
    on foetus 309
  high-risk misusers 34, 35, 51
  methadone maintenance clients 212
  prescribing 149
  prison inmates 245, 248
  withdrawal 57, 182, 187, 281, 282
Biopsychosocial model of dependency
  179, 207
Bleach cleaning 165
Body language 262
'Breaking denial', stage of 191–3
Breastfeeding 309

Brief interventions 188–9, 207, 209, 265
British Journal of Addiction 185, 216
British National Curriculum Council
  102
Buprenorphine 183

Cannabis 5, 6, 7, 20
  dependent drug misusers 59
  effects of 144, 282
  high-risk misusers 34, 35, 36
  ISDD 1995 Audit 10
  misuse in US 11
  police
    cautions 315
    seizures 8
  prison inmates 84, 86, 245, 246, 248,
    249
  progression from cigarettes 96
  recreational use 19, 22
  statistics 8, 9, 11
Care plans 258, 273, 284–6
  aftercare 284, 287, 289
  completed plan 285
  misusers with children 310
Case studies
  community drug team intervention
    222–9
  drug agency syringe exchange 155–8
  recreational drug prevention
    116–21
  Twelve Step programme day centre
    229–35
Catchment areas, agency-based
  syringe exchanges 158
Categories of misuse xiii
  assessment of 272, 273–7
    different categories at different
      times 274
    two categories at once 274
  change over time xiv, 96–8
Cause and effect of misuse
  confusion about 49, 57, 60, 64,
    69–70, 90, 208–9, 277
  respondents' views 46, 67
  criminal activity 314
  problems contributing and arising
    from misuse 278–9
  see also Reasons for misuse
Caution and diversion schemes 10,
  314–15

International Journal of Addiction 217
Intervention, types of xiii–xiv
Intimidation in prison 246, 250
Intoxication, high-risk misusers 37

Juvenile offenders 314–15

Knowledge, transfer of 104

Language development, infants of
    maternal misusers 139
Learning difficulties 141
Legal risks 44–5, 72–3
Licensing laws 17–18
Life skills programmes 104–5
Lifeline Project, Manchester 102
Lifestyle 12–13, 140, 150
  changing 217
  dependent drug misusers 74
  developing safer environment
    292–4
  dissatisfaction with 271
  high-risk drug misusers 34–6
Long-term effects, drug misuse 144–5,
  280–4
Longitudinal studies 105, 106, 115
LSD 7, 8, 10, 11, 20, 34
  effects of 144, 283–4
  police seizures 8
  use at raves and nightclubs 19, 22

Magic mushrooms 144, 283–4
Matching clients to help and treat-
    ments 213–14, 257, 273–7
Materials, *see* Resources
Maternal
  drug misuse 138–9, 308–10
  transmission, HIV 133, 309
MDMA, *see* Ecstasy
Menstruation, cessation of 308
Mental health workers 306–7
Mentally ill misusers 140, 141
Methadone 9, 136, 177, 182–3, 272, 308
  with alcohol 71, 281
  for harm minimization 145–6
  HIV risk clients 205
  with other drugs 212
  prescribing practice 303–4
  respondants' statements 50, 58–9,
    60, 72, 75

Methodology
  drug education and prevention
    103–5
  research studies 13–16
Minnesota Method 178, 189, 190–1,
  195, 196, 221, 301–2
  day centre case study 229–35
Monitoring progress 273, 297
  post-treatment 295
  self-monitoring 186, 290
Mood control strategies 292
Moral explanations 63
Motivation 265–70
  need for client 255, 257–8
  to change 266–70
  to seek help 265–6
Motivational interviewing 215, 215–16,
  240, 266, 272
  limitations for recreational
    misusers 269–70
  uses of 268–9
Multidisciplinary team approach 307,
  316–17
Music at raves 18

Naloxone 144
Naltrexone Anonymous 178, 180, 190,
  191, 192, 214, 220, 229, 235
  'disease' model 129, 177, 178, 189
  meetings 234
  role of counsellors 233
National Institute of Drug Misuse
  (NIDA), US 11, 155
Needles, *see* Sharing needles and
  syringes; Syringe and needle
  exchanges
Neonatal withdrawal syndrome 309
Nightclubs, misuse at 17–19, 28–30, 34,
  36, 109
  'Peanut Pete' campaign 102
  Project 'Pitfall' 116–21
  qualitative data 20–7
  quantitative review 19–29
  views of non-misusers 27–8
Non-directive, client-centred
  counselling 261, 264
Non-drug users 95
  at raves 27–8
  prison 95
Non-professional staff 302